Charles Seale-Hayne Library

University of Plymouth

(01752) 588 588

LibraryandITenquiries@plymouth.ac.uk

Springer Tracts in Modern Physics
Volume 166

Springer
Berlin
Heidelberg
New York
Barcelona
Hong Kong
London
Milan
Paris
Singapore
Tokyo

Physics and Astronomy

ONLINE LIBRARY

http://www.springer.de/phys/

Springer Tracts in Modern Physics

Springer Tracts in Modern Physics provides comprehensive and critical reviews of topics of current interest in physics. The following fields are emphasized: elementary particle physics, solid-state physics, complex systems, and fundamental astrophysics.

Suitable reviews of other fields can also be accepted. The editors encourage prospective authors to correspond with them in advance of submitting an article. For reviews of topics belonging to the above mentioned fields, they should address the responsible editor, otherwise the managing editor.

See also http://www.springer.de/phys/books/stmp.html

Managing Editor

Gerhard Höhler

Institut für Theoretische Teilchenphysik
Universität Karlsruhe
Postfach 69 80
76128 Karlsruhe, Germany
Phone: +49 (7 21) 6 08 33 75
Fax: +49 (7 21) 37 07 26
Email: gerhard.hoehler@physik.uni-karlsruhe.de
http://www-ttp.physik.uni-karlsruhe.de/

Elementary Particle Physics, Editors

Johann H. Kühn

Institut für Theoretische Teilchenphysik
Universität Karlsruhe
Postfach 69 80
76128 Karlsruhe, Germany
Phone: +49 (7 21) 6 08 33 72
Fax: +49 (7 21) 37 07 26
Email: johann.kuehn@physik.uni-karlsruhe.de
http://www-ttp.physik.uni-karlsruhe.de/~jk

Thomas Müller

Institut für Experimentelle Kernphysik
Fakultät für Physik
Universität Karlsruhe
Postfach 69 80
76128 Karlsruhe, Germany
Phone: +49 (7 21) 6 08 35 24
Fax: +49 (7 21) 6 07 26 21
Email: thomas.muller@physik.uni-karlsruhe.de
http://www-ekp.physik.uni-karlsruhe.de

Fundamental Astrophysics, Editor

Joachim Trümper

Max-Planck-Institut für Extraterrestrische Physik
Postfach 16 03
85740 Garching, Germany
Phone: +49 (89) 32 99 35 59
Fax: +49 (89) 32 99 35 69
Email: jtrumper@mpe-garching.mpg.de
http://www.mpe-garching.mpg.de/index.html

Solid-State Physics, Editors

Andrei Ruckenstein
Editor for The Americas

Department of Physics and Astronomy
Rutgers, The State University of New Jersey
136 Frelinghuysen Road
Piscataway, NJ 08854-8019, USA
Phone: +1 (732) 445 43 29
Fax: +1 (732) 445-43 43
Email: andreir@physics.rutgers.edu
http://www.physics.rutgers.edu/people/pips/
Ruckenstein.html

Peter Wölfle

Institut für Theorie der Kondensierten Materie
Universität Karlsruhe
Postfach 69 80
76128 Karlsruhe, Germany
Phone: +49 (7 21) 6 08 35 90
Fax: +49 (7 21) 69 81 50
Email: woelfle@tkm.physik.uni-karlsruhe.de
http://www-tkm.physik.uni-karlsruhe.de

Complex Systems, Editor

Frank Steiner

Abteilung Theoretische Physik
Universität Ulm
Albert-Einstein-Allee 11
89069 Ulm, Germany
Phone: +49 (7 31) 5 02 29 10
Fax: +49 (7 31) 5 02 29 24
Email: steiner@physik.uni-ulm.de
http://www.physik.uni-ulm.de/theo/theophys.html

Walter Dittrich   Holger Gies

Probing the Quantum Vacuum

Pertubative Effective Action Approach
in Quantum Electrodynamics
and its Application

With 16 Figures

 Springer

Professor Walter Dittrich
Dr. Holger Gies

University of Tuebingen
Institute of Theoretical Physics
Auf der Morgenstelle 14
72076 Tuebingen, Germany

E-mail: walter.dittrich@uni-tuebingen.de
E-mail: holger.gies@uni-tuebingen.de

Physics and Astronomy Classification Scheme (PACS): 12.20.-m, 11.10.E, 11.10.Wx, 13.40.-f, 41.20.Jb

ISSN 0081-3869
ISBN 3-540-67428-4 Springer-Verlag Berlin Heidelberg New York

Library of Congress Cataloging-in-Publication Data.

Dittrich, Walter. Probing the quantum vacuum: perturbative effective action approach in quantum electrodynamics and its application / Walter Dittrich, Holger Gies. p.cm. – (Springer tracts in modern physics, ISSN 0081-3869; v. 166). Includes bibliographical references and index. ISBN 3540674284 (alk. paper). 1. Quantum electrodynamics. I. Gies, Holger, 1972– . II. Title. III. Springer tracts in modern physics; 166.   QC1.S797 vol. 166   [QC680]   539 s–dc21 [530.14'33]   00-030078

Springer-Verlag Berlin Heidelberg New York
a member of BertelsmannSpringer Science+Business Media GmbH

© Springer-Verlag Berlin Heidelberg 2000
Printed in Germany

The use of general descriptive names, registered names, trademarks, etc. in this publication does not imply, even in the absence of a specific statement, that such names are exempt from the relevant protective laws and regulations and therefore free for general use.

Typesetting: Camera-ready copy by the authors using a Springer LaTeX macro package
Cover design: design& production GmbH, Heidelberg

Printed on acid-free paper   SPIN: 10767117   56/3144/tr   5 4 3 2 1 0

This volume is dedicated to the memory of Julian Schwinger,
great physicist, teacher and friend

Preface

This book is devoted to an investigation of the vacuum of quantum electrodynamics (QED), relying on the perturbative effective-action approach. If the vacuum is probed with external perturbations, the response of the system can be analyzed after averaging over the high-energy degrees of freedom. This results in an effective description of the properties of the vacuum, which are comparable to the properties of a classical medium.

We concentrate primarily on the physics of slowly varying fields or soft photons by integrating out the high-energy degrees of freedom, i.e. the electrons, employing Schwinger's proper-time method. We derive a new representation of the one-loop photon polarization tensor, coupling to all orders to an arbitrary constant electromagnetic field, fully maintaining the dependence on the complete set of invariants.

On the basis of effective Lagrangians, we derive the light cone condition for low-frequency photons propagating in strong fields. Our formalism can be extended to various external perturbations, such as temperature and Casimir situations. We give a proof of the "unified formula" for low-energy phenomena that describes the refractive indices of various perturbed quantum vacua. In the high-energy domain, we observe similarities between a vacuum with a superstrong magnetic field and a magnetized plasma. The question of measurability of the various effects is addressed; a violation of causality is not found.

Furthermore, the QED one-loop effective Lagrangian in the presence of an arbitrary constant electromagnetic background field at finite temperature is derived. We show that the requirement for gauge invariance induces an additional invariant on the quantum level, which is physically related to a chemical potential and to Debye screening. Further applications are presented.

In our treatment of 2+1-dimensional QED, we discuss several unfamiliar features of the effective-action approach, such as the formation of a magnetically induced chiral condensate, the perturbative generation of a parity-odd Chern–Simons term and a derivative expansion of the QED_{2+1} effective action for inhomogeneous fields.

During the course of writing the manuscript, we have benefited from discussions and correspondence with many colleagues. In particular, we wish to thank Professor Martin Reuter for numerous discussions and various useful suggestions. We are indebted to Dr. Klaus Scharnhorst for his helpful comments, criticism and encouragement.

We are also grateful to Mrs. Virginia Dittrich, who helped us to put this work into correct English.

Tübingen,   *Walter Dittrich*
April 2000   *Holger Gies*

Contents

1. Introduction

During the development of quantum field theory, it has become a popular point of view to consider the quantum vacuum as a medium. The manifold phenomena which arise from the presence (or the provoked absence) of *virtual* particles in the vacuum, such as the Lamb shift or the Casimir effect, tempt one to assign *real* properties to the vacuum. On the other hand, a physicist can hardly accept an establishment of this kind of "modern ether" without reservations; it is, at least, unsatisfactory to formulate a fundamental theory with the aid of some ingredients which elude direct measurement.

In the present volume, we do not insist on either point of view as the one and only foundation of physics, but intend to demonstrate that it can be pragmatically appropriate and useful to understand certain quantum effects as peculiarities of the vacuum – both intuitively and formally.

The formal concept of relating the full quantum theory to properties of the vacuum is given by the effective action. By integrating out the high-energy degrees of freedom of the exact theory, one arrives at an effective description of the low-energy degrees of freedom which are relevant to the physics of the vacuum.

This procedure is particularly successful for theories such as quantum electrodynamics (QED), where it is easy to identify and separate the different degrees of freedom. In QED, it is the electron, which is separated from the low-energy photons by its mass, that represents the fundamental scale in the theory. In strongly coupled (gauge) systems such as quantum chromodynamics (QCD), one has to rely on some more or less justified approximations in order to establish the effective-action approach.

The effective-action approach has also proved useful when additional energy scales are introduced which can be varied over a wide range compared with the high-energy scale. These additional scales may represent perturbations of the vacuum, probing its response under external influences. The (pragmatic) idea then is to compare the response of the vacuum with the response of a medium under the influence of such a perturbation. The properties of a medium are finally assigned to the vacuum itself in order to complete the comparison.

In the present work, we apply this philosophy to QED, which is perturbatively accessible. We primarily employ external electromagnetic fields as the perturbation of the vacuum, i.e. as the additional energy scale. Further examples of perturbation are given by coupling the system to an external heat bath or limiting its spatial extension by means of perfectly conducting (Casimir) plates. Also imaginable are gravitational perturbations and perturbations that arise from space–time embeddings with nontrivial topology.

Technically speaking, we integrate out the fluctuating particles which are coupled to the external perturbation, to all orders if possible. The result is then compared with the unperturbed vacuum, and the effects that arise from the discrepancies are investigated. Since the remaining low-energy degrees of freedom of QED are electromagnetic fields, we study the properties of the modified vacua by, in particular, searching for a modification of the propagation of low-frequency light in these modified vacua. It is proposed that the deformation of the light cone can be used as a detector of the various characteristics of the vacuum. Our intention is that this theoretical work can be related to experiments, currently in preparation, that are aimed at proving directly the quantum-induced nonlinear extensions of classical electrodynamics for the first time [167].

These investigations are presented from two different starting points. In the first part, we start with the complete quantum theory and study the polarization tensor and induced current in an electromagnetic field in detail, whereas in the second part, we take the effective action as the primary object. Once the effective action is deduced from the quantum theory, we can employ it as the definition of a new classical theory which *effectively* incorporates the quantum physics.

In the perturbative approach, the proper-time method as developed by Schwinger is extremely useful, since it not only represents a powerful regularization technique, but also supports a detailed investigation of gauge invariance. Moreover, it provides for a clear formulation and solution of the technical difficulties.

Another important tool, which we use throughout these investigations of QED, derives from the fact that the effective Lagrangian is a scalar (in all respects) and can therefore depend only on the invariants of all allowed symmetry transformations of the theory (gauge and Lorentz symmetry). Identifying the complete set of invariants at the beginning of the investigations allows us to formulate the calculations in full generality, and simplifies the calculations in most cases. The simplifications arise from the fact that the computations of the dynamics of a system are separate from the algebraic computations referring to the symmetries of the theory.

2. Nonlinear Electrodynamics: Quantum Theory

2.1 Introduction to Proper-Time Methods

We give a brief survey of the proper-time technique as formulated by Schwinger, in order to deal with quantized fermionic fields interacting with an external electromagnetic field. The relation to one-particle quantum mechanics is emphasized, and questions of gauge invariance are studied. We describe three different methods for computing the proper-time transition amplitude, which represents the central object of this technique. The derivation of the fermionic Green's function and the Heisenberg–Euler effective Lagrangian of QED in Schwinger's representation is sketched.

The main purpose of this section is to establish conventions and notation, as well as to lay the foundations for subsequent investigations.

One of the most important tools for the analysis of quantum electrodynamics is the proper-time method introduced by Fock [79] and Schwinger [148]. It has been proved extremely useful in studying propagators and effective Lagrangians, especially for the case of additional external fields.

Schwinger's main intention was to circumvent problems associated with gauge invariance. The proper-time method explicitly deals with objects that transform gauge-covariantly, thereby ensuring the invariance of the procedure and its results. In the literature, the proper-time method is in most cases simply used as a regularization procedure, because it is able to isolate the divergences of a calculation with respect to the proper time. Lorentz and gauge invariance of this isolation process is maintained, since the proper-time parameter is not related to a particular reference frame or choice of gauge.

But more than being a regularization procedure, the proper-time technique relates the field-theoretic problem of how a particle field interacts with an external ("c-number") electromagnetic field to the description of a particle's motion with respect to an additional evolution parameter, the proper time (the "fifth parameter"). In this sense the technique falls back on the principles of classical and one-particle quantum mechanics.

In the following, we shall give a brief introduction to the proper-time method, aiming especially at the construction of the Green's function of a Dirac field and at the calculation of the effective Lagrangian. The electromagnetic field is considered to be an *external* one, i.e. it is treated as a c-number;

hence, radiative corrections are neglected and we are, by definition, dealing with the one-loop approximation.

We begin by stating that the vacuum-to-vacuum persistence amplitude in the presence of an external electromagnetic field A_μ is related to the effective action (effective Lagrangian) by

$$\langle 0_+ | 0_- \rangle^A = e^{iW^{(1)}[A]} = e^{i \int d^4x \, \mathcal{L}^{(1)}(x)}. \tag{2.1}$$

The superscript (1) indicates the one-loop character of the quantity to which it is attached. $W^{(1)}$ is defined so that it generates the vacuum expectation value of the current j^μ upon differentiating with respect to the external field A_μ:

$$\frac{\delta W^{(1)}[A]}{\delta A_\mu(x)} = \langle 0| j^\mu(x) |0\rangle^A. \tag{2.2}$$

According to Schwinger, the correct definition of the current operator is based on the explicitly symmetric treatment of the oppositely charged Dirac fields:[1]

$$j^\mu = \frac{e}{2} [\bar\psi, \gamma^\mu \psi]. \tag{2.3}$$

When the desired vacuum expectation value of the current operator is calculated, the charge symmetrization translates into a time symmetrization, and we finally recover the definition of the propagator of a Dirac particle:

$$\langle 0| j^\mu(x) |0\rangle^A = -e \lim_{x' \to x} \gamma^\mu \langle 0| T\, \psi(x')\bar\psi(x) |0\rangle^A$$
$$= ie\,\mathrm{tr}[\gamma^\mu\, G(x, x|A)], \tag{2.4}$$

where the limit has to be taken symmetrically with respect to the time coordinate, and T denotes the time-ordering operator. The propagator G satisfies the Green's function equation of a Dirac particle:

$$[(\gamma^\mu \Pi_\mu) + m]\, G(x, x'|A) = \delta(x - x'), \tag{2.5}$$

with $\Pi_\mu = -i\partial_\mu - eA_\mu$. Equation (2.5) can be interpreted as the representation of an operator equation in configuration space ($G(x, x'|A) = \langle x|G[A]|x'\rangle$):

$$(\gamma\Pi + m)\, G[A] = 1. \tag{2.6}$$

We introduce the proper-time representation of the formal solution of the operator equation (2.6),

$$G[A] = \frac{1}{\gamma\Pi + m} = \frac{\gamma\Pi - m}{(\gamma\Pi)^2 - m^2} = (m - \gamma\Pi)\,i\int_0^\infty ds\, e^{-is[m^2 - (\gamma\Pi)^2]}. \tag{2.7}$$

Obviously, the proper-time representation simply expresses an inverse operator in terms of a convenient exponential function. Convergence of the proper-time integral at infinity is ensured by the implicit prescription $m^2 \to m^2 - i\epsilon$.

---

[1] Spinor indices will be suppressed in the calculation.

We are interested in the proper-time representation of the effective action $W^{(1)}$. Hence, we have to solve the equation (cf. (2.2) and (2.4))

$$i\frac{\delta W^{(1)}[A]}{\delta A_\mu(x)} = -e\,\mathrm{tr}_\gamma[\gamma^\mu\,G(x,x|A)] \tag{2.8}$$

for $W^{(1)}$. We now show that the ansatz

$$iW^{(1)} \equiv i\int \mathrm{d}^4x\,\mathcal{L}^{(1)} = -\frac{1}{2}\int\limits_0^\infty \frac{\mathrm{d}s}{s}\,\mathrm{e}^{-ism^2}\,\mathrm{tr}_{x,\gamma}\left[\mathrm{e}^{is(\gamma\Pi)^2}\right] \tag{2.9}$$

fulfills (2.8) and thereby gives $W^{(1)}$ to within a constant. This constant has to be chosen so that the action satisfies the boundary condition that it vanishes for vanishing external field. For the calculation of the functional derivative of (2.9), we note that $\delta\Pi_\alpha(x) = -e\,\delta A_\alpha(x)$:

$$i\frac{\delta W^{(1)}[A]}{\delta A_\alpha(x)} = \frac{e}{2}\int\limits_0^\infty \frac{\mathrm{d}s}{s}\,\mathrm{e}^{-ism^2}\,\mathrm{tr}\int \mathrm{d}^4y\,is\,\gamma_\mu\gamma_\nu\,\frac{\delta[\Pi^\mu(y)\Pi^\nu(y)]}{\delta\Pi_\alpha(x)}\,\mathrm{e}^{is\,\gamma_\mu\gamma_\nu\,\Pi^\mu\,\Pi^\nu}$$

$$= ie\int\limits_0^\infty \mathrm{d}s\,\mathrm{e}^{-ism^2}\,\mathrm{tr}\left[\gamma^\alpha\,\langle x|\gamma\Pi\,\mathrm{e}^{is(\gamma\Pi)^2}|x\rangle\right]$$

$$= -e\,\mathrm{tr}\left[\gamma^\alpha\,\langle x|(m-\gamma\Pi)\,i\int\limits_0^\infty \mathrm{d}s\,\mathrm{e}^{-is[m^2-(\gamma\Pi)^2]}|x\rangle\right]$$

$$\overset{(2.7)}{=} -e\,\mathrm{tr}[\gamma^\alpha\,G(x,x|A)],$$

where in the third line we have made use of the fact that the trace of an odd number of γ's vanishes.

In this way, we have demonstrated the validity of (2.9), and so we can write for the unrenormalized Lagrangian

$$\mathcal{L}^{(1)} = \frac{i}{2}\mathrm{tr}_\gamma\int\limits_0^\infty \frac{\mathrm{d}s}{s}\,\mathrm{e}^{-ism^2}\,\langle x|\mathrm{e}^{is(\gamma\Pi)^2}|x\rangle. \tag{2.10}$$

For the propagator $G(x,x'|A)$, as well as for the effective Lagrangian, we need to evaluate the object

$$K(x,x';s|A) = \langle x|\mathrm{e}^{is(\gamma\Pi)^2}|x'\rangle. \tag{2.11}$$

For the remainder of this section, we shall mainly be concerned with the various methods of finding an explicit representation in configuration or momentum space.

The first method that we shall discuss was proposed by Schwinger. It is based on the interpretation of $K(x,x';s|A)$ as the coordinate representation of the proper-time "evolution operator"

$$U(s) = e^{-iHs}, \tag{2.12}$$

where we have introduced the "Hamiltonian"

$$H = -(\gamma \Pi)^2 = \Pi^2 - \frac{e}{2}\sigma_{\mu\nu}F^{\mu\nu}, \qquad \Pi_\mu = p_\mu - eA_\mu, \tag{2.13}$$

and $\sigma_{\mu\nu} = i/2\,[\gamma_\mu, \gamma_\nu]$. In this sense, the desired object $K(x, x'; s|A)$ is equal to a transformation amplitude $\langle x|\,U(s)\,|x'\rangle = \langle x, s|x', 0\rangle$, while the evolution operator satisfies a "Schrödinger equation"

$$i\partial_s U(s) = H U(s). \tag{2.14}$$

To obtain the transformation amplitude, we have to solve the dynamical problem formulated by the following one-particle equations of (proper-time) motion of Heisenberg type:

$$\frac{\mathrm{d}x_\mu}{\mathrm{d}s} = -i[x_\mu, H] = 2\Pi_\mu,$$

$$\frac{\mathrm{d}\Pi_\mu}{\mathrm{d}s} = -i[\Pi_\mu, H] = e\left(F_{\mu\nu}\Pi^\nu + \Pi^\nu F_{\mu\nu}\right) + \frac{e}{2}\sigma_{\lambda\nu}\,\partial_\mu F^{\lambda\nu}$$

$$= 2e\,F_{\mu\nu}\Pi^\nu - ie\,\partial^\nu F_{\mu\nu} + \frac{e}{2}\sigma_{\lambda\nu}\,\partial_\mu F^{\lambda\nu}. \tag{2.15}$$

Our strategy is as follows: first, solve the (operator) equations of motion (2.15) for $\Pi(s)$ as a function of $x(s)$ and $x(0)$; secondly, insert these findings into the Hamiltonian; and thirdly, solve the Schrödinger equation (2.14) in the coordinate representation, in which the Hamiltonian is now diagonal (for a pedagogical review, see e.g. [106]).

Of course, this procedure can only be performed analytically for a couple of simple external-field configurations, e.g. for constant and plane wave fields [148]. In Sect. 2.3, a detailed example of such a type of calculation is presented.

The second method to be discussed is based on the classical WKB approximation (see e.g. [147]). Here, we want to compute the transformation amplitude by means of

$$\langle x', s|x'', 0\rangle = (2\pi i)^{-d/2}\sqrt{D}\,e^{iS}, \tag{2.16}$$

where $d = 4$ is the dimension and $\hbar = 1$, and the Van Vleck determinant is

$$D(x', x''; s) \stackrel{d=4}{=} (-1)^4 \det\left(\frac{\partial^2 S}{\partial x'^\mu \partial x''^\nu}\right); \tag{2.17}$$

S denotes the classical action. For the constant-field case, this was treated in [57], and also in [97] and [155] using the heat kernel method in QED (with not necessarily constant fields). As a side remark, we note that the transformation amplitude can also be expressed by means of the path integral representation:

$$\langle x', s|x'', 0\rangle = K(x', x''; s) = \int\limits_{x(0)=x'', x(s)=x'} \mathcal{D}x(\lambda)e^{iS[x(\lambda)]}, \tag{2.18}$$

where

$$S = \int_0^s d\lambda\, L[x(\lambda), \dot{x}(\lambda)]. \tag{2.19}$$

The Lagrangian associated with the Hamiltonian of (2.13) is given by

$$L = \frac{1}{4}\dot{x}_\mu \dot{x}^\mu + eA^\mu \dot{x}_\mu + \frac{e}{2}\sigma^{\mu\nu} F_{\mu\nu}. \tag{2.20}$$

Now, we have to calculate the classical action (2.19). This can be achieved by first solving the equations of motion (2.15), which, for constant fields, reads $(\Pi = \dot{x}/2 = p - eA)$

$$\ddot{x}(\lambda) = 2eF\dot{x}(\lambda), \tag{2.21}$$

where we have used matrix notation, i.e. $F\dot{x} \equiv F^\mu{}_\nu \dot{x}^\nu$. This equation can be solved with the ansatz

$$\dot{x}(\lambda) = e^{2eF\lambda}\dot{x}(0), \tag{2.22}$$

and a further integration, together with the initial conditions $x(\lambda = 0) = x''$, $x(\lambda = s) = x'$, yields

$$\dot{x}(0) = \frac{1}{e^{2eFs} - 1}\, 2eF(x' - x''), \quad x(\lambda) - x(0) = \frac{e^{2eF\lambda} - 1}{e^{2eFs} - 1}(x' - x''). \tag{2.23}$$

This information is sufficient to compute the classical action, which turns out to be

$$S(x', x''; s) = e \int_{x''}^{x'} d\xi_\mu\, A^\mu(\xi)$$

$$+\frac{1}{4}(x' - x'')^\alpha eF_\alpha{}^\beta \left[\coth(eFs)\right]_\beta{}^\gamma (x' - x'')_\gamma + \frac{e}{2}\sigma_{\mu\nu} F^{\mu\nu} s.$$

The last step on the way to evaluating (2.16) involves two partial derivatives of S with respect to x' and x'' (cf. (2.17)), which at last brings us to the desired quantity,

$$\sqrt{D} = \frac{i}{4s^2} \exp\left(-\frac{1}{2}\operatorname{tr}\ln\frac{\sinh eFs}{eFs}\right).$$

According to (2.16), the complete transformation amplitude for constant fields is given by the result

$$\langle x', s | x'', 0 \rangle = -\frac{i}{16\pi^2} \frac{1}{s^2} \exp\left(-\frac{1}{2}\operatorname{tr}\ln\frac{\sinh eFs}{eFs}\right) \exp\left(ie \int_{x''}^{x'} d\xi_\mu A^\mu(\xi)\right)$$

$$\times \exp\left[\frac{i}{4}(x' - x'')(eF \coth eFs)(x' - x'')\right] e^{i(e/2)\sigma Fs}, \tag{2.24}$$

where the entire gauge dependence has been isolated in the form of the holonomy factor (counter-gauge factor) $\exp\left(ie \int_{x''}^{x'} d\xi_\mu A^\mu(\xi)\right)$, and the integration is along the straight path connecting x' and x''.

A useful result is given by (cf. (2.18))

$$\int_{x(0)=x'', x(s)=x'} \mathcal{D}x(\lambda) \exp\left[i \int_0^s d\lambda \left(\frac{1}{4}\dot{x}_\mu \dot{x}^\mu + eA^\mu \dot{x}_\mu\right)\right]$$

$$= e^{ie \int_{x''}^{x'} d\xi_\mu A''(\xi)} \frac{(-i)}{16\pi^2} \frac{1}{s^2} \exp\left(-\frac{1}{2}\text{tr} \ln \frac{\sinh eFs}{eFs}\right)$$

$$\times \exp\left[\frac{i}{4}(x' - x'')(eF \coth eFs)(x' - x'')\right]. \tag{2.25}$$

Finally, we shall present a third method to evaluate the transformation amplitude $K(x, x'; s|A)$ that, on the one hand, represents a more pedestrian way of solving the problem but, on the other hand, supplies us with the Fourier transform of $K(x, x'; s|A)$. The latter turns out to be extremely useful for finite-temperature calculations.

In the following, we confine ourselves again to constant-field configurations where $F^{\mu\nu} = \text{const.}$ and work in the Schwinger–Fock gauge

$$A^\mu_{\text{SF}} = -\frac{1}{2}F^{\mu\nu}(x - x')_\nu. \tag{2.26}$$

First, we move one step backwards and consider the object

$$\Delta(x, x'|A) = i \int_0^\infty ds\, e^{-im^2 s} K(x, x'; s|A), \tag{2.27}$$

which is contained in the propagator of a Dirac particle (in coordinate representation) (2.7). From the Green's function equation (2.5) for the Dirac particle, we can easily read off the corresponding Green's function equation for $\Delta(x, x'|A_{\text{SF}})$ in the Schwinger–Fock gauge:

$$\delta(x - x') \overset{(2.5)}{=} (\gamma \Pi + m) G(x, x'|A_{\text{SF}}) \overset{(2.7)}{=} [m^2 - (\gamma \Pi)^2] \Delta(x, x'|A_{\text{SF}})$$

$$= \left[m^2 - \partial^2 - \frac{e^2}{4}(x - x')_\mu (F^{\mu\lambda} F^\nu_\lambda)(x - x')_\nu \right. \tag{2.28}$$

$$\left. - \frac{e}{2}\sigma F + \frac{ie}{2}F^{\mu\nu}(x_\mu \partial_\nu - x_\nu \partial_\mu)\right] \Delta(x, x'|A_{\text{SF}}).$$

Translational invariance reduces the coordinate dependence of Δ: $\Delta = \Delta(x - x'|A_{\text{SF}})$. The differential equation (2.28) is further simplified by observing that any term except for the last one in the equation is invariant under (generalized) Lorentz rotations. Since the differential operator acting on $\Delta(x - x'|A_{\text{SF}})$ contains the generator of these rotations in this last term, namely $(x_\mu \partial_\nu - x_\nu \partial_\mu)$, $\Delta(x - x'|A_{SF})$ must also be invariant. We conclude that

$$\frac{ie}{2} F^{\mu\nu} (x_\mu \partial_\nu - x_\nu \partial_\mu) \, \Delta(x - x'|A_{SF}) = 0. \tag{2.29}$$

This leads us to the problem of solving the Green's function equation

$$\left[-\partial^2 + m^2 - \frac{e}{2}\sigma_{\mu\nu}F^{\mu\nu} - \frac{e^2}{4}(x-x')_\mu F^{\mu\lambda}F^\nu_\lambda(x-x')_\nu \right] \Delta(x - x'|A_{SF})$$
$$= \delta(x-x'). \tag{2.30}$$

It is helpful to study this equation in Fourier space by introducing the momentum description

$$\Delta(x - x'|A_{SF}) = \int \frac{d^4p}{(2\pi)^4} \, e^{ip(x-x')} \, \Delta(p|A_{SF}). \tag{2.31}$$

The differential equation (2.30) can then be written as

$$\left(p^2 + \kappa^2 + \frac{e^2}{4} \frac{\partial}{\partial p^\mu} F^{\mu\lambda} F^\nu_\lambda \frac{\partial}{\partial p^\nu} \right) \Delta(p|A_{SF}) = 1, \tag{2.32}$$

where we have used the short form $\kappa^2 = m^2 - (e/2)\sigma F$.

The solution is given by [55]

$$\Delta(p|A_{SF}) = i \int_0^\infty ds \, e^{-i\kappa^2 s} e^{-M(is)} = \int_0^{i\infty} ds \, e^{-\kappa^2 s} e^{-M(s)}, \tag{2.33}$$

where, in the last step, we have performed the substitution $is \to s$ (no contour rotation!), and $M(s)$ can be decomposed according to

$$M(s) = p_\alpha X^{\alpha\beta}(s) p_\beta + Y(s). \tag{2.34}$$

The quantities X and Y depend additionally on the field strength and are given by

$$Y(s) = \frac{1}{2}\text{tr}\,\ln[\cos(eFs)], \tag{2.35}$$

$$X(s) = \frac{\tan(eFs)}{eF}, \tag{2.36}$$

where we have used matrix notation, e.g. $F^{\mu\nu} \equiv (F)^{\mu\nu}$.

From the definition of $\Delta(x - x'|A_{SF})$ ((2.27) and (2.31)), we can read off the expression for the transformation amplitude $K(x - x'; s|A_{SF})$:

$$K(x - x'; s|A_{SF}) = \int \frac{d^4p}{(2\pi)^4} e^{ip(x-x')} e^{i(e/2)\sigma Fs} e^{-Y(is)} e^{-pX(is)p}. \tag{2.37}$$

For the purpose of obtaining the Lagrangian, we are especially interested in the coincidence limit $x' \to x$. We obtain

$$K(s|A_{\rm SF}) \equiv K(0;s|A_{\rm SF})$$

$$= \int \frac{{\rm d}^4 p}{(2\pi)^4}\, {\rm e}^{{\rm i}(e/2)\sigma Fs}\, {\rm e}^{-Y({\rm i}s)}\, {\rm e}^{-pX({\rm i}s)p} \tag{2.38}$$

$$= \frac{{\rm i}}{(4\pi)^2}\, {\rm e}^{{\rm i}(e/2)\sigma Fs}\, {\rm e}^{-Y({\rm i}s)}\, [\det X({\rm i}s)]^{-1/2}$$

$$\overset{(2.35)}{=} \frac{{\rm i}}{(4\pi)^2}\, {\rm e}^{{\rm i}(e/2)\sigma Fs}\, \exp\left[-\frac{1}{2}{\rm tr}\,\ln\cosh(eFs)\right]$$

$$\times \frac{1}{({\rm i}s)^2}\, \exp\left(-\frac{1}{2}{\rm tr}\,\ln\frac{X}{{\rm i}s}\right)$$

$$\overset{(2.36)}{=} -\frac{{\rm i}}{(4\pi s)^2}\, {\rm e}^{{\rm i}(e/2)\sigma Fs}\, \exp\left(-\frac{1}{2}{\rm tr}\,\ln\frac{\sinh(eFs)}{eFs}\right) \tag{2.39}$$

$$\equiv \langle x|{\rm e}^{{\rm i}(\gamma\Pi)^2 s}|x\rangle,$$

which is the result we obtained in (2.24) for $x' = x'' = x$. To complete our study of the proper-time transition amplitude, the question of gauge dependence has to be discussed. Again, we start with the Green's function equation (2.5) in coordinate space for an arbitrary choice of gauge:

$$\left[\gamma(-{\rm i}\partial - eA) + m\right] G(x,x'|A) = \delta(x - x'). \tag{2.40}$$

Replacing A_μ by a different gauge $A'_\mu = A_\mu - \partial_\mu\Lambda$ (where $\Lambda(x)$ is continuously differentiable), this equation reads

$$\delta(x - x') = \left[\gamma(-{\rm i}\partial - eA + e\partial\Lambda) + m\right] G(x,x'|A - \partial\Lambda)$$

$$= {\rm e}^{-{\rm i}e\Lambda(x)}\left[\gamma(-{\rm i}\partial - eA) + m\right]{\rm e}^{{\rm i}e\Lambda(x)}G(x,x'|A - \partial\Lambda)$$

$$= {\rm e}^{-{\rm i}e\Lambda(x')}\left[\gamma(-{\rm i}\partial - eA) + m\right]{\rm e}^{{\rm i}e\Lambda(x)}G(x,x'|A - \partial\Lambda)$$

$$= \left[\gamma(-{\rm i}\partial - eA) + m\right]{\rm e}^{{\rm i}e\left[\Lambda(x)-\Lambda(x')\right]}G(x,x'|A - \partial\Lambda), \tag{2.41}$$

where, in the third line, we have made use of the properties of the δ function. Comparing (2.41) with (2.40), we can identify the gauge transformation property of the Green's function:

$$G(x,x'|A - \partial\Lambda) = {\rm e}^{-{\rm i}e\left[\Lambda(x)-\Lambda(x')\right]}G(x,x'|A). \tag{2.42}$$

Since $\Lambda(x)$ can be viewed as a "potential" for the difference between two equivalent choices of gauge, we may write for (2.42)

$$G(x,x'|A') = {\rm e}^{{\rm i}e\int_{x'}^{x}{\rm d}\xi_\mu\left[A'^\mu(\xi)-A^\mu(\xi)\right]}G(x,x'|A). \tag{2.43}$$

Since $\Delta(x,x'|A)$ and $K(x,x';s|A)$ are linearly related to $G(x,x'|A)$, they behave equivalently under gauge transformations.

Taking these considerations into account, the proper-time transition amplitude in an arbitrary gauge can be obtained from that in the Schwinger–Fock gauge according to

$$K(x, x'; s|A) = \Phi(x, x'|A) \, K(x, x'; s|A_{SF}),\tag{2.44}$$

where we have introduced

$$\Phi(x, x'|A) = \exp\left[ie \int_{x'}^{x} d\xi_\mu \left(A^\mu(\xi) + \frac{1}{2}F^{\mu\nu}(\xi - x')_\nu\right)\right].\tag{2.45}$$

Since $\Phi(x, x'|A)$ represents a mapping from a path in coordinate space onto the gauge group $U(1)$, it is called the *holonomy factor* (or simply the holonomy). Note that $\Phi(x, x'|A)$ is invariant under continuous deformations of the integration path, because the integrand is curl-free.[2] In the coincidence limit $x' \to x$, the holonomy factor simply reduces to 1, which explicitly demonstrates the gauge invariance of the effective action.

This completes our search for an appropriate expression for the transformation amplitude $K(x - x'; s|A)$ or its coincidence limit $K(s|A)$. In (2.39), we have rediscovered the findings of Schwinger [148]. The results of (2.37) and (2.39) are sufficient to finally evaluate the propagator of a Dirac particle and the effective Lagrangian according to the formulas (2.7) and (2.10).

We conclude this section by stating the final expressions for the desired objects. The Green's function for a Dirac particle (the electron propagator) in a constant external electromagnetic field (in an arbitrary gauge) reads

$$G(x, x'|A) = \Phi(x, x'|A)\frac{1}{(4\pi)^2}\int_0^\infty \frac{ds}{s^2}\left[m - \frac{1}{2}\gamma^\mu[f(s) + eF]_{\mu\nu}(x - x')^\nu\right]$$

$$\times e^{-im^2 s - L(s)}\exp\left[\frac{i}{4}(x - x')f(s)(x - x')\right]e^{i(e/2)\sigma F s},\tag{2.46}$$

where we have used the abbreviations

$$f(s) = eF\coth(eFs) \quad\text{and}\quad L(s) = \frac{1}{2}\text{tr}\ln\frac{\sinh(eFs)}{eFs},\tag{2.47}$$

and

$$\Phi(x, x'|A) = e^{ie\int_{x'}^{x} d\xi_\mu[A^\mu(\xi) + (1/2)F^{\mu\nu}(\xi - x')_\nu]}\tag{2.48}$$

carries completely the gauge dependence of the propagator.

According to (2.10) and (2.39), the one-loop effective Lagrangian for constant external fields can be written as

---

[2] These statements are not strictly tied to the use of the Schwinger–Fock gauge.

$$\mathcal{L}^{(1)} = \frac{1}{8\pi^2} \int\limits_0^\infty \frac{ds}{s^3} e^{-im^2 s} \left\{ (es)^2 |\mathcal{G}| \cot\left[es(\sqrt{\mathcal{F}^2 + \mathcal{G}^2} + \mathcal{F})^{1/2}\right] \right. \tag{2.49}$$

$$\left. \times \coth\left[es(\sqrt{\mathcal{F}^2 + \mathcal{G}^2} - \mathcal{F})^{1/2}\right] + \frac{2}{3}(es)^2 \mathcal{F} - 1 \right\},$$

where we have introduced the gauge and Lorentz invariants of the electromagnetic field,

$$\mathcal{F} = \frac{1}{4} F_{\mu\nu} F^{\mu\nu}, \tag{2.50}$$

$$\mathcal{G} = \frac{1}{4} F_{\mu\nu} {}^\star F^{\mu\nu}, \qquad {}^\star F^{\mu\nu} = \frac{1}{2} \epsilon^{\mu\nu\lambda\sigma} F_{\lambda\sigma}. \tag{2.51}$$

The last two terms in the curly brackets of (2.49) stem from a field strength and charge renormalization.

An expansion of the integral for small values of e corresponds to a weak-field approximation $F^{\mu\nu} \ll m^2$ and yields

$$\mathcal{L}^{(1)} = \frac{8}{45} \frac{\alpha^2}{m^4} \mathcal{F}^2 + \frac{14}{45} \frac{\alpha^2}{m^4} \mathcal{G}^2 \cdots$$

$$= \frac{2\alpha^2}{45 m^4} \left[(\boldsymbol{E}^2 - \boldsymbol{B}^2)^2 + 7(\boldsymbol{E} \cdot \boldsymbol{B})^2 \right]. \tag{2.52}$$

Euler, Kockel and Heisenberg [75, 98] and, independently, Weisskopf [164] were the first to discuss effective Lagrangians of this type.

Equation (2.52) can be interpreted as the first nonlinear correction term to the Maxwell Lagrangian, which defines the linear theory of classical electrodynamics.

This concludes our introductory remarks about the proper-time method, which turns out to be a powerful tool with which to analyze the underlying (fundamental) gauge field theory of quantum electrodynamics.

2.2 Polarization Tensor in External Fields

We derive a new representation of the one-loop photon polarization tensor coupling to all orders to an arbitrary constant electromagnetic field, fully maintaining the dependence on the complete set of invariants. The problem of extensive Dirac algebraic computations is avoided by constructing a one-to-one mapping from the special case of parallel electric and magnetic fields, as treated by Urrutia, to the general case. The polarization tensor is decomposed into its eigenvalues and eigenspace projectors. We work out explicitly certain limiting cases, e.g. purely magnetic fields, crossed fields and arbitrary weak electromagnetic fields. Finally, the modification of light propagation in external fields is discussed.

The polarization tensor $\Pi^{\mu\nu}(k)$, also known as the photon self-energy, can be viewed as a modification of the (classically trivial) photon structure by virtual processes of vacuum polarization. To one-loop order, it is an electron–positron pair which comes into existence for a certain time and annihilates itself during the propagation of the photon.

On a heuristic level, the physical consequences of these vacuum polarization effects can be illustrated by an intuitive picture: the virtual transitions, e.g. the electron–positron loop, confer the properties of the participant particles on the photon. By this means, the photon can therefore acquire, for instance, a "size" (of the order of the Compton wavelength of the participant particle) or a "charge distribution".

In the *trivial* vacuum, i.e. in the absence of external perturbations, these internal photon properties are hidden; the photon propagates with the speed of light on the *trivial light cone*, which is defined by $k^2 = -(k^0)^2 + (\boldsymbol{k})^2 = 0$, where k^μ denotes the wave vector of a photon of arbitrary frequency.

This situation changes substantially in the presence of various external modifications of the vacuum, even if an interaction with the propagating light is classically forbidden. These classically forbidden processes are, of course, the most interesting ones, and will be elucidated in this and the following sections. Within our heuristic picture, the vacuum modifications can interact with those properties of the photon which are caused by the virtual particles.

In the present section, we consider constant, but otherwise arbitrary, electromagnetic fields as the external perturbation. The study of the polarization tensor $\Pi^{\mu\nu}(k|A)$ will therefore serve as an investigation of the properties of the QED vacuum with external fields. The derivation of this kind of polarization tensor has a rather long history. The properties of the tensor in a constant magnetic field have been studied by many authors [14,45,47,124,152,159]; focusing on light propagation, a comprehensive investigation was performed by Tsai and Erber [160]. Another special case, the crossed-fields configuration, has been investigated by Narozhnyi [126] and Ritus [139]. However, a generalization to arbitrary constant electromagnetic fields in a straightforward manner is associated with a substantial increase in calculational difficulties; detailed results like those for the purely magnetic case have still not been worked out. Although Batalin and Shabad had already obtained a representation in proper-time form for the polarization tensor in an electromagnetic background in 1971 [22], their extensive result was only brought into a practical form almost 20 years later [11]. A calculation employing special operator techniques has been performed by Baier et al. [15]. Within the world-line formalism, the polarization tensor has recently been calculated by Schubert [146] with comparably little computational effort.

In the following, we shall derive the polarization tensor in an arbitrary constant electromagnetic field in a simplified manner; our approach is based on the findings of Urrutia [162], who solved the problem for the special case of parallel electric and magnetic fields. We shall demonstrate that the general case can be deduced from this particular one by an appropriate Lorentz transformation. Applications will be described afterwards.

2.2.1 Derivation of $\Pi^{\mu\nu}$

To begin with, we shall outline the direct calculation of $\Pi^{\mu\nu}$ and point out where one encounters technical difficulties. According to the usual rules of perturbation theory, the polarization tensor can be calculated from

$$\Pi_{\mu\nu}(k|A) = -\mathrm{i}e^2 \operatorname{tr}_\gamma \int \frac{\mathrm{d}^4 p}{(2\pi)^4} \gamma_\mu\, g(p)\, \gamma_\nu\, g(p-k), \qquad (2.53)$$

where we have chosen to work in momentum space. Therefore, $g(p)$ denotes the Fourier transform of the Dirac fermion propagator, which, according to (2.7) and (2.27), can be written generally as

$$G(x, x'|A') = \phi(x, x'|A' - A)\, (m - \not{I}\!\!\not{I})\, \Delta(x, x'|A). \qquad (2.54)$$

In the following, we shall use exclusively the Schwinger–Fock gauge for constant fields, $A_{\mathrm{SF}}^\mu = -(1/2)F^{\mu\nu}(x-x')_\nu$, which allows us to express the gauge potential in terms of the field strength as well as to forget about the holonomy factor $\phi(x, x'|A'-A)$. Introducing the Fourier representation $\Delta(p|A_{\mathrm{SF}})$ of $\Delta(x, x'|A_{\mathrm{SF}})$, we can easily derive the Fourier representation of $G(x, x'|A_{\mathrm{SF}})$:

$$G(x, x'|A_{\mathrm{SF}}) = \int \frac{\mathrm{d}^4 p}{(2\pi)^4}\, \mathrm{e}^{\mathrm{i}p(x-x')} \left[m - \gamma_\alpha \left(p^\alpha + \frac{\mathrm{i}e}{2} F^{\alpha\kappa} \frac{\partial}{\partial p^\kappa} \right) \right] \Delta(p|A_{\mathrm{SF}})$$

$$= \int \frac{\mathrm{d}^4 p}{(2\pi)^4}\, \mathrm{e}^{\mathrm{i}p(x-x')}\, g(p). \qquad (2.55)$$

In (2.33)–(2.36), we obtained the explicit form of $\Delta(p|A_{\mathrm{SF}})$; inserting these results into $g(p)$ and performing the momentum differentiation leads us to

$$g(p) = \mathrm{i} \int_0^\infty \mathrm{d}s\, \mathrm{e}^{-\mathrm{i}m^2 s} \mathrm{e}^{-Y(\mathrm{i}s)} \left[m - \gamma_\alpha \left(p - \mathrm{i}e\mathsf{F}\mathsf{X}p \right)^\alpha \right] \mathrm{e}^{-p\mathsf{X}p} \mathrm{e}^{\mathrm{i}(e/2)s\sigma F}, (2.56)$$

where we have employed matrix notation, and the quantities Y and X are defined in (2.35) and (2.36). After insertion of (2.56) into (2.53), the remaining problems are some Gaussian integrations and various Dirac traces; in fact, the Dirac traces can be looked up in Appendix C ((C.25)–(C.31)); these obviously demonstrate that the direct computation will be more than tedious. But the main problem is the reorganization of the terms at the end of the calculation, especially the simplification of the trigonometric and hyperbolic functions. The latter would even prohibit successful use of algebraic-manipulation programs, at least for higher-order calculations.

As already mentioned, the calculation becomes enormously simplified for a selected class of field configurations, such as the case of parallel magnetic and electric fields. Since space–time can effectively be decomposed into a transverse and a longitudinal subspace with respect to the field axis, the four-dimensional tensor analysis reduces to two-dimensional. If we choose the field to point along the z axis, without loss of generality, a vector decomposes according to

$$k^\mu = k_\parallel^\mu + k_\perp^\mu, \qquad k_\parallel^\mu = (k^0, 0, 0, k^3), \qquad k_\perp^\mu = (0, k^1, k^2, 0). \qquad (2.57)$$

In the same manner, tensors can be decomposed, e.g. $g^{\mu\nu} = g_\parallel^{\mu\nu} + g_\perp^{\mu\nu}$. With respect to each subspace, we easily find the unique vector orthogonal to a given vector:

$$\tilde{k}_\parallel^\mu = (k^3, 0, 0, k^0), \qquad\qquad \tilde{k}_\perp^\mu = (0, k^2, -k^1, 0). \qquad (2.58)$$

The polarization tensor for this special field configuration has been calculated in [162] in the way outlined above. We simply state the result here:

$$\Pi^{\mu\nu}(k|A) = \frac{\alpha}{2\pi} \int\limits_0^\infty \frac{ds}{s} \int\limits_{-1}^1 \frac{d\nu}{2} \Bigg\{ e^{-is\phi_0} \frac{zz'}{\sin z \sinh z'}$$

$$\left[\left(g^{\mu\nu}k^2 - k^\mu k^\nu\right)N_0 + \left(g_\parallel^{\mu\nu}k_\parallel^2 - k_\parallel^\mu k_\parallel^\nu\right)N_1 \right. \qquad (2.59)$$

$$\left. + \left(g_\perp^{\mu\nu}k_\perp^2 - k_\perp^\mu k_\perp^\nu\right)N_2 - \left(\tilde{k}_\perp^\mu \tilde{k}_\parallel^\nu + \tilde{k}_\parallel^\mu \tilde{k}_\perp^\nu\right)N_3 \right] + \text{c.t.} \Bigg\},$$

where "c.t." means "contact term". The electric and magnetic field strengths E, B are contained in the variables $z = eBs$ and $z' = eEs$. The exponent ϕ_0 is given by[3]

$$\phi_0 = m^2 + \frac{k_\parallel^2}{2}\frac{\cosh z' - \cosh \nu z'}{z' \sinh z'} + \frac{k_\perp^2}{2}\frac{\cos \nu z - \cos z}{z \sin z}. \qquad (2.60)$$

The functions N_i are

$$N_0 = \cosh \nu z' \cos \nu z - \sinh \nu z' \sin \nu z \cot z \coth z',$$

$$N_1 = 2 \cos z \frac{\cosh z' - \cosh \nu z'}{\sinh^2 z'} - N_0 = \tilde{N}_1 - N_0,$$

$$N_2 = 2 \cosh z' \frac{\cos \nu z - \cos z}{\sin^2 z} - N_0 = \tilde{N}_2 - N_0, \qquad (2.61)$$

$$N_3 = \frac{1 - \cos z \cos \nu z}{\sin z} \frac{\cosh \nu z' \cosh z' - 1}{\sinh z'} - \sin \nu z \sinh \nu z',$$

where we have also defined the functions $\tilde{N}_{1,2}$ for later use. The contact term is, finally, determined by the requirement that $\Pi^{\mu\nu}$ should vanish in the limit of vanishing fields and for wave vectors on the mass shell (light cone), $k^2 = 0$. This corresponds to a charge and field strength renormalization and ensures that the real photon will remain massless. We obtain

$$\text{c.t.} = -e^{-im^2 s}(1 - \nu^2)\left(g^{\mu\nu}k^2 - k^\mu k^\nu\right). \qquad (2.62)$$

Now, our claim is that the Lorentz-invariant form of the polarization tensor for an arbitrary constant electromagnetic field can be completely reconstructed from the special form given above for parallel electric and magnetic

---

[3] This formula has been misprinted in [162].

fields. The situation should be familiar from the simpler case of a problem where it is known that the solution depends on the field invariants only. In that case it is sufficient to solve the problem for the special case of parallel E and B fields, i.e. to go to a Lorentz frame in which $\boldsymbol{E}\| \pm \boldsymbol{B}$, and reconstruct the invariants later on according to $\mathcal{F} = (1/2)\left(B^2 - E^2\right)$ and $\mathcal{G} = -EB$. This procedure leads to the correct results, since there is a one-to-one correspondence between the two sets of linear independent variables $E, B \to \mathcal{F}, \mathcal{G}$.

Returning to the present problem, we have to establish such a one-to-one correspondence between the dynamical variables contained in (2.59) and a complete set of linearly independent Lorentz and gauge invariants, as well as a one-to-one correspondence between the tensor structure of (2.59) and a complete tensor basis for the polarization tensor in arbitrary fields.

Let us begin with the scalar quantities. In the present problem, the basic building blocks for the construction of gauge- and Lorentz-invariant scalars or pseudoscalars are given by the field strength tensors $F^{\mu\nu}$ and $^\star F^{\mu\nu}$ and the wave vector k^μ. The number of linearly independent invariants is four, as can be proved by a simple group-theoretical argument (cf. Sect. 3.5). Here it is convenient to choose the secular invariants a, b as defined in Appendix B for the purely field-dependent invariants, since they reduce to $a \to B$ and $b \to -E$ in the case of antiparallel electric and magnetic fields. Note that we shall assume $E < 0$ in the following without loss of generality to obtain a simple invariant structure (cf. Appendix B). A convenient choice for the remaining invariants is given by

$$z_k = \left(k_\alpha F^{\alpha\kappa}\right)\left(k_\beta F^\beta{}_\kappa\right), \tag{2.63}$$
$$k^2 = k^\mu k_\mu.$$

In fact, there are exactly four dynamical variables which $\Pi^{\mu\nu}$ in (2.59) depends on: E, B, $k_\|^2$ and $k_\perp^2$. The map from these variables onto the set of invariants is simply given by

$$
\begin{aligned}
a &= B, & b &= -E, \\
z_k &= -E^2 k_\|^2 + B^2 k_\perp^2, & k^2 &= k_\|^2 + k_\perp^2,
\end{aligned}
\tag{2.64}
$$

which is valid in a system where $\boldsymbol{B}\|-\boldsymbol{E}$. The inverse map is obtained by a simple calculation; the nontrivial relations are

$$
k_\|^2 = \frac{a^2 k^2 - z_k}{a^2 + b^2}, \qquad k_\perp^2 = \frac{b^2 k^2 + z_k}{a^2 + b^2}. \tag{2.65}
$$

Hence, the one-to-one correspondence between the scalar quantities is established.

Let us turn to the tensor structure of $\Pi^{\mu\nu}$. There are exactly four linearly independent vectors which we can construct from k^μ and $F^{\mu\nu}$; we shall employ the choice

$$
k^\mu, \quad Fk^\mu \equiv F^{\mu\alpha}k_\alpha, \quad F^2 k^\mu \equiv F^{\mu\alpha}F_{\alpha\beta}k^\beta, \quad {}^\star F k^\mu \equiv {}^\star F^{\mu\alpha}k_\alpha. \tag{2.66}
$$

(Note that vectors containing higher powers of the field strength tensors can be decomposed with the aid of the fundamental algebraic relations (B.3), (B.4).) Since $\Pi^{\mu\nu}$ is a symmetric second-rank tensor by construction, it can only possess ten independent components; indeed, we can only construct ten independent symmetric Lorentz second-rank tensors from the four vectors given above.

Next, we take advantage of Furry's theorem, which in this case tells us that the polarization tensor can only contain an even number of powers of $F^{\mu\nu}$ in each term. This is because, at each external line, a field strength tensor is attached to the loop, and there is only an even number of external lines. Since the invariants, as chosen above for example, only involve even numbers of $F^{\mu\nu}$ (note that a and b are defined via $F^{\mu\nu}F_{\mu\nu}$ and $F^{\mu\nu}\,{}^{\star}F_{\mu\nu}$), the possible Lorentz tensors must also contain only even numbers of $F^{\mu\nu}$. This leads to the precept that we are only allowed to combine the four vectors in (2.66) in such a way that the tensor product contains even powers of $F^{\mu\nu}$. So, the number of possible tensors reduces to six.

Finally, we employ the Ward identity as a consequence of gauge symmetry, which demands that $\Pi^{\mu\nu}k_\nu = 0$. This reduces the number of basic tensors to four. In other words, there are only four linearly independent tensors that generate the tensor structure of $\Pi^{\mu\nu}(k|A)$. In fact, the polarization tensor for the special case (2.59) is constructed from four linearly independent tensor quantities which respect all of these symmetries:

$$g^{\mu\nu}k^2 - k^\mu k^\nu, \quad g_\parallel^{\mu\nu}k_\parallel^2 - k_\parallel^\mu k_\parallel^\nu, \quad g_\perp^{\mu\nu}k_\perp^2 - k_\perp^\mu k_\perp^\nu, \quad \tilde{k}_\perp^\mu \tilde{k}_\perp^\nu + \tilde{k}_\parallel^\mu \tilde{k}_\parallel^\nu. \tag{2.67}$$

Our final task is to establish a one-to-one correspondence between these tensors and an appropriate set of Lorentz-invariant tensors constructed from the vectors (2.66) in the special Lorentz frame where $\boldsymbol{B} \| - \boldsymbol{E}$.

As a first step, it is useful to note the following identities:

$$g_\parallel^{\mu\nu}k_\parallel^2 - k_\parallel^\mu k_\parallel^\nu \equiv -\tilde{k}_\parallel^\mu \tilde{k}_\parallel^\nu,$$
$$g_\perp^{\mu\nu}k_\perp^2 - k_\perp^\mu k_\perp^\nu \equiv \tilde{k}_\perp^\mu \tilde{k}_\perp^\nu, \tag{2.68}$$

which can be verified by a straightforward calculation employing the definitions (2.57) and (2.58). Next, we represent the Lorentz-invariant vectors (2.66) by $k_{\parallel,\perp}^\mu$ and $\tilde{k}_{\parallel,\perp}^\mu$ in the special Lorentz frame. By simple matrix multiplication, we find

$$Fk^\mu = E\,\tilde{k}_\parallel^\mu + B\,\tilde{k}_\perp^\mu = -b\,\tilde{k}_\parallel^\mu + a\,\tilde{k}_\perp^\mu,$$
$${}^{\star}Fk^\mu = B\,\tilde{k}_\parallel^\mu - E\,\tilde{k}_\perp^\mu = a\,\tilde{k}_\parallel^\mu + b\,\tilde{k}_\perp^\mu,$$
$$F^2k^\mu = E^2\,k_\parallel^\mu - B^2\,k_\perp^\mu = b^2\,k_\parallel^\mu - a^2\,k_\perp^\mu,$$
$$k^\mu = k_\parallel^\mu + k_\perp^\mu. \tag{2.69}$$

These equations can easily be inverted:

$$\tilde{k}^\mu_\| = \frac{1}{a^2+b^2}\left(a\,{}^\star Fk^\mu - b\,Fk^\mu\right), \qquad k^\mu_\| = \frac{1}{a^2+b^2}\left(F^2 k^\mu + a^2\,k^\mu\right),$$

$$\tilde{k}^\mu_\perp = \frac{1}{a^2+b^2}\left(b\,{}^\star Fk^\mu + a\,Fk^\mu\right), \qquad k^\mu_\perp = \frac{1}{a^2+b^2}\left(-F^2 k^\mu + b^2\,k^\mu\right), \quad (2.70)$$

which are valid in the special system. However, the right-hand sides of (2.70) are written in a Lorentz-invariant form and can therefore be transformed into any Lorentz system. For instance, let us define the Lorentz transformations of $\tilde{k}^\mu_{\|,\perp}$ as

$$\tilde{k}^\mu_\| \quad \to \quad v^\mu_\| = \frac{1}{a^2+b^2}\left(a\,{}^\star Fk^\mu - b\,Fk^\mu\right),$$

$$\tilde{k}^\mu_\perp \quad \to \quad v^\mu_\perp = \frac{1}{a^2+b^2}\left(b\,{}^\star Fk^\mu + a\,Fk^\mu\right), \qquad (2.71)$$

where the subscripts $\|$ and $\perp$ are intended to remind us of the meaning of $v_\|$ and $v_\perp$ in the special Lorentz system (the longitudinal and transverse parts of $\tilde{k}$). As a cross-check, we can prove the following identities in a general system; they are, of course, also valid for $\tilde{k}^\mu_{\|,\perp}$ in the special system:

$$v^2_\| = v^\mu_\| v_{\|\mu} = -\frac{a^2 k^2 - z_k}{a^2+b^2} \quad \left(\equiv \tilde{k}^2_\| = -k^2_\|\right),$$

$$v^2_\perp = v^\mu_\perp v_{\perp\mu} = \frac{b^2 k^2 + z_k}{a^2+b^2} \quad \left(\equiv \tilde{k}^2_\perp = k^2_\perp\right), \qquad (2.72)$$

$$v^\mu_\| v_{\perp\mu} = 0 \qquad \left(\equiv \tilde{k}^\mu_\| \tilde{k}_{\perp\mu}\right).$$

Inserting the Lorentzian generalizations of $k^\mu_{\|,\perp}$ and $\tilde{k}^\mu_{\|,\perp}$ given in (2.70) into the tensor quantities of the special system (2.67), we find a Lorentz-invariant tensor basis for the polarization tensor in an arbitrary constant electromagnetic field:

$$g^{\mu\nu}_\| k^2_\| - k^\mu_\| k^\nu_\| \to -v^\mu_\| v^\nu_\|,$$

$$g^{\mu\nu}_\perp k^2_\perp - k^\mu_\perp k^\nu_\perp \to v^\mu_\perp v^\nu_\perp,$$

$$\tilde{k}^\mu_\perp \tilde{k}^\nu_\| + \tilde{k}^\mu_\| \tilde{k}^\nu_\perp \to v^\mu_\perp v^\nu_\| + v^\mu_\| v^\nu_\perp, \qquad (2.73)$$

$$g^{\mu\nu} k^2 - k^\mu k^\nu \to k^2 \left(P^{\mu\nu}_0 + P^{\mu\nu}_\| + P^{\mu\nu}_\perp\right),$$

where the projectors $P^{\mu\nu}_{0,\|,\perp}$ are defined by

$$P^{\mu\nu}_0 = \frac{1}{k^2\left[2\mathcal{F}z_k/k^2 + \mathcal{G}^2 - (z_k/k^2)^2\right]}\left(F^2 k^\mu + \frac{z_k}{k^2}\,k^\mu\right)\left(F^2 k^\nu + \frac{z_k}{k^2}\,k^\nu\right),$$

$$P^{\mu\nu}_\| = \frac{v^\mu_\| v^\nu_\|}{v^2_\|}, \qquad\qquad P^{\mu\nu}_\perp = \frac{v^\mu_\perp v^\nu_\perp}{v^2_\perp}, \qquad (2.74)$$

and satisfy the usual projector identities $P^2_{0,\|,\perp} = P_{0,\|,\perp}$, $P_{0,\|,\perp}{}^\mu{}_\mu = 1$. Moreover, the projectors are orthogonal to each other: $P^{\mu\nu}_i P_{j\,\nu}{}^\lambda = 0$ for

$i \neq j$ and $i, j = 0, \|, \perp$. We may furthermore introduce the short form $Q^{\mu\nu} = v_\|^\mu v_\perp^\nu + v_\perp^\mu v_\|^\nu$, which is neither a projector nor orthogonal to the $P_{\|,\perp}^{\mu\nu}$'s but is orthogonal to $P_0^{\mu\nu}$.

We are finally in a position to transform the polarization tensor for the parallel field configuration into its generalized form for arbitrary constant electromagnetic fields :

$$\Pi^{\mu\nu}(k|A) = \Pi_0 \, P_0^{\mu\nu} + \Pi_\| \, P_\|^{\mu\nu} + \Pi_\perp \, P_\perp^{\mu\nu} + \Theta \, Q^{\mu\nu}, \tag{2.75}$$

where $\Pi_{0,\|,\perp}$ and Θ are functions of the invariants and are given by

$$
\begin{Bmatrix} \Pi_0 \\ \Pi_\| \\ \Pi_\perp \\ \Theta \end{Bmatrix}
= \frac{\alpha}{2\pi} \int_0^\infty \frac{ds}{s} \int_{-1}^1 \frac{d\nu}{2}
\left(e^{-is\phi_0} \frac{eas\,ebs}{\sin eas \sinh ebs}
\begin{Bmatrix} k^2 N_0 \\ N_0 v_\perp^2 - \tilde{N}_1 v_\|^2 \\ \tilde{N}_2 v_\perp^2 - N_0 v_\|^2 \\ -N_3 \end{Bmatrix} + \text{c.t.} \right).
\tag{2.76}
$$

Substituting the invariants into (2.60) and (2.61), the functions N_i and ϕ_0 yield

$$
\phi_0 = m^2 - \frac{v_\|^2}{2} \frac{\cosh ebs - \cosh \nu ebs}{ebs \sinh ebs} + \frac{v_\perp^2}{2} \frac{\cos \nu eas - \cos eas}{eas \sin eas},
$$
$$
N_0 = \cosh \nu ebs \, \cos \nu eas - \sinh \nu ebs \, \sin \nu eas \, \cot eas \coth ebs,
$$
$$
\tilde{N}_1 = 2 \cos eas \frac{\cosh ebs - \cosh \nu ebs}{\sinh^2 ebs},
$$
$$
\tilde{N}_2 = 2 \cosh ebs \frac{\cos \nu eas - \cos eas}{\sin^2 eas}, \tag{2.77}
$$
$$
N_3 = -\frac{1 - \cos eas \, \cos \nu eas}{\sin eas} \frac{\cosh \nu ebs \, \cosh ebs - 1}{\sinh ebs} + \sin \nu eas \, \sinh \nu ebs.
$$

The scalars $v_{\|,\perp}^2$ are given by certain combinations of the invariants and can be found in (2.71). The contact term given in (2.62) contributes equally to the Π_i's,

$$\text{c.t.} = -e^{-im^2 s} k^2 (1 - \nu^2), \tag{2.78}$$

but does not modify the function Θ, which is already finite.

Although (2.75) is the most convenient representation, a physical meaning can only be ascribed to the particular terms in a diagonalized version. While $P_0^{\mu\nu}$ indeed projects onto an eigenspace of $\Pi^{\mu\nu}$ with eigenvalue Π_0, this is generally not the case for the projectors $P_{\|,\perp}^{\mu\nu}$, owing to the mixed term $\Theta \, Q^{\mu\nu}$. However, the function Θ vanishes in those cases where either a or b vanishes, i.e. $\boldsymbol{E} \cdot \boldsymbol{B} = 0$, since $N_3(a = 0 \text{ or } b = 0) = 0$. Then, the vectors $v_{\|,\perp}^\mu$ indeed represent eigenvectors of $\Pi^{\mu\nu}$ with eigenspace projectors $P_{\|,\perp}^{\mu\nu}$ and eigenvalues $\Pi_{\|,\perp}$; in other cases, the eigenvectors are formed by linear combinations of $v_\|^\mu$ and $v_\perp^\mu$. These can be found by a direct computation, which we shall encounter in the next subsections.

The fourth eigenvalue of $\Pi^{\mu\nu}$ is, of course, zero, since the remaining eigenspace projector $P_k^{\mu\nu} \equiv g^{\mu\nu} - P_0^{\mu\nu} - P_\parallel^{\mu\nu} - P_\perp^{\mu\nu} = k^\mu k^\nu / k^2$ satisfies the Ward identity $\Pi_{\mu\nu}(k) P_k^{\nu\lambda} = 0$.

In view of the technical problems associated with a general, direct computation of $\Pi^{\mu\nu}$ for arbitrary constant field configurations, the simple tensor formalism developed so far reduces the calculational efforts considerably. Moreover, the use of the one-to-one correspondences between the invariants and the basic tensors established here is not limited to the one-loop level, but remains applicable to all orders in perturbation theory.

2.2.2 Applications

This representation of the polarization tensor in an arbitrary constant electromagnetic field can be reduced to various well-known representations for the limiting cases of certain field configurations. As a demonstration of the physical information contained in $\Pi^{\mu\nu}$, we shall apply our findings to the question of how the propagation of light is modified in an electromagnetic background. Here, we mainly focus on low-frequency photons in order to make contact with the effective-action approach in chapter 3. Nevertheless, the polarization tensor contains information about light propagation at any frequency scale.

Having integrated out the fermions (electrons and positrons), the effects of vacuum polarization can be included in the effective action for the electromagnetic field by writing

$$
W_{\text{eff}} = \int d^4x \, \mathcal{L}_{\text{eff}} = -\int d^4x \, \frac{1}{4} F^{\mu\nu}(x) F_{\mu\nu}(x) \bigg|_{A+a}
$$
$$
- \frac{1}{2} \int d^4x \, d^4x' \, a_\mu(x) \Pi^{\mu\nu}(x, x'|A) a_\nu(x'). \qquad (2.79)
$$

In view of the physical situation under consideration, we have separated the electromagnetic field into a constant background part A_μ and a plane wave field $f^{\mu\nu}$ with potential $a_\mu(k)$ and wave vector k^μ. Deriving the equation of motion for the plane wave field in momentum space leads us to the light cone condition for the propagating wave:

$$
\left[k^2 g^{\mu\nu} - k^\mu k^\nu + \Pi^{\mu\nu}(k|A) \right] a_\nu(k) = 0. \qquad (2.80)
$$

Employing our knowledge about the tensor structure of $\Pi^{\mu\nu}$, the light cone condition simplifies if the polarization of the plane wave lies entirely in an eigenspace of $\Pi^{\mu\nu}$. For example, if $a^\mu(k) = P_0^{\mu\nu} a_\nu(k)$, i.e. $a^\mu(k)$ is proportional to $F^2 k^\mu + (z_k/k^2) k^\mu$, then we obtain the light cone condition for a "longitudinal" photon,

$$
k^2 + \Pi_0 = 0, \qquad \text{for} \quad a^\mu(k) = P_0^{\mu\nu} a_\nu(k). \qquad (2.81)
$$

Since $\Pi_0 \sim k^2$ (cf. (2.76)), (2.81) can always be solved trivially by $k^2 = 0$. From this, we may infer that there are no longitudinal photons. The interesting question is whether there are further, nontrivial solutions to (2.81) which hint at the existence of longitudinal photons and a new mode of propagation. Different points of view on this subject can be found in [47], [152] and [161].

The existence of non-trivial solutions to the light cone condition should be expected if the polarization of the plane wave lies in the plane spanned by $v_{\parallel,\perp}^{\mu}$, since these polarizations are continuously related to the transverse ones in the trivial vacuum. The light cone condition then reads

$$\left[\left(k^2 + \Pi_{\parallel}\right)P_{\parallel}^{\mu\nu} + \left(k^2 + \Pi_{\perp}\right)P_{\perp}^{\mu\nu} + \Theta\, Q^{\mu\nu}\right] a_{\nu}(k) = 0,$$
$$\text{for} \quad a^{\mu}(k) = \left(P_{\parallel}^{\mu\nu} + P_{\perp}^{\mu\nu}\right)a_{\nu}(k), \tag{2.82}$$

which we shall refer to as the light cone condition for transverse photons. Of course, the physical polarization modes still have to be identified by diagonalization of (2.82).

For the following applications, it is useful to be aware of some identities that relate the Lorentz-covariant objects to their Galilean counterparts in a certain reference frame; these identities can be checked by simple matrix algebra (cf. Appendix A):

$$k^{\mu} = (\omega, \boldsymbol{k}) \equiv \mathrm{k}(v, \hat{\boldsymbol{k}}),$$
$$Fk^{\mu} = \mathrm{k}\left(\hat{\boldsymbol{k}} \cdot \boldsymbol{E},\, \hat{\boldsymbol{k}} \times \boldsymbol{B} + v\boldsymbol{E}\right),$$
$${}^{*}Fk^{\mu} = \mathrm{k}\left(\hat{\boldsymbol{k}} \cdot \boldsymbol{B},\, -\hat{\boldsymbol{k}} \times \boldsymbol{E} + v\boldsymbol{B}\right), \tag{2.83}$$
$$F^2 k^{\mu} = \mathrm{k}\left[\boldsymbol{E} \cdot (\hat{\boldsymbol{k}} \times \boldsymbol{B}) + v\boldsymbol{E}^2,\, (\hat{\boldsymbol{k}} \cdot \boldsymbol{B})\boldsymbol{B} + (\hat{\boldsymbol{k}} \cdot \boldsymbol{E})\boldsymbol{E} - B^2\hat{\boldsymbol{k}} + v\boldsymbol{E} \times \boldsymbol{B}\right],$$
$$z_k = \mathrm{k}^2\left[B^2 \sin^2\theta_B + E^2 \sin^2\theta_E - (1 - v^2)E^2 + 2v\boldsymbol{E} \cdot (\hat{\boldsymbol{k}} \times \boldsymbol{B})\right].$$

In the first line, we have introduced the phase velocity of the plane wave, defined by $v = \omega/|\boldsymbol{k}|$; we have furthermore employed the short form $\mathrm{k} = |\boldsymbol{k}|$ (the Lorentz scalar k^2 should not be confused with the $O(3)$ scalar $\mathrm{k}^2 \equiv |\boldsymbol{k}|^2$). In the last line, we have defined the angles θ_B and θ_E between the magnetic and electric fields and the wave vector $\boldsymbol{k}$ via $B^2 \sin^2\theta_B = B^2 - (\boldsymbol{B} \cdot \hat{\boldsymbol{k}})^2$ and similarly for θ_E. In terms of the phase velocity, a light cone condition of the form $k^2 + \Pi = 0$ can be written as

$$v^2 = 1 + \frac{\Pi}{\mathrm{k}^2}, \tag{2.84}$$

where Π generally still depends on v^2.

Magnetic Field

The problem of evaluating the polarization tensor becomes considerably simplified if at least one of the invariants vanishes. This is the case, for example, if a reference frame exists in which the electric field vanishes. We are left with a purely magnetic field, and the invariant structure can then be written as

$$b = 0, \quad a = B, \quad z_k = \mathrm{k}^2 B^2 \sin^2 \theta_B, \quad \mathrm{k}^2 = \mathrm{k}^2(1 - v^2). \tag{2.85}$$

It is therefore useful to rewrite the problem in terms of B, k, v instead of a, z_k, k^2; for convenience we shall omit the subscript B on the angle θ_B. Furthermore the functions N_i and ϕ_0 reduce to

$$N_0 = \cos v e B s - v \sin v e B s \cot e B s,$$

$$\tilde{N}_1 = (1 - v^2) \cos e B s, \qquad \tilde{N}_2 = 2 \frac{\cos v e B s - \cos e B s}{\sin^2 e B s}, \tag{2.86}$$

$$\phi_0 = m^2 - \mathrm{k}^2 \left\{ [\sin^2 \theta - (1 - v^2)] \frac{1 - v^2}{4} + \frac{\sin^2 \theta}{2} \frac{\cos v e B s - \cos e B s}{e B s \sin e B s} \right\}.$$

Since N_3 and, consequently, Θ vanish, $\Pi_{\|,\perp}$ become eigenvalues of the polarization tensor, and the transverse light cone condition (2.82) decouples. For the different modes, we obtain

$$\mathrm{k}^2 + \Pi_\| = 0, \qquad \| \text{ mode}: \quad a^\mu \sim v_\|^\mu,$$

$$\mathrm{k}^2 + \Pi_\perp = 0, \qquad \perp \text{ mode}: \quad a^\mu \sim v_\perp^\mu. \tag{2.87}$$

The eigenvectors for purely magnetic fields read:

$$v_\|^\mu = \mathrm{k}\big[(\hat{\boldsymbol{k}} \cdot \hat{\boldsymbol{B}}), v\,\hat{\boldsymbol{B}}\big],$$

$$v_\perp^\mu = \mathrm{k}\big(0, \hat{\boldsymbol{k}} \times \hat{\boldsymbol{B}}\big). \tag{2.88}$$

Referring to (2.87), the vectors $v_{\|,\perp}^\mu$ determine the polarization directions of the plane wave which diagonalize $\Pi^{\mu\nu}$. The direction of the electric field vector of the propagating plane wave $\hat{\boldsymbol{e}}$ is related to these vectors via $e_i = f^{0i} = k^0 a^i - k^i a^0 \sim \omega\, v_{\|,\perp}^i - v_{\|,\perp}^0 k^i$. Hence, the $\perp$ mode is polarized perpendicular to the plane spanned by the magnetic field $\hat{\boldsymbol{B}}$ and the propagation direction $\hat{\boldsymbol{k}}$, while the $\|$ mode is polarized parallel to this plane:

$$e_\| \sim v^2 \hat{\boldsymbol{B}} - (\hat{\boldsymbol{k}} \cdot \hat{\boldsymbol{B}})\,\hat{\boldsymbol{k}},$$

$$e_\perp \sim \hat{\boldsymbol{k}} \times \hat{\boldsymbol{B}}. \tag{2.89}$$

Note that the $\|$ mode is only "almost" transverse, owing to the factor of v^2.

Now we arrive at a subtle point with a long tradition in the literature: since we are dealing with a one-loop calculation, we expect the shift of the phase velocity to be small compared with the vacuum velocity. This assumption should at least hold for moderate magnetic fields $0 \leq B \leq \mathcal{O}(B_{\mathrm{cr}} = m^2/e)$ and moderate frequencies $0 \leq \omega \leq \mathcal{O}(m)$. In this case, we may set $v = 1$ and $\mathrm{k} = \omega$ in the polarization tensor, because a small deviation would lead to only a next-to-leading order correction to the velocity shift. Note that this assumption must be confirmed by consistent results. The first consequence is that the longitudinal light cone condition is already satisfied, since $v = 1 \Leftrightarrow \mathrm{k}^2 = 0$; in this regime, there is no longitudinal photon. The contact terms also vanish. The function ϕ_0 reduces to

$$\phi_0 = m^2 - \omega^2 \sin^2 \theta \left(\frac{1-\nu^2}{4} - \frac{1}{2} \frac{\cos \nu eBs - \cos eBs}{eBs \sin eBs} \right). \tag{2.90}$$

The scalars $v_{\parallel,\perp}^2$ simply become $v_{\parallel,\perp}^2 = \omega^2 \sin^2 \theta$. The relevant components of the polarization tensor can then be written as

$$\Pi_{\parallel,\perp} = \omega^2 \sin^2 \theta \frac{\alpha}{2\pi} \int\limits_0^\infty \frac{ds}{s} \int\limits_{-1}^1 \frac{d\nu}{2} e^{-is\phi_0} N_{\parallel,\perp}, \tag{2.91}$$

where

$$N_{\parallel} = \frac{eBs \cos \nu eBs}{\sin eBs} - eBs \cot eBs \left(1 - \nu^2 + \nu \frac{\sin \nu eBs}{\sin eBs} \right), \tag{2.92}$$

$$N_{\perp} = -\frac{eBs \cos \nu eBs}{\sin eBs} + \frac{\nu eBs \sin \nu eBs \cot eBs}{\sin eBs}$$
$$+ \frac{2eBs(\cos \nu eBs - \cos eBs)}{\sin^3 eBs}.$$

Here, we have arrived at the representation of the polarization tensor found by Tsai and Erber [160], who employed this expression to obtain refractive indices for a magnetic background over a wide range of frequency and field strength. We shall confine ourselves to the case of low-frequency photons, $\omega \ll m$, which allows us to approximate the function ϕ_0 in the exponent by $\Pi_0 \to m^2$. Bearing in mind that we assume a small velocity shift, we may write for the phase velocity

$$v = 1 - \frac{\alpha}{4\pi} \sin^2 \theta \, \eta_{\parallel,\perp}, \tag{2.93}$$

where $\eta_{\parallel,\perp}$ are functions only of the field strength:

$$\eta_{\parallel,\perp} = -\int\limits_0^\infty \frac{ds}{s} \int\limits_{-1}^1 \frac{d\nu}{2} e^{-im^2 s} N_{\parallel,\perp}. \tag{2.94}$$

The necessary integrations can be performed analytically; details may be looked up in Appendix D (see (D.28) and (D.29)). We find

$$\eta_{\parallel}(B) = 8\zeta' \left(-1, \frac{B_{cr}}{2B} \right) - 4 \frac{B_{cr}}{2B} \zeta' \left(0, \frac{B_{cr}}{2B} \right) - \frac{2}{3} \Psi \left(1 + \frac{B_{cr}}{2B} \right)$$
$$+ 2 \frac{B_{cr}}{2B} \ln \frac{B_{cr}}{2B} - 2 \left(\frac{B_{cr}}{2B} \right)^2 + \frac{2B}{3B_{cr}} - \frac{1}{3},$$

$$\eta_{\perp}(B) = -4 \frac{B_{cr}}{2B} \zeta' \left(0, \frac{B_{cr}}{2B} \right) + 4 \left(\frac{B_{cr}}{2B} \right)^2 \Psi \left(1 + \frac{B_{cr}}{2B} \right)$$
$$- 2 \frac{B_{cr}}{2B} \ln \frac{B_{cr}}{2B} - 2 \frac{B_{cr}}{2B} - 4 \left(\frac{B_{cr}}{2B} \right)^2 + \frac{2}{3}. \tag{2.95}$$

Our representation employs the first derivative of the Hurwitz zeta function $\zeta'(s, q)$. With the aid of the identities of Appendix D (see (D.40)), its equivalence to the representation of [160], which involves a generalized gamma function, can be proved.

In the limit of weak fields, $B \ll B_{cr}$, we recover the well-known result for the velocity shift:

$$v_\parallel \simeq 1 - \frac{14}{45} \frac{\alpha^2}{m^4} B^2 \sin^2 \theta + \mathcal{O}\left[\left(\frac{B}{B_{cr}}\right)^4\right],$$

$$v_\perp \simeq 1 - \frac{8}{45} \frac{\alpha^2}{m^4} B^2 \sin^2 \theta + \mathcal{O}\left[\left(\frac{B}{B_{cr}}\right)^4\right]. \tag{2.96}$$

The limit of strong fields can be studied by expanding (2.95) in terms of $B_{cr}/B \ll 1$. We obtain

$$v_\parallel \simeq 1 - \frac{\alpha}{4\pi} \sin^2 \theta \left[\frac{2}{3} \frac{B}{B_{cr}} + \left(\frac{1}{3} + \frac{2\gamma}{3} - 8L_1\right) + \mathcal{O}\left(\frac{B_{cr}}{2B} \ln \frac{B_{cr}}{2B}\right)\right],$$

$$v_\perp \simeq 1 - \frac{\alpha}{4\pi} \sin^2 \theta \left[\frac{2}{3} + \mathcal{O}\left(\frac{B_{cr}}{2B} \ln \frac{B_{cr}}{2B}\right)\right], \tag{2.97}$$

where γ denotes Euler's constant and $L_1 = 0.248754477\ldots$. Note that the phase velocity for the $\perp$ mode remains close to one, even for strong magnetic fields. The maximum velocity shift amounts to $\delta v_{\perp \text{max}} = -\alpha/(6\pi) \simeq -3.8 \times 10^{-4}$. This justifies the assumption regarding the smallness of the velocity shift at the beginning of the calculation for the $\perp$ mode.

In contrast, the velocity shift for the $\parallel$ mode increases linearly with B for strong fields and seems to be unbounded. This, of course, contradicts the assumption that the velocity shift should be small; hence, (2.97) for the $\parallel$ mode is only valid as long as $-\delta v_\perp \ll 1$. This restricts the magnetic field strength so as to obey $B/B_{cr} \lesssim \pi/\alpha \simeq 430$.

To obtain the correct strong-field behavior in the limit $B \to \infty$, one must evaluate the integrals without a restriction to small velocity shifts. In Sect. 3, we shall obtain the correct strong-field behaviour within the effective-action approach. In this approach, it is not necessary to perform tedious integrations.

Let us finally remark that these phase velocities for the low-frequency limit do not explicitly depend on the frequency; hence the group velocity coincides with the phase velocity in this limit. It will turn out that this is a general feature of the low-frequency (= soft-photon) limit.

Absorption in a Magnetic Field

As an example of a high-frequency process, we investigate the absorption rate for a propagating photon in a magnetic field due to $e^+ e^-$ pair production, following the work of Tsai and Erber [159]. By reference to the optical theorem, the absorption coefficients for the different polarization modes are related to the imaginary part of the polarization tensor in a magnetic field (2.91) by

$$\kappa_{\|,\perp}(\omega) = -\frac{1}{\omega} \operatorname{Im} \Pi_{\|,\perp}. \tag{2.98}$$

Of course, it is hardly possible to evaluate the imaginary parts of $\Pi_{\|,\perp}$ exactly; therefore, we shall confine ourselves to the energy regimes relevant to the present situation.

In particular, the exponential factor $e^{-i\phi_0 s}$ has to be treated with great care. Substituting for the s integration by means of $eBs = z$, the exponent reads

$$\phi_0 s = \frac{m^2}{eB} \left[z + z\frac{\omega^2}{m^2} \sin^2 \theta \left(\frac{1}{2} \frac{\cos \nu z - \cos z}{z \sin z} - \frac{1 - \nu^2}{4} \right) \right]. \tag{2.99}$$

First, note that the term in the round brackets has the following expansion for small z:

$$\left(\frac{1}{2} \frac{\cos \nu z - \cos z}{z \sin z} - \frac{1 - \nu^2}{4} \right) = \frac{1}{48}(1 - \nu^2)^2 z^2 + \mathcal{O}(z^4). \tag{2.100}$$

We are allowed to insert (2.100) into (2.99) if the principal contribution to the z integral stems from the lower bound, where $z \ll 1$. Regarding (2.99), this is the case if $m^2/eB \gg 1$ and $\omega/m \sin \theta \gg 1$, i.e. for weak fields and high frequencies.[4] For photon absorption, the latter condition exactly matches our requirements, since pair production can only occur beyond the two-particle threshold, $\omega \gg 2m$.

Therefore, we approximate the exponential factor by

$$e^{-i\phi_0 s} \simeq e^{-i\Xi}, \tag{2.101}$$

where Ξ is defined as

$$\begin{aligned} \Xi &= \frac{m^2}{eB} \left(z + \frac{\omega^2}{m^2} \sin^2 \theta \frac{(1 - \nu^2)^2}{48} z^3 \right) \\ &= \frac{3}{2} \xi \left(y + \frac{1}{3} y^3 \right). \end{aligned} \tag{2.102}$$

In the last step, we have performed the substitution

$$y = \frac{1 - \nu^2}{4} \frac{\omega}{m} \sin \theta \, z, \qquad \xi = \frac{8}{3} \frac{m^2}{eB} \frac{m}{\omega} \frac{1}{\sin \theta} \frac{1}{1 - \nu^2}. \tag{2.103}$$

For our further purposes, it is useful to introduce another parameter:

$$\lambda = \frac{3}{2} \frac{eB}{m^2} \frac{\omega}{m} \sin \theta, \quad \text{and hence} \quad \xi = \frac{4}{\lambda} \frac{1}{1 - \nu^2}. \tag{2.104}$$

---

[4] Throughout these considerations, it is understood that the z integral contour lies slightly below the positive real axis.

Note that the requirements $m^2/eB \gg 1$ and $(\omega/m)\sin\theta \gg 1$ do not restrict λ to a particular interval; λ can become large for sufficiently high frequencies or small for weak magnetic fields.

Since only small values of z are relevant for the integral, it suffices to approximate the functions $N_{\parallel,\perp}$ by the first terms of their Taylor expansions:

$$N_{\parallel} = \frac{1}{2}(1-\nu^2)\left(1-\frac{1}{3}\nu^2\right)z^2,$$

$$N_{\perp} = \frac{1}{2}(1-\nu^2)\left(\frac{1}{2}+\frac{1}{6}\nu^2\right)z^2. \tag{2.105}$$

Finally, the absorption coefficients can be written as

$$\kappa_{\parallel,\perp}(\omega) = -\frac{\alpha}{2\pi}\omega\sin^2\theta\,\mathrm{Im}\int_0^\infty\frac{dy}{y}\int_0^1 d\nu\,e^{-i\Xi}\,N_{\parallel,\perp}$$

$$= \frac{4\alpha}{\pi}\frac{m^2}{\omega}\int_0^1 d\nu\,(1-\nu^2)^{-1}\left[\left(1-\frac{\nu^2}{3}\right)_{\parallel},\left(\frac{1}{2}+\frac{\nu^2}{6}\right)_{\perp}\right]\int_0^\infty dy\,y\sin\Xi$$

$$= \frac{1}{2}\alpha\sin\theta\frac{eB}{m}\frac{4\sqrt{3}}{\pi\lambda} \tag{2.106}$$

$$\times\int_0^1 d\nu\,(1-\nu^2)^{-1}\left[\left(1-\frac{\nu^2}{3}\right)_{\parallel},\left(\frac{1}{2}+\frac{\nu^2}{6}\right)_{\perp}\right]K_{2/3}\left(\frac{4}{\lambda}\frac{1}{1-\nu^2}\right),$$

where the representation of the modified Bessel function (Airy integral)

$$\frac{1}{\sqrt{3}}K_{2/3}(\xi) = \int_0^\infty dy\,y\sin\left[\frac{3}{2}\xi\left(y+\frac{1}{3}y^3\right)\right] \tag{2.107}$$

has been employed. Equation (2.106) demonstrates that the absorption coefficients are proportional to the Larmor frequency $\omega_L = eB/m$. Apart from the angle dependence $\sin\theta$, the remaining formula is a pure number depending on the parameter λ as defined in (2.104).

In the various limits of λ, one can make use of the usual asymptotic expansions of the Bessel function. For example, in the region where $\lambda \gg 1$, i.e. sufficiently high frequency, we find, following [159],

$$\kappa_{\parallel,\perp}(\omega) \simeq \frac{1}{2}\alpha\sin\theta\frac{eB}{m}\lambda^{-1/3}\frac{2^{1/3}\sqrt{3}}{\pi}\Gamma(3/2)$$

$$\times\int_0^1 d\nu\,(1-\nu^2)^{-1/3}\left[\left(1-\frac{\nu^2}{3}\right)_{\parallel},\left(\frac{1}{2}+\frac{\nu^2}{6}\right)_{\perp}\right]$$

$$= \frac{1}{2}\alpha\sin\theta\frac{eB}{m}\lambda^{-1/3}\frac{2^{1/3}\sqrt{3}}{7\sqrt{\pi}}\frac{\Gamma^2(2/3)}{\Gamma(7/6)}\left[(3)_{\parallel},(2)_{\perp}\right]$$

$$\simeq \frac{1}{2}\alpha \sin\theta \frac{eB}{m}\lambda^{-1/3}\left[(1.04)_\parallel,(0.69)_\perp\right]. \tag{2.108}$$

For small values of $\lambda \ll 1$, i.e. sufficiently weak magnetic fields, we obtain

$$\kappa_{\parallel,\perp}(\omega) \simeq \alpha \sin\theta \frac{eB}{m}\sqrt{\frac{3}{2\pi\lambda}}$$

$$\times \int_0^1 d\nu\,(1-\nu^2)^{-1/2}e^{-[4/(\lambda(1-\nu^2))]}\left[\left(1-\frac{\nu^2}{3}\right)_\parallel,\left(\frac{1}{2}+\frac{\nu^2}{6}\right)_\perp\right]$$

$$= \frac{1}{2}\alpha \sin\theta \frac{eB}{m}\sqrt{\frac{3}{2\pi\lambda}}$$

$$\times \int_1^\infty du\,(u-1)^{-1/2}\left[\left(\frac{2}{3u}+\frac{1}{3u^2}\right)_\parallel,\left(\frac{2}{3u}-\frac{1}{6u^2}\right)_\perp\right]e^{-4u/\lambda},$$

$$= \frac{1}{2}\alpha \sin\theta \frac{eB}{m}e^{-4/\lambda}\sqrt{\frac{3}{2}}\left[\left(\frac{1}{2}\right)_\parallel,\left(\frac{1}{4}\right)_\perp\right]. \tag{2.109}$$

These findings coincide with the expressions found in [14, 74, 116, 117, 135, 158, 166].

Note that (2.108) and (2.109) describe the asymptotic absorptive behavior of the magnetic field. Both absorption coefficients tend to zero for $\lambda \to 0$ or $\lambda \to \infty$. In between, a maximum exists at $\lambda \simeq 16$ for $\kappa_\parallel/\omega_L$ and at $\lambda \simeq 19$ for $\kappa_\perp/\omega_L$. These features are exhibited in more detail in Fig. 2.1.

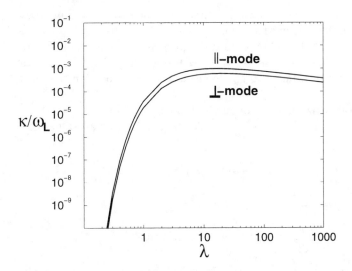

Fig. 2.1. Absorption coefficients $\kappa_{\parallel,\perp}$ in units of the Larmor frequency $\omega_L = eB/m$ versus the parameter λ as defined in (2.104); the propagation direction is assumed to be orthogonal to the magnetic field, i.e. $\theta = \pi/2$, $\sin\theta = 1$

Crossed Fields

A particularly interesting field configuration is provided by so-called "crossed" fields, where E and B are not only orthogonal but also of equal modulus: $E \perp B$ and $E^2 = B^2$. This causes the vanishing of the two invariants $\mathcal{F} = (1/2)(B^2 - E^2) = (1/2)(a^2 - b^2)$ and $\mathcal{G} = -E \cdot B = ab$. The field dependence is therefore completely carried by the invariant $z_k = (k_\mu F^{\mu\nu})^2$.

The structure of the polarization tensor in such a background was originally studied by Narozhnyi [126] and Ritus [139]. In the present work, the crossed-field configuration serves as an instructive example of the question of how to take the correct limits for the invariants; it also represents a touchstone for our formalism, since we derived our representation for $\Pi^{\mu\nu}$ from a completely different field configuration, which cannot be trivially related to the crossed-field situation by a smooth limiting process.

Indeed, taking the limit $a, b \to 0$ requires special care. Problems arise from the fact that the choice $a = b = 0$ does not uniquely determine a class of Lorentz-equivalent field configurations; e.g. $F^{\mu\nu} = 0$ implies $a, b = 0$, but $a, b = 0$ obviously does not force $F^{\mu\nu}$ to vanish. Thus, we must ensure that our limiting procedure of taking $a, b \to 0$ does not automatically lead us to the trivial vacuum $F^{\mu\nu} = 0$. The latter would be the case if, for example, we first set b to zero and then took the limit $a \to 0$. Since $b = 0$ corresponds to the existence of a reference frame in which the field is purely magnetic, $a = B$, taking the limit $a \to 0$ would then definitely lead us to the trivial vacuum. The same undesired result would be achieved by interchanging the roles of a and b.

The correct procedure is to set $a = b$ first, and then take the limit $(a = b) \to 0$. In terms of field strengths, we first choose $|E| = |B|$ and afterwards rotate the direction of either E or B into the perpendicular position $E \perp B$. By construction, taking the limit $(a = b) \to 0$ does not affect the electric or magnetic field strength but changes their orientation in a given reference frame.

To obtain the polarization tensor in the limit of vanishing a and b, it is sufficient to consider the functions ϕ_0 and N_i in (2.77) for small values of a and b:

$$\phi_0 = m^2 + \frac{1}{4}(1 - \nu^2)\,k^2 + \frac{(es)^2}{48}(1 - \nu^2)^2\,z_k,$$

$$N_0 = (1 - \nu^2) - \frac{(es)^2}{6}\nu^2(1 - \nu^2)\,(a^2 - b^2),$$

$$\tilde{N}_1 = (1 - \nu^2) + \frac{(es)^2}{12}(1 - \nu^2)(-3b^2 + \nu^2 b^2 - 6a^2),$$

$$\tilde{N}_2 = (1 - \nu^2) - \frac{(es)^2}{12}(1 - \nu^2)(-3a^2 + \nu^2 a^2 - 6b^2),$$

$$N_3 = -\frac{(es)^2}{4}(1 - \nu^2)^2\,ab. \tag{2.110}$$

Employing (2.72), we find the following identity in the limit $(a = b) \to 0$:

$$\tilde{N}_2 v_\perp^2 - N_0 v_\parallel^2 = k^2(1 - v^2) + \frac{(es)^2}{24}(1 - v^2)(9 - v^2) z_k$$

$$\equiv N_0 v_\perp^2 - \tilde{N}_1 v_\parallel^2, \qquad \text{for} \quad (a = b) \to 0. \qquad (2.111)$$

Referring to (2.76), we are able to conclude that the contributions to the polarization tensor $\Pi_\parallel$ and $\Pi_\perp$ are equal for crossed fields. Nevertheless, the polarization tensor is not degenerate, since the term $\Theta\, Q^{\mu\nu}$ does not vanish for crossed fields.[5] Therefore, the projectors $P_{\parallel,\perp}^{\mu\nu}$ are not related to the eigenspaces of $\Pi^{\mu\nu}$. In order to diagonalize the polarization tensor, it is useful to note the following relation, which is valid for crossed fields ($\Pi_\parallel = \Pi_\perp \equiv \Pi_{\parallel,\perp}$):

$$\Pi_\parallel P_\parallel^{\mu\nu} + \Pi_\perp P_\perp^{\mu\nu} = \Pi_{\parallel,\perp} \left(\frac{FkFk}{z_k} + \frac{{}^\star Fk^\star Fk}{z_k} \right)^{\mu\nu}, \qquad (2.112)$$

where we have made use of the explicit form of the projectors $P_{\parallel,\perp}^{\mu\nu}$ as defined in (2.71), (2.72) and (2.74). The contribution $\Theta\, Q^{\mu\nu}$ can similarly be represented; we find for the combination $-N_3 Q^{\mu\nu}$ in (2.76)

$$-N_3 Q^{\mu\nu} = \frac{(es)^2}{8}(1 - v^2)^2\, z_k \left(-\frac{FkFk}{z_k} + \frac{{}^\star Fk^\star Fk}{z_k} \right)^{\mu\nu}. \qquad (2.113)$$

In fact, the tensors $Fk^\mu Fk^\nu / z_k$ and ${}^\star Fk^\mu {}^\star Fk^\nu / z_k$ are orthonormal projectors in the case of crossed fields and they are orthogonal to the eigenspace projector $P_0^{\mu\nu}$. Hence, we arrive at the diagonalized form of the one-loop polarization tensor in crossed fields :

$$\Pi^{\mu\nu} = \Pi_0\, P_0^{\mu\nu} + \Pi_1\, \frac{{}^\star Fk^\mu {}^\star Fk^\nu}{z_k} + \Pi_2 \frac{Fk^\mu Fk^\nu}{z_k}, \qquad (2.114)$$

where

$$\begin{Bmatrix} \Pi_0 \\ \Pi_1 \\ \Pi_2 \end{Bmatrix} = \frac{\alpha}{2\pi} \int_0^\infty \frac{ds}{s} \int_{-1}^1 \frac{dv}{2}(1 - v^2)\, e^{-im^2 s}$$

$$\times \left[e^{-i\phi_1 s} \begin{Bmatrix} k^2 + \frac{(es)^2}{3} z_k + \frac{(es)^2}{6}(1 - v^2) z_k \\ k^2 \\ k^2 + \frac{(es)^2}{3} z_k - \frac{(es)^2}{12}(1 - v^2) z_k \end{Bmatrix} - k^2 \right],$$

$$\phi_1 = \frac{1}{4}(1 - v^2)\, k^2 + \frac{(es)^2}{48}(1 - v^2)^2\, z_k. \qquad (2.115)$$

For the derivation of (2.115), we have employed (2.111)–(2.113), which we have inserted into (2.76), taking the limit $(a = b) \to 0$.

---

[5] At first sight, the situation appears to be paradoxical: $\Theta Q^{\mu\nu} \to 0$ for either $a \to 0$ or $b \to 0$, but $\Theta Q^{\mu\nu}$ remains finite for $(a = b) \to 0$. This demonstrates how important the correct treatment of the limiting process is.

The dynamical content of the polarization tensor (2.114) is completely determined by the two dynamical variables k^2 and z_k, while the orientation of the eigenspaces can be read off from (2.83) in the limit $|\boldsymbol{E}| = |\boldsymbol{B}|$ and $\boldsymbol{E} \cdot \boldsymbol{B} = 0$.

A different representation of $\Pi^{\mu\nu}$ for the case of crossed fields, in terms of integrals over Airy functions, can be found in [139]; the corresponding formulas for the functions Π_i are particularly useful for studying the analytical properties of $\Pi^{\mu\nu}$ in the complex k^2 plane.

Finally, we apply these findings to the question of light propagation in a crossed-fields background; for convenience, we confine ourselves to the soft-photon limit $\omega \ll m$ and consider weak fields only. Regarding z_k in (2.83), the weak-field condition can be summarized as the requirement for z_k/k^2 to be smaller than m^4: $z_k/k^2 \ll m^4$. Again, we have to impose the additional assumption that the velocity shift remains small in order to set $k^2 = 0$ in the polarization tensor. This implies that we can neglect the function ϕ_1 in the exponent of (2.115): $e^{-is\phi_1} \to 1$. As an initial consequence, we observe that Π_0 vanishes in this limit; this is not astonishing, since new longitudinal modes of photon propagation are not expected in the classical limit of weak fields. The remaining integrals in $\Pi_{1,2}$ can be carried out elementarily, leading to the following simple result to first order in z_k:

$$\left\{ \begin{matrix} \Pi_1 \\ \Pi_2 \end{matrix} \right\} = -2 \left\{ \begin{matrix} 14/45 \\ 8/45 \end{matrix} \right\} \frac{\alpha^2}{m^4} z_k. \tag{2.116}$$

Upon insertion of this into the light cone condition (2.84), we find for the phase and group velocities of the respective polarization modes

$$v_1 = 1 - \frac{14}{45} \frac{\alpha^2}{m^4} \frac{z_k}{k^2},$$

$$v_2 = 1 - \frac{8}{45} \frac{\alpha^2}{m^4} \frac{z_k}{k^2}. \tag{2.117}$$

Equations (2.117) are identical to the findings of Narozhnyi [126], who also studied the strong-field case.

From the form of the eigenvectors Fk^μ and $^\star Fk^\mu$ as given in (2.83), we can conclude, similarly to (2.89), that the polarization directions of the electric field in the two modes are ($v^2 \simeq 1$)

$$e_1 \sim -\hat{\boldsymbol{k}} \times \boldsymbol{E} + \boldsymbol{B} - (\boldsymbol{B} \cdot \hat{\boldsymbol{k}})\boldsymbol{B},$$

$$e_2 \sim \hat{\boldsymbol{k}} \times \boldsymbol{B} + \boldsymbol{E} - (\boldsymbol{E} \cdot \hat{\boldsymbol{k}})\boldsymbol{E}. \tag{2.118}$$

It is also useful to consider z_k/k in a given reference frame; for this, we also retain the assumption $v \simeq 1$, and conclude from the last line of (2.83) that

$$\frac{z_k}{k^2} = B^2 \sin^2 \theta_B + E^2 \sin^2 \theta_E + 2\hat{\boldsymbol{k}} \cdot (\boldsymbol{B} \times \boldsymbol{E})$$

$$= B^2 \sin^2 \theta_B + E^2 \sin^2 \theta_E - 2\hat{\boldsymbol{k}} \cdot \boldsymbol{S} \tag{2.119}$$

$$= \mathcal{E}^2 \left(1 + (\hat{\boldsymbol{k}} \cdot \hat{\boldsymbol{n}})\right)^2. \tag{2.120}$$

Equation (2.119) expresses z_k/k^2 in terms of the transverse electric and magnetic fields and the Poynting vector $\boldsymbol{S} = \boldsymbol{E} \times \boldsymbol{B}$. Equation (2.120) explicitly exploits the properties of the crossed-fields configuration: we have defined a variable $\mathcal{E}$ for the modulus of the field strengths $\mathcal{E} = E = B$ and introduced the vector $\hat{n}$ normal to the plane spanned by $\boldsymbol{B}$ and $\boldsymbol{E}$: $\hat{n} = \hat{\boldsymbol{B}} \times \hat{\boldsymbol{E}}$. From this representation, we can read off that a propagation direction in the crossed-fields case exists which renders no modifications: if $\hat{\boldsymbol{k}} \cdot \hat{n} = -1$, then $z_k = 0 = \delta v_{1,2}$. In other words, the propagation of light is not affected by weak crossed background fields if $(\hat{\boldsymbol{k}}, \hat{\boldsymbol{E}}, \hat{\boldsymbol{B}})$ form the basis of an orthonormalized coordinate system.

Arbitrary Constant Weak Fields

For weak fields, we recognized in the previous sections that the assumption $v \simeq 1$ or $k^2 \simeq 0$ is consistent with the results for the velocity shifts in the low-frequency limit. To the same order of accuracy, we are now interested in the form of the polarization tensor for an arbitrary constant field configuration in the weak-field limit. As usual, the contribution proportional to $\Pi_0\, P_0^{\mu\nu}$ can be neglected. The physically more interesting part of $\Pi^{\mu\nu}$, which exists in the $\|\perp$ subspace spanned by $v_{\|,\perp}^\mu$, can be written as a matrix:

$$
\Pi^{\mu\nu} = \begin{pmatrix} \Pi_\| & \Theta\sqrt{v_\|^2 v_\perp^2} \\ \Theta\sqrt{v_\|^2 v_\perp^2} & \Pi_\perp \end{pmatrix}. \tag{2.121}
$$

Here we have employed the fact that $v_\|^2 v_\perp^2 = z_k^2/(a^2+b^2)^2 + \mathcal{O}(k^2) > 0$, and $\sqrt{v_\|^2 v_\perp^2}$ is the normalization of the tensor $Q^{\mu\nu}$. Diagonalization of this matrix is straightforward; we obtain the eigenvalues

$$
\lambda_{1,2} = \frac{1}{2}\left(\Pi_\| + \Pi_\perp \mp \sqrt{\left(\Pi_\| - \Pi_\perp\right)^2 + 4\Theta^2 v_\|^2 v_\perp^2}\right). \tag{2.122}
$$

With the aid of the weak-field approximations (2.110) to the functions N_i and setting $\phi_0 = m^2$ to our order of approximation, the functions $\Pi_{\|,\perp}$ and Θ can easily be evaluated:

$$
\Pi_\| = -2\frac{\alpha^2}{m^4}\frac{z_k}{a^2+b^2}\left(\frac{14}{45}a^2 + \frac{8}{45}b^2\right),
$$

$$
\Pi_\perp = -2\frac{\alpha^2}{m^4}\frac{z_k}{a^2+b^2}\left(\frac{14}{45}b^2 + \frac{8}{45}a^2\right), \tag{2.123}
$$

$$
\Theta = 2 - \frac{\alpha^2}{m^4}\,ab\,\frac{6}{45}.
$$

Inserting these equations into (2.122), we find, surprisingly, that the eigenvalues do not depend on the invariants a and b any more:

$$\lambda_{1,2} = -2\frac{\alpha^2}{m^4}\, z_k \left(\frac{11}{45} \pm \frac{3}{45}\right).$$
(2.124)

Hence, we obtain exactly the same eigenvalues for the $\|, \perp$ subspace of the polarization tensor as in the crossed-fields case in the weak-field limit: $\Pi_{1,2} = \lambda_{1,2}$. Obviously, this leads to the same velocity shifts as obtained in (2.117):

$$v_{1,2} = 1 - \left(\frac{11}{45} \pm \frac{3}{45}\right) \frac{\alpha^2}{m^4} \frac{z_k}{k^2}.$$
(2.125)

The eigenvectors $u_{1,2}^\mu$ of the matrix (2.121) are constructed from certain linear combinations of $v_\|^\mu$ and $v_\perp^\mu$:

$$u_{1,2}^\mu = m_{1,2}^\| \frac{v_\|^\mu}{\sqrt{v_\|^2}} + m_{1,2}^\perp \frac{v_\perp^\mu}{\sqrt{v_\perp^2}}.$$
(2.126)

The coefficients can immediately be calculated:

$$m_1^\| = \frac{a}{\sqrt{a^2 + b^2}}, \qquad m_1^\perp = \frac{b}{\sqrt{a^2 + b^2}},$$

$$m_2^\| = -\frac{b}{\sqrt{a^2 + b^2}} \qquad m_2^\perp = \frac{a}{\sqrt{a^2 + b^2}}.$$
(2.127)

Inserting (2.127) into (2.126), the eigenvectors $u_{1,2}^\mu$ reduce to

$$u_1^\mu = \frac{{}^*Fk^\mu}{\sqrt{z_k}}, \qquad u_2^\mu = \frac{Fk^\mu}{\sqrt{z_k}},$$
(2.128)

so the directions of the polarization modes are given by

$$e_1 \sim -\hat{k} \times E + B - (B \cdot \hat{k})B,$$
$$e_2 \sim \hat{k} \times B + E - (E \cdot \hat{k})E.$$
(2.129)

This is the same result as the one that we found for the crossed-fields case. In the weak-field limit, the polarization tensor near the light cone does not depend on the complete set of invariants k^2, a, b, z_k, but loses its dependence on the pure field invariants a and b. Furthermore, the remaining influence of the field via

$$z_k = k^2 \left[B^2 \sin^2 \theta_B + E^2 \sin^2 \theta_E + 2\hat{k} \cdot (B \times E)\right]$$
(2.130)

is strongly dependent on the orientation of k, E and B.

2.3 Induced Electromagnetic Current

The calculation of the electromagnetic current induced by an external constant magnetic field is comprehensively performed in all technical details. For this, we solve explicitly the proper-time equations of motion, mapping Lorentz matrix computations onto simple Pauli spin algebra. The outcome agrees with Adler's result, and coincides with the findings obtained via the polarization tensor in the previous section.

As an example of a proper-time calculation, we investigate here the electro-magnetic current induced by an external c-number field. For this, we employ the Hamiltonian formalism for a quantum mechanical one-particle problem as outlined in Sect. 2.1. In particular, we review the approach of Adler [4], whose calculation of the effects of photon splitting and vacuum birefringence is based on this technique. Adler's work, in turn, was inspired by the ideas and calculations of Minguzzi [124]. A generalization of the results for vacuum birefringence was later worked out by Daugherty and Lerche [51].

As our physical starting point, consider the effective field equation of a wave field a_μ propagating in an external (constant magnetic) field:

$$\Box^2 a_\mu = -\langle j_\mu \rangle.$$

The appearance of the expectation value of the current on the right-hand side accounts for the fact that an external field modulated by a propagating wave field generally induces vacuum polarization, which can be diagrammatically understood in terms of virtual loop processes. Hence, the expectation value of the current depends, in general, nonlinearly on the external field as well as on the wave field. In a classical sense, the linear part with respect to the wave field can then be reinterpreted as an effective contribution to the index of refraction for the propagation of light.

To be precise, if there is a contribution to the current which is linear in a_μ,

$$\langle j_\mu \rangle \simeq -f(\omega, B)\, a_\mu, \qquad f(\omega, B) \quad \text{arbitrary},$$

then the field equation is solved by a plane wave field

$$a_\mu \sim \mathrm{e}^{\mathrm{i}\omega(-t + n\hat{\boldsymbol{n}}\cdot\boldsymbol{x})},$$

where the refractive index is given by

$$n^2 = 1 - \frac{f(\omega, B)}{\omega^2}.$$

Bearing this concept in mind, we begin with the expression for the current given in (2.4):

$$\langle 0|\, j^\mu(x)\, |0 \rangle^A = \mathrm{i} e \,\mathrm{tr}[\gamma^\mu\, G(x, x|A)]. \tag{2.131}$$

Representing the Green's function as an operator in configuration space, $G(x, x|A) = \langle x|(\gamma \Pi + m)^{-1}|x\rangle$, and treating the inverse operator expression symmetrically leads us to

$$\langle 0|j^\mu(x)|0\rangle^A = -\frac{\mathrm{i} e}{2}\, \mathrm{tr}\left[\gamma^\mu \langle x|(\gamma \Pi) \frac{1}{m^2 - (\gamma \Pi)^2} + \frac{1}{m^2 - (\gamma \Pi)^2}(\gamma \Pi)|x\rangle \right], \tag{2.132}$$

where we have omitted two mass terms which contain odd numbers of γ matrices. With the aid of the proper-time integral representation (2.7) of the inverse squared Dirac operator, we may write the current in the following way:

$$\langle 0|j^{\mu}(x)|0\rangle^{A} = \frac{e}{2}\int_{0}^{\infty}ds\, e^{-im^{2}s}\, \text{tr}\left[\gamma^{\mu}\gamma^{\nu}\langle x|\Pi_{\nu}U(s)|x\rangle + \gamma^{\nu}\gamma^{\mu}\langle x|U(s)\Pi_{\nu}|x\rangle\right],$$

$$(2.133)$$

where the proper-time evolution operator $U(s)$ was defined in (2.12).

Introducing the proper-time definitions

$$\begin{aligned} |x\rangle &\equiv |x(0)\rangle, & \langle x(s)| &= \langle x(0)|\,U(s), \\ x &\equiv x(0), & \Pi &\equiv \Pi(0), \\ x(s) &= U^{-1}(s)\,x(0)\,U(s), & \Pi(s) &= U^{-1}(s)\,\Pi(0)\,U(s), \end{aligned}$$

$$(2.134)$$

and employing the identity $\gamma^{\mu}\gamma^{\nu} = -g^{\mu\nu} - i\sigma^{\mu\nu}$ (cf. Appendix A), we obtain our final representation of the current for arbitrary external electromagnetic fields:

$$\langle j^{\mu}(x)\rangle = -\frac{e}{2}\int_{0}^{\infty}ds\, e^{-im^{2}s}\, \text{tr}\left[\langle x(s)|\Pi^{\mu}(s) + \Pi^{\mu}(0)|x(0)\rangle\right.$$

$$\left. + i\sigma^{\mu\nu}\langle x(s)|\Pi_{\nu}(s) - \Pi_{\nu}(0)|x(0)\rangle\right]. \quad (2.135)$$

To evaluate this expression further, one has to diagonalize the operators $\Pi(s) \pm \Pi(0)$ in configuration space to obtain c-number expressions, i.e. one has to represent $\Pi(s) \pm \Pi(0)$ as functions of $x(s)$ and $x(0)$. This can be achieved by solving the one-particle equations of proper-time motion as given in (2.15).

Obviously, analytical solutions to these equations for arbitrary electromagnetic fields are difficult to obtain. Therefore, we specialize to the situation under consideration in which a plane wave field $f^{\mu\nu}$ propagates in a strong magnetic background field B pointing along the z axis. In particular, we make the replacements

$$F_{\mu\nu} \quad \to \quad F_{\mu\nu} + f_{\mu\nu}(\xi), \quad \xi = n \cdot x, \quad (2.136)$$

where $n^{\mu} = (1, \hat{n})$ defines the direction of propagation of the plane wave and

$$F^{\mu\nu} \equiv F_{\mu\nu} \,\hat{=}\, \mathsf{F} = \begin{pmatrix} 0 & 0 & 0 & 0 \\ 0 & 0 & B & 0 \\ 0 & -B & 0 & 0 \\ 0 & 0 & 0 & 0 \end{pmatrix} = i\,B\,\sigma_{t2}. \quad (2.137)$$

The index "t" (transverse) on the second Pauli spin matrix specifies the Lorentz subspace in which $\mu, \nu = 1, 2$. The remaining tensor space will be labeled by the index "l" (longitudinal). Introducing a potential $a_{\mu} = \epsilon_{\mu}\exp(i\omega\xi)$ for the plane wave field, with a polarization amplitude ϵ_{μ}, we may write

$$f_{\mu\nu} = \partial_\mu a_\nu - \partial_\nu a_\mu = i\omega(n_\mu \epsilon_\nu - n_\nu \epsilon_\mu)e^{i\omega\xi}. \tag{2.138}$$

An appropriate expression for $\Pi(s) - \Pi(0)$ is immediately obtainable by integrating the equations of motion (2.15) once. Considering the second line of (2.15), we obtain

$$\frac{d\Pi_\mu(s)}{ds} = e\left[\{F_{\mu\nu} + f_{\mu\nu}[\xi(s)]\}\Pi^\nu \right.$$

$$\left. + \Pi^\nu(s)\{F_{\mu\nu} + f_{\mu\nu}[\xi(s)]\} + \frac{1}{2}\sigma(s)_{\lambda\nu}\,\partial_\mu f^{\lambda\nu}[\xi(s)]\right]$$

$$= 2e\,\Pi^\nu(s)F_{\mu\nu} + \Phi_\mu(s), \tag{2.139}$$

where

$$\Phi_\mu(s) = e\left(\Pi^\nu(s)\,f_{\mu\nu}[\xi(s)] + f_{\mu\nu}[\xi(s)]\,\Pi^\nu(s) + \frac{1}{2}\sigma_{\lambda\nu}(s)\partial_\mu f^{\lambda\nu}[\xi(s)]\right). \tag{2.140}$$

Substituting the equation of motion (2.15) for $x_\mu(s)$ into (2.139), we can perform a proper-time integration to find

$$\Pi_\mu(s) - \Pi_\mu(0) = \int_0^s \frac{d\Pi_\mu(t)}{dt}\,dt \stackrel{(2.139)}{=} \int_0^s dt\left(e\frac{dx^\nu(t)}{dt}F_{\mu\nu} + \Phi_\mu(t)\right)$$

$$= C_\mu^{(-)\,\nu}\left[x_\nu(s) - x_\nu(0)\right] + \int_0^s \Phi_\mu(t)\,dt, \tag{2.141}$$

where we have introduced the short form $C_\mu^{(-)\,\nu} = eF_\mu{}^\nu$.

Some more effort is needed to find an expression for $\Pi(s) + \Pi(0)$. Clearly, we have to extract the information from an integration of the proper-time equation of motion for the coordinate variable x_μ. For this, we need an integrable expression for the momentum $\Pi_\mu(s)$, which can be obtained by solving the momentum equation of motion (this time without inserting the coordinate equation of motion):

$$\frac{d\Pi_\mu(s)}{ds} = 2e\,C_\mu^{(-)\,\nu}\Pi^\nu(s) + \Phi_\mu(s)$$

$$\Longrightarrow\quad \Pi_\mu(s) = \left(e^{2C^{(-)}s}\right)_\mu{}^\nu \Pi_\nu(0) + \int_0^s \left(e^{2C^{(-)}(s-t)}\right)_\mu{}^\nu \Phi_\nu(t)\,dt. \tag{2.142}$$

In the following, the 4×4 matrix calculations can be simplified by employing Pauli spin algebra, since it is useful to note that

$$2C^{(-)}s \stackrel{(2.137)}{=} 2ieBs\,\sigma_{t2}$$

$$\Longrightarrow\quad e^{2C^{(-)}s} = \mathbb{1}_l + \mathbb{1}_t\,\cos 2eBs + i\sigma_{t2}\,\sin 2eBs, \tag{2.143}$$

where we have made use of the notation $\mathbb{1}_l = \mathrm{diag}(1,0,0,1)$ and $\mathbb{1}_t = \mathrm{diag}(0,1,1,0)$. Owing to the absence of an electric component in the external field, the Lorentz tensor structure completely decomposes into purely transverse and longitudinal tensors which commute, i.e. $[P_l, Q_t] = 0$ for arbitrary tensors P and Q.

Substituting (2.143) into (2.142), we find the desired expression for $\Pi_\mu(s)$, which can now be inserted into the equation of motion for $x_\mu(s)$. Upon integration, we obtain

$$x_\mu(s) - x_\mu(0) = \int_0^s \frac{\mathrm{d}x_\mu(t)}{\mathrm{d}t}\,\mathrm{d}t = 2\int_0^s \Pi_\mu(t)\,\mathrm{d}t$$

$$= 2\Pi_\nu(0)\int_0^s \mathrm{d}t\,(\mathbb{1}_l + \mathbb{1}_t \cos 2eBt + i\sigma_{t2}\,\sin 2eBt)_\mu{}^\nu$$

$$+2\int_0^s \mathrm{d}r\,(\mathbb{1}_l + \mathbb{1}_t \cos 2eBr + i\sigma_{t2}\,\sin 2eBr)_\mu{}^\lambda$$

$$\times \int_0^r \mathrm{d}t\,(\mathbb{1}_l + \mathbb{1}_t \cos 2eBt - i\sigma_{t2}\,\sin 2eBt)_\lambda{}^\nu \Phi_\nu(t).$$

Integrating the r integral in the second term by parts leads us to

$$eB\left[x_\mu(s) - x_\mu(0)\right]$$
$$= \left[2eBs\,\mathbb{1}_l + \mathbb{1}_t \sin 2eBs + i\sigma_{t2}\,(1 - \cos 2eBs)\right]_\mu{}^\nu \Pi_\nu(0)$$

$$+ \int_0^s \mathrm{d}t\,\left[2eBs\,\mathbb{1}_l + \mathbb{1}_t \sin 2eB(s-t) - i\sigma_{t2}\,\cos 2eB(s-t)\right]_\mu{}^\nu \Phi_\nu(t)$$

$$- \int_0^s \mathrm{d}t\,\left(2eBs\,\mathbb{1}_l - i\sigma_{t2}\right)_\mu{}^\nu \Phi_\nu(t). \tag{2.144}$$

From this equation, we can determine $\Pi_\mu(0)$ completely in terms of x_μ and Φ_μ by multiplying by the inverse of the matrix

$$\mathsf{M} = 2eBs\,\mathbb{1}_l + \mathbb{1}_t \sin 2eBs + i\sigma_{t2}\,(1 - \cos 2eBs). \tag{2.145}$$

As can be easily checked by direct matrix multiplication, the inverse M^{-1} is given by

$$\mathsf{M}^{-1} = \frac{1}{2eBs}\,\mathbb{1}_l + \frac{1 + \cos 2eBs}{2\sin 2eBs}\,\mathbb{1}_t - \frac{i}{2}\,\sigma_{t2}. \tag{2.146}$$

Multiplication of (2.144) from the left yields

$$\Pi_\lambda(0)$$
$$= \left(\mathsf{M}^{-1}\right)_\lambda{}^\mu eB\left[x_\mu(s) - x_\mu(0)\right] \tag{2.147}$$

$$+ \int_0^s dt \left[\frac{t}{s} \mathbb{1}_l - \frac{1}{2} \mathbb{1}_t - i\sigma_{t2} \left(\frac{1 + \cos 2eBs}{2\sin 2eBs} \right) \right]_\lambda{}^\nu \Phi_\nu(t)$$

$$- \int_0^s dt \left[\mathbb{1}_l + \mathbb{1}_t \left(\frac{(1 + \cos 2eBs)\sin 2eB(s-t)}{2\sin 2eBs} - \frac{\cos 2eB(s-t)}{2} \right) \right.$$

$$\left. - i\sigma_{t2} \left(\frac{(1+\cos 2eBs)\cos 2eB(s-t)}{2\sin 2eBs} + \frac{\sin 2eB(s-t)}{2} \right) \right]_\lambda{}^\nu \Phi_\nu(t).$$

For the moment, let us go back to the integrated momentum equation of motion (2.142), use the representation (2.143) and add $\Pi_\mu(0)$ on both sides:

$$\Pi_\mu(s) + \Pi_\mu(0)$$
$$= \left[2\mathbb{1}_l + \mathbb{1}_t (1 + \cos 2eBs) + i\sigma_{t2} \sin 2eBs \right]_\mu{}^\nu \Pi_\nu(0) \qquad (2.148)$$

$$+ \int_0^s dt \left[\mathbb{1}_l + \mathbb{1}_t \cos 2eB(s-t) + i\sigma_{t2} \sin 2eB(s-t) \right]_\mu{}^\nu \Phi_\nu(t).$$

Finally, we can insert the representation of $\Pi_\nu(0)$ found in (2.147) into the right-hand side of (2.148) to obtain

$$\Pi_\mu(s) + \Pi_\mu(0) = C^{(+)\nu}_\mu \left[x_\nu(s) - x_\nu(0) \right] + \int_0^s dt\, T(s,t)_\mu{}^\nu \Phi_\nu(t), \quad (2.149)$$

where, after a straightforward calculation, the tensors $C^{(+)}$ and $T(s,t)$ are found to be given by

$$C^{(+)} = \left[2\mathbb{1}_l + \mathbb{1}_t (1 + \cos 2eBs) + i\sigma_{t2} \sin 2eBs \right] M^{-1} eB,$$

$$T = v\,\mathbb{1}_l + C(s,t)\,\mathbb{1}_t + S(s,t)\,i\sigma_{t2} = \begin{pmatrix} v & 0 & 0 & 0 \\ 0 & C(s,t) & S(s,t) & 0 \\ 0 & -S(s,t) & C(s,t) & 0 \\ 0 & 0 & 0 & v \end{pmatrix}. \quad (2.150)$$

The functions in the last line of (2.150) are computed as

$$v = \frac{2t}{s} - 1, \quad C(s,t) = \frac{\sin eBvs}{\sin eBs}, \quad S(s,t) = \frac{\cos eBvs - \cos eBs}{\sin eBs}. \quad (2.151)$$

Now we are in a position to rewrite the current for this special field configuration. We insert the expressions for $\Pi(s) \pm \Pi(0)$ found in (2.141) and (2.149) into the representation for a general current, (2.135). Noticing that terms proportional to $x(s) - x(0)$ vanish, since $\langle x(s)|x_\nu(s) - x_\nu(0)|x(0)\rangle = x_\nu - x_\nu = 0$, we find the induced current to be

$$\langle j_\mu(x) \rangle = -\frac{e}{2} \int_0^\infty ds\, e^{-im^2 s} \int_0^s dt\, \mathrm{tr}\left\{ \left[T(s,t)_\mu{}^\nu + i\sigma_\mu{}^\nu \right] \langle x(s)|\Phi_\nu(t)|x(0)\rangle \right\}.$$

$$(2.152)$$

It should be pointed out that this result is exact for an external field configuration of a constant magnetic background field modulated by a plane wave field where both are treated without taking radiative corrections into account. Diagrammatically, we are dealing with an electron loop with infinitely many external photons of both origins (constant and plane wave fields) attached to it.

However, c-number results can only be gained by performing further approximations. The physical circumstances of a weak plane wave field propagating in a strong background field suggest expanding the current in powers of the plane wave amplitude. This can be achieved in a systematic way by going over to an interaction picture in which the "free" evolution operator contains only the constant magnetic field, and the plane wave field is treated as a perturbation. We therefore need the exact solution of the zeroth-order approximation, i.e. the case of a magnetic field alone – a problem that was originally solved by Schwinger [148].

In the following review of this solution, the zeroth-order quantities will be marked by a superscript $^{(0)}$. In particular, we employ the definitions

$$U^{(0)}(s) = e^{-iH^{(0)}s}, \qquad\qquad H^{(0)} = \Pi^{(0)2} - \frac{e}{2}\sigma F,$$

$$\Pi^{(0)}(0) \equiv \Pi^{(0)}, \qquad\qquad \Pi^{(0)}(s) = U(0)^{-1}(s)\,\Pi^{(0)}(0)\,U^{(0)}(s),$$

$$x^{(0)}(0) \equiv x^{(0)}, \qquad\qquad x^{(0)}(s) = U(0)^{-1}(s)\,x^{(0)}(0)\,U^{(0)}(s),$$

$$[x_\mu^{(0)}, \Pi_\nu^{(0)}] = i\,g_{\mu\nu}, \tag{2.153}$$

$$[\Pi_\mu^{(0)}, \Pi_\nu^{(0)}] = ie\,F_{\mu\nu}, \qquad (F)_{\mu\nu} = i\,B\sigma_{t2}.$$

It is evident that the equations of motion in this case simplify to

$$\frac{dx_\mu^{(0)}(s)}{ds} = 2\Pi_\mu^{(0)}(s),$$

$$\frac{d\Pi_\mu^{(0)}(s)}{ds} = 2e\,F_{\mu\nu}\Pi^{(0)\nu}(s), \tag{2.154}$$

since $[F_{\mu\nu}, \Pi_\rho^{(0)}] = 0$. Solving these equations is a simple task of integration. In particular, the momentum equation yields

$$\Pi_\mu^{(0)}(s) = \left(e^{2eFs}\right)_\mu{}^\nu \Pi_\nu^{(0)}(0)$$

$$= \left(\mathbb{1}_1 + \mathbb{1}_t \cos 2eBs + i\sigma_{t2}\sin 2eBs\right)_\mu{}^\nu \Pi_\nu^{(0)}(0)$$

$$= R(s)_\mu{}^\nu \Pi_\nu^{(0)}(0), \tag{2.155}$$

where we have defined

$$R(s)_\mu{}^\nu \,\hat{=}\, R(s) \equiv e^{2C^{(-)}s} = \begin{pmatrix} 1 & 0 & 0 & 0 \\ 0 & \cos 2eBs & \sin 2eBs & 0 \\ 0 & -\sin 2eBs & \cos 2eBs & 0 \\ 0 & 0 & 0 & 1 \end{pmatrix}. \tag{2.156}$$

Inserting (2.155) into the coordinate equation of motion (2.154), the latter can immediately be integrated, with the result

$$x_\mu^{(0)}(s) = x_\mu^{(0)}(0) + 2\left(\mathbb{1}_1 s + \mathbb{1}_t \frac{\sin 2eBs}{2eB} - i\sigma_{t2}\frac{\cos 2eBs - 1}{2eB}\right)_\mu^{\ \nu} \Pi_\nu^{(0)}(0)$$

$$= x_\mu^{(0)}(0) + I(s)_\mu^{\ \nu}\, \Pi_\nu^{(0)}(0), \tag{2.157}$$

where

$$I(s)_\mu^{\ \nu} \ \hat{=} \ \mathsf{I}(s)$$

$$= \begin{pmatrix} 2s & 0 & 0 & 0 \\ 0 & (\sin 2eBs)/(eB) & (1 - \cos 2eBs)/(eB) & 0 \\ 0 & (\cos 2eBs - 1)/(eB) & (\sin 2eBs)/(eB) & 0 \\ 0 & 0 & 0 & 2s \end{pmatrix}. \tag{2.158}$$

One further necessary ingredient for the following expansion is the zeroth-order evolution operator in coordinate space $\langle x^{(0)}(s)|x^{(0)}(0)\rangle$, which has already been found in (2.39) of Sect. 2.1. In the case of a purely magnetic field, the expression reduces to

$$\langle x^{(0)}(s)|x^{(0)}(0)\rangle \equiv \langle x^{(0)}| U^{(0)}(s) |x^{(0)}\rangle$$

$$= -\frac{i}{(4\pi s)^2}\frac{eBs}{\sin eBs}\, e^{i(e/2)\sigma Fs}. \tag{2.159}$$

This completes our investigation of the zeroth order, i.e. the constant-field part of the problem.

Our next intermediate step is characterized by finding the relations between the zeroth-order and exact quantities. While the exact and zeroth-order coordinates at zero proper time coincide and thus can be viewed as a boundary condition, the corresponding canonical momenta differ by the plane wave amplitude:

$$x_\mu(0) = x_\mu^{(0)}(0), \quad \Pi_\mu(0) = p_\mu - eA_\mu = \Pi_\mu^{(0)}(0) - ea_\mu[\xi^{(0)}(0)], \tag{2.160}$$

whereby $\xi^{(0)} = n \cdot x^{(0)}$. By insertion of (2.160), the exact Hamiltonian can be related to the zeroth-order Hamiltonian:

$$H = \Pi^2 - \frac{e}{2}\sigma(F + f) = \{\Pi_\mu^{(0)}(0) - ea_\mu[\xi^{(0)}(0)]\}^2 - \frac{e}{2}\sigma(F + f)$$

$$= H^{(0)} - e\left[a_\mu[\xi^{(0)}(0)]\,\Pi^{(0)\mu} + \Pi^{(0)\mu}\, a_\mu[\xi^{(0)}(0)]\right. \tag{2.161}$$

$$\left. - e\, a_\mu[\xi^{(0)}(0)]a^\mu[\xi^{(0)}(0)] + \frac{1}{2}\sigma_{\mu\nu}f^{\mu\nu}[\xi^{(0)}(0)]\right].$$

In order to divide the complete evolution operator into the "free" (zeroth-order) and interaction parts,

$$U(s) = U^{(0)}(s)\, U_{\mathrm{I}}(s) \quad \Longleftrightarrow \quad e^{-iHs} = e^{-iH^{(0)}s}\, U_{\mathrm{I}}(s), \tag{2.162}$$

we employ the operator identity

$$e^{A+B} = e^A \, \mathrm{T} \exp \left(\int_0^1 dt \, e^{At} B e^{At} \right), \tag{2.163}$$

where $A = -iH^{(0)}s$ and $B = -i(H - H^{(0)})s$. Obviously, the interaction evolution operator can be computed as

$$U_{\mathrm{I}}(s) = \mathrm{T} \exp \left(-i \int_0^1 dt \, U^{(0)^{-1}}(st) \, (H - H^{(0)}) s \, U^{(0)}(st) \right)$$

$$= \mathrm{T} \exp \left(-i \int_0^s du \, U^{(0)^{-1}}(u) \, (H - H^{(0)}) \, U^{(0)}(u) \right). \tag{2.164}$$

With the help of (2.161), the integrand in the exponent of (2.164) can be written as

$$-iU^{(0)^{-1}}(u) \, (H - H^{(0)}) \, U^{(0)}(u)$$

$$= ie \Big[a_\mu[\xi^{(0)}(u)] \, \Pi^{(0)\mu}(u) + \Pi^{(0)\mu}(0) \, a_\mu[\xi^{(0)}(u)] \tag{2.165}$$

$$- e \, a_\mu[\xi^{(0)}(u)] a^\mu[\xi^{(0)}(u)] + \frac{1}{2} \sigma_{\mu\nu}^{(0)}(u) f^{\mu\nu}[\xi^{(0)}(u)] \Big],$$

where we have repeatedly inserted factors of $U^{(0)}(u)U^{(0)-1}(u)$. Incidentally, the proper-time-dependent σ matrix in the last term is given by

$$\sigma_{\mu\nu}^{(0)}(u) = U^{(0)-1}(u) \, \sigma_{\mu\nu} \, U^{(0)}(u) = e^{-i(e/2)u\sigma F} \, \sigma_{\mu\nu} \, e^{i(e/2)u\sigma F}, \tag{2.166}$$

since $[\Pi_\rho^{(0)}, \sigma_{\mu\nu}] = 0$.

Substituting (2.165) into (2.164) produces

$$U_{\mathrm{I}} = \mathrm{T} \exp \left\{ ie \int_0^s du \Big[a_\mu[\xi^{(0)}(u)] \, \Pi^{(0)\mu}(u) + \Pi^{(0)\mu}(0) \, a_\mu[\xi^{(0)}(u)] \right. \tag{2.167}$$

$$\left. - e \, a_\mu[\xi^{(0)}(u)] a^\mu(\xi^{(0)}(u)) + \frac{1}{2} \sigma_{\mu\nu}^{(0)}(u) f^{\mu\nu}[\xi^{(0)}(u)] \Big] \right\}.$$

The collection of the tools required for the interaction picture is now completed, and we can turn to the remaining problem of rephrasing the object $\langle x(s)|\Phi_\nu(t)|x(0)\rangle$, which is contained in the equation for the the current (2.152):

$$\langle x(s)|\Phi_\nu(t)|x(0)\rangle$$

$$= \langle x(0)|U(s)U^{-1}(t)\Phi_\nu(0)U(t)|x(0)\rangle$$

$$= \langle x(0)|U^{(0)}(s)U_{\mathrm{I}}(s)U_{\mathrm{I}}^{-1}(t)U^{(0)-1}(t)\Phi_\nu(0)U^{(0)}(t)U_{\mathrm{I}}(t)|x(0)\rangle$$

$$= \langle x^{(0)}(s)|U_{\mathrm{I}}(s)U_{\mathrm{I}}^{-1}(t)\Phi_\nu^{(0)}(t)U_{\mathrm{I}}(t)|x(0)^{(0)}\rangle. \tag{2.168}$$

We may give a more explicit expression for $\Phi_\nu^{(0)}(t)$ in the last line by employing its definition in (2.140) and noting that $\Pi^\nu = \Pi^{(0)\nu} - ea^\nu$:

$$\Phi_\nu^{(0)}(t) = e\Big(\Pi^{(0)\rho}(t)\, f_{\nu\rho}[\xi^{(0)}(t)] + f_{\nu\rho}[\xi^{(0)}(t)]\, \Pi^{(0)\rho}(t)$$

$$-2e\, a^\rho[\xi^{(0)}(t)] f_{\nu\rho}[\xi^{(0)}(t)] + \frac{1}{2}\sigma_{\lambda\kappa}^{(0)}(t)\partial_\nu f^{\lambda\kappa}[\xi^{(0)}(t)] \Big). \quad (2.169)$$

Inserting the last line back into (2.168), which finally enters (2.152), we obtain the desired expression for the current in the interaction picture:

$$\langle j_\mu(x)\rangle$$

$$= -\frac{e^2}{2}\int\limits_0^\infty ds\, e^{-im^2 s}\int\limits_0^s dt\, \mathrm{tr}\Big\{ [T(s,t)_\mu{}^\nu + i\sigma_\mu{}^\nu] \qquad (2.170)$$

$$\times \langle x^{(0)}(s)|U_I(s)U_I^{-1}(t)\Big[\Pi^{(0)\rho}(t)\, f_{\nu\rho}[\xi^{(0)}(t)] + f_{\nu\rho}[\xi^{(0)}(t)]\, \Pi^{(0)\rho}(t)$$

$$-2ea^\rho[\xi^{(0)}(t)] f_{\nu\rho}[\xi^{(0)}(t)] + \frac{1}{2}\sigma_{\lambda\kappa}^{(0)}(t)\partial_\nu f^{\lambda\kappa}[\xi^{(0)}(t)]\Big]U_I(t)|x^{(0)}(0)\rangle\Big\}.$$

To study the birefringence properties of the vacuum, it is sufficient to consider the first-order approximation of the current, which describes the lowest-order interaction of the plane wave field with the magnetically induced currents. Since the operator in the large square brackets in (2.170) is at least of first order in the plane wave field, the interaction evolution operators U_I have to be approximated simply by $U_I \simeq 1$. Furthermore, the third term in the large square brackets of (2.170) has to be omitted, because it is already of second order in the plane wave field. The resulting first-order approximation of the current then reads

$$\langle j_\mu(x)\rangle = -\frac{e^2}{2}\int\limits_0^\infty ds\, e^{-im^2 s}\int\limits_0^s dt\, \mathrm{tr}\Big\{ [T(s,t)_\mu{}^\nu + i\sigma_\mu{}^\nu] \qquad (2.171)$$

$$\times \langle x(s)|\Big[\Pi^\rho(t)\, f_{\nu\rho}[\xi(t)]$$

$$+ f_{\nu\rho}[\xi(t)]\, \Pi^\rho(t) + \frac{1}{2}\sigma_{\lambda\kappa}(t)\partial_\nu f^{\lambda\kappa}[\xi(t)]\Big]|x(0)\rangle\Big\}.$$

For convenience, we have omitted the superscript $^{(0)}$ in (2.171) and shall continue to suppress it in the following, since it should be clear that from now on we are only dealing with zeroth-order quantities.

Our final task is to evaluate the matrix elements of the operators in the large square brackets in (2.171) in coordinate representation and then take the Dirac trace to obtain an explicit c-number expression. With the aid of (2.155), and inverting (2.157), we obtain a coordinate representation for the momentum operator:

$$\Pi(t) = \mathsf{R}(t) \cdot \mathsf{I}^{-1}(s) \cdot [x(s) - x(0)], \tag{2.172}$$

where we have used matrix notation and the dot denotes matrix multiplication. Similarly, we can represent the plane wave phase by

$$\exp[i\omega\xi(t)] = \exp[i\omega\, n \cdot x(t)] \overset{(2.157)}{=} \exp\{i\omega\, n \cdot [x(0) + \mathsf{I}(t) \cdot \Pi(0)]\}$$
$$= \exp\big\{A[x(s)] + B[x(0)]\big\}, \tag{2.173}$$

where we have defined

$$A[x(s)] = i\omega\, n \cdot \mathsf{I}(t) \cdot \mathsf{I}^{-1}(s) \cdot x(s),$$
$$B[x(0)] = i\omega\, n \cdot \big[1 - \mathsf{I}(t) \cdot \mathsf{I}^{-1}(s)\big] \cdot x(0). \tag{2.174}$$

Since the commutator of $x(s)$ and $x(0)$ is nonvanishing,

$$[x(s)_\mu, x(0)_\nu] \overset{(2.157)}{=} -i\, I_{\mu\nu}(s), \tag{2.175}$$

we find for the commutator of A and B

$$R(s,t) = -\frac{1}{2\omega^2}[A[x(s)], B[x(0)]]$$
$$= -\frac{i}{2}[n \cdot \mathsf{I}(t)] \cdot \Big\{n \cdot \big[1 - \mathsf{I}(t) \cdot \mathsf{I}^{-1}(s)\big]\Big\} \tag{2.176}$$

which is c-number valued. Employing the Baker–Campbell–Hausdorff formula,

$$e^{A+B} = e^A e^B e^{-\frac{1}{2}[A,B]},$$

we may write for (2.173)

$$e^{i\omega\xi(t)} = e^{A[x(s)]} e^{B[x(0)]} e^{\omega^2 R(s,t)}. \tag{2.177}$$

Regarding the products of $\Pi^\rho(t)$ and $f_{\nu\rho}[\xi(t)]$ in (2.171), we need to order the operators in such a way that all the $x(s)$'s stand to the left of all the $x(0)$'s. In this way, they can be pulled out of the $\langle x(s)|$ and $|x(0)\rangle$ brackets. For this, we make use of the operator identity

$$e^a b = b e^a + [a, b] e^a,$$

which we apply to

$$x^\alpha(0)\, e^{A[x(s)]} = e^{A[x(s)]}\, x^\alpha(0) - \omega\big[n \cdot \mathsf{I}(t)\big]^\alpha e^{A[x(s)]} \tag{2.178}$$

and

$$e^{B[x(0)]}\, x(s)^\alpha = x^\alpha(s)\, e^{B[x(0)]}$$
$$+ \omega\left(\mathsf{I}(s) \cdot \Big\{n \cdot \big[\mathsf{I}(t) \cdot \mathsf{I}^{-1}(s) - 1\big]\Big\}\right)^\alpha e^{B[x(0)]}. \tag{2.179}$$

With the help of (2.172)–(2.179), we can rewrite the equation for the current (2.171):

$$\langle j_\mu(x)\rangle = -\frac{e^2}{2}\omega \int_0^\infty ds\, e^{-im^2 s} \int_0^s dt\, e^{\omega^2 R(s,t)} \mathrm{tr}\left\{ \left(T_\mu{}^\nu + i\sigma_\mu{}^\nu\right)\langle x(s)|x(0)\rangle\right.$$

$$\times \left[\left(f\cdot R(t)\cdot\left\{I^{-1}(s)\cdot[n\cdot I(t)]+n\cdot I(t)\cdot I^{-1}(s)-n\right\}\right)_\nu\right.\tag{2.180}$$

$$\left.\left.+\frac{i}{2}n_\mu f^{\lambda\kappa}\sigma(t)_{\lambda\kappa}\right]\right\},$$

where the transition amplitude $\langle x(s)|x(0)\rangle \equiv K(s)$ is given in (2.159). Here it should be pointed out that the object $\sigma(t)_{\lambda\kappa}$ can only be pulled out of the sandwiching bracket to the right, since $\langle x(s)|$ possesses a nontrivial Dirac structure, while $|x(0)\rangle$ does not.

In principle, the Dirac traces, as well as the matrix manipulations, could be performed in full generality, but for reasons of lucidity we shall make use of the simplifying specifications of our particular configuration. The magnetic background field points along the positive z axis; we consider the plane wave as propagating along the positive x direction: $n^\mu = (1,1,0,0)$. Consequently, we denote the two possible polarization vectors as $\epsilon_\perp^\mu = (0,0,1,0)$ and $\epsilon_\parallel^\mu = (0,0,0,1)$, specifying the directions perpendicular ($\perp$) and parallel ($\parallel$) to the plane spanned by the magnetic field and the propagation direction.[6] This choice also normalizes the plane wave potential, $|a_{\parallel,\perp}|^2 = a_{\parallel,\perp\mu}^* a_{\parallel,\perp}^\mu$.

The matrix multiplications can be performed straightforwardly; $I(s)$ is given in (2.158), and its inverse reads

$$I^{-1}(s) = \frac{1}{2s}\mathbb{1}_l + \frac{eB\sin 2eBs}{2(1-\cos 2eBs)}\mathbb{1}_t - \frac{i}{2}eB\,\sigma_{t2},\tag{2.181}$$

which enables us to evaluate the function $R(s,t)$ defined in (2.176):

$$R(s,t) = \frac{i}{2}\left[\frac{s}{2}(1-v^2) - \frac{\cos eBvs - \cos eBs}{eB\sin eBs}\right].\tag{2.182}$$

Using (2.156), the plane-wave-independent part of the first term in the second line of (2.180) is now computed as

$$R(t)\cdot\left\{I^{-1}(s)\cdot[n\cdot I(t)]+n\cdot I(t)\cdot I^{-1}(s)-n\right\}\alpha = \begin{pmatrix} -v \\ C(s,t) \\ S(s,t) \\ 0 \end{pmatrix}_\alpha$$

$$\equiv \left(n\cdot T\right)_\alpha,\tag{2.183}$$

where the functions $S(s,t)$ and $C(s,t)$ are defined in (2.151) and the matrix T is given in (2.150).

---

[6] Besides the use of different metric conventions, note that Adler's definition of perpendicular and parallel components is opposite to ours, since he refers to the magnetic field vector of the photon field. In the present work, the polarization vector designates the direction of the electric field vector of the photon field.

Substituting (2.183) and (2.159) into (2.180) leads us to

$$\langle j^\mu(x) \rangle$$

$$= \omega \frac{\alpha}{2\pi} i \int\limits_0^\infty \frac{ds}{s^2} e^{-im^2 s} \int\limits_0^s dt\, e^{\omega^2 R(s,t)} \left(\frac{1}{4} \frac{eBs}{\sin eBs} \right) \tag{2.184}$$

$$\times \left\{ [\mathbf{T} \cdot \mathbf{f} \cdot (\mathbf{n} \cdot \mathbf{T})]_\mu \, \mathrm{tr} \left[e^{i(e/2)s\sigma F} \right] + i[\mathbf{f} \cdot (\mathbf{n} \cdot \mathbf{T})]_\nu \, \mathrm{tr} \left[\sigma_\mu{}^\nu e^{i(e/2)s\sigma F} \right] \right.$$

$$+ \frac{i}{2} (\mathbf{T} \cdot \mathbf{n}) f^{\lambda \kappa} \, \mathrm{tr} \left[e^{i(e/2)s\sigma F} \sigma_{\lambda \kappa}(t) \right]$$

$$\left. - \frac{1}{2} n_\nu f^{\lambda \kappa} \, \mathrm{tr} \left[\sigma_\mu{}^\nu e^{i(e/2)s\sigma F} \sigma_{\lambda \kappa}(t) \right] \right\}.$$

The simplest way to, on the one hand, calculate the Dirac traces and, on the other hand, extract the desired information is to project the current onto the polarization directions by multiplying by $a^*_{\|,\perp}{}^\mu$. Incidentally, the current can only be constructed as a Lorentz vector from the polarization vectors, because gauge invariance implies transversality.

For the reduced case of a purely magnetic field, the exponential function of σF can be conveniently represented by Pauli spin matrices (cf. Appendix C):

$$e^{i(e/2)s\sigma F} = \mathbb{1} \cos eBs + i \sigma_3 \sin eBs, \tag{2.185}$$

where it is understood that the (2×2) Pauli matrices are doubled blockwise in Dirac space. We can immediately read off that

$$\mathrm{tr} \left[e^{i(e/2)s\sigma F} \right] = 4 \cos eBs, \tag{2.186}$$

and, in conjunction with the identities listed in Appendix C, we also find

$$\mathrm{tr} \left[\sigma_{\mu\nu} e^{i(e/2)s\sigma F} \right] = -4 \sin eBs \left(\sigma_{t2} \right)_{\mu\nu}. \tag{2.187}$$

The same relation holds for the third trace in (2.184). For the last trace, we observe that, by employing the contraction of the current with $a^*_{\|,\perp}{}^\mu$, this object can be written as

$$\mathrm{tr} \left[\left(f^*_{\|,\perp} \sigma \right) e^{i(e/2)(s-t)\sigma F} \left(f_{\|,\perp} \sigma \right) e^{i(e/2)t\sigma F} \right]$$

$$= -\omega^2 (\pm 16)_{\|,\perp} (\cos eBvs - \cos eBs), \tag{2.188}$$

which involves the definition of $\sigma_{\mu\nu}(t)$ in (2.166). (Details can be looked up in Appendix C.)

After inserting these traces into (2.184) and computing the scalar products by projecting onto the polarization directions, we can summarize our findings for the induced current regarding this particular field configuration as

$$\langle j^\mu_{\|,\perp} \rangle = -\omega^2 A_{\|,\perp}(\omega, B) \, a^\mu_{\|,\perp}, \tag{2.189}$$

where $A_{\parallel,\perp}$ is given by

$$A_{\parallel,\perp}(\omega, B) = \frac{\alpha}{2\pi} \int\limits_0^\infty \frac{ds}{s^2} e^{im^2 s} \int\limits_0^s dt\, e^{\omega^2 R(s,t)} \, N_{\parallel,\perp}(s,t), \qquad (2.190)$$

and the functions $N_{\parallel,\perp}$ are constructed from simple combinations of trigonometric functions:

$$N_\parallel = \frac{eBs \cos eBvs}{\sin eBs} - eBs \cot eBs \left(1 - v^2 + v \frac{\sin eBvs}{\sin eBs}\right), \qquad (2.191)$$

$$N_\perp = -\frac{eBs \cos eBvs}{\sin eBs} + \frac{veBs \sin eBvs \cot eBs}{\sin eBs}$$
$$+ \frac{2eBs(\cos eBvs - \cos eBs)}{\sin^3 eBs}. \qquad (2.192)$$

An expression for $R(s,t)$ has been obtained in (2.182).

These results are completely consistent with the findings of the previous section, in which the same situation was discussed via the polarization tensor. To be precise, the induced current as calculated here for a plane wave field moving in an external constant magnetic field is related to the polarization tensor given in (2.91) according to

$$\langle j^\mu_{\parallel,\perp} \rangle = -\Pi^{\mu\nu}(\omega n)\, a_\nu(\omega n), \qquad (2.193)$$

which simply reflects the implications of the fundamental principle of linear response as described by (2.2). At this point, we want to stress that the present calculation has been simplified in two ways compared with the previous section: first, we assumed a specific propagation of the plane wave field, namely, perpendicular to the B field; however, the general case can be immediately restored by inserting a factor of $\sin^2 \theta$ into the function $R(s,t)$ in (2.182) and as an overall factor in front of (2.190) (θ denotes the angle between the magnetic field and the propagation direction). Secondly, we assumed during the whole calculation that the plane wave field propagates on the light cone by requiring that n^μ is a null vector (cf. (2.136)). This implies that the refractive index for the plane wave field remains close to 1, which is, of course, not true in the strong-field limit ($B/B_{\mathrm{cr}} > \pi/\alpha$). As has been demonstrated by Daugherty and Lerche [51], it is not essential to maintain this assumption during the calculation; indeed, from the details given above, it is obvious that it only enters the final matrix manipulations, which simplify slightly for this case. Of course, the strong-field limit has to be analyzed without any *a priori* assumptions about the wave vector (cf. Sect. 3.2).

As mentioned above, the technique outlined here can also be applied to the calculation of photon-splitting matrix elements [4]. For this, the induced current has to be evaluated to second order in the plane wave field.

Finally, let us mention that a generalization of the present calculation to the case of parallel electric and magnetic fields was treated by Daugherty and Lerche [51].

2.4 Schwinger's Equivalence Theorem and the Axial-Vector Anomaly

Schwinger's equivalence theorem, which claims equivalence between the pseudoscalar and pseudovector interactions of a pseudoscalar field with electrically charged fermions at low energies, is derived. It is shown that there are no non-perturbative effects arising from the all-order coupling between the external electromagnetic field and the fermion loop.

The relation between the equivalence theorem and the axial-vector anomaly is outlined, proving their coincidence for small photon momenta only. As a corollary, we find that there are no nonperturbative contributions to the axial-vector anomaly to one loop in the domain of small photon momenta.

Besides being a powerful tool for investigating QED, proper-time techniques provide for a detailed study of axial couplings between pseudoscalar and fermionic fields in the presence of external electromagnetic fields. In particular, problems related to gauge invariance can be investigated thoroughly.

In his seminal work, Schwinger [148] proved the *equivalence theorem*, which states that, in the low-energy regime, a *pseudoscalar* interaction between a spinless neutral meson and a fermion field leads to the same result for the decay of the meson into two photons as a *pseudovector* interaction.

In the following, we shall reinvestigate Schwinger's equivalence theorem with special emphasis on the nonperturbative domain, i.e. the coupling of the external electromagnetic field to the fermion loop to all orders. Furthermore, we shall point out in what respect the equivalence theorem is related to the axial-vector anomaly as discovered by Adler, Bell and Jackiw [2, 25].

2.4.1 Equivalence Theorem

Let us first describe Schwinger's view on the two-photon decay of the neutral pion, which was inspired by a paper by Steinberger [156] on the question of the equivalence of various interaction Lagrangians. In fact, it will turn out that the present considerations cannot be immediately applied to pion decay, owing to a mismatch of energy–momentum regimes; let us, however, stick to the historical viewpoint for the time being and later translate our findings into modern language.

Pseudoscalar Interaction

For the *pseudoscalar* interaction between a neutral meson field and a fermionic field, we begin with the Lagrangian:

$$\mathcal{L}^{\mathrm{PS}} = -\mathrm{i}g\,\phi(x)\,\frac{1}{2}\left[\bar{\psi}(x), \gamma_5\psi(x)\right], \qquad (2.194)$$

where g denotes a dimensionless coupling constant, and ϕ represents the spinless meson (pion). The fermion was identified with the proton; nowadays, ψ should be associated with a quark appearing in three colors.

In order to describe the decay of the pion into two photons, Schwinger replaced the fermion fields by their vacuum expectation value in the presence of an external electromagnetic field:

$$\mathcal{L}^{PS} \to \mathcal{L}^{PS}_{eff} = -ig\,\phi(x)\,\frac{1}{2}\langle[\bar\psi(x),\gamma_5\psi(x)]\rangle^A$$
$$= g\,\phi(x)\,\mathrm{tr}\,\gamma_5 G(x,x|A), \tag{2.195}$$

where taking the vacuum expectation value in the first line translates into a time-ordering prescription via the point splitting procedure:

$$i\,\mathrm{tr}\,\gamma_5 G(x,x|A) = \lim_{x' \overset{S}{\to} x} \gamma_5\langle 0|T\,\bar\psi(x')\psi(x)|0\rangle^A. \tag{2.196}$$

In (2.196), the limit has to be performed with respect to the time-like components, and the invariant distance $(x-x')^2$ should be space-like, i.e. > 0.

Equation (2.195) can be diagrammatically represented as in Fig. 2.2,

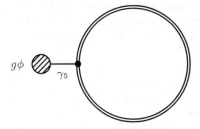

Fig. 2.2. Diagrammatic representation of (2.195)

where the double line represents the coupling to the external electromagnetic field to all orders. Obviously, taking the vacuum expectation value with respect to the external field corresponds to integrating out the fermions to one loop.

Inserting the proper-time representation of the Greens function G (cf. (2.7)) into (2.195), we obtain

$$\mathcal{L}^{PS}_{eff} = g\,\phi(x)\,\mathrm{tr}\left(\gamma_5\,\langle x|(M-\gamma\varPi)i\int_0^\infty ds\,e^{-is[M^2-(\gamma\varPi)^2]}|x\rangle\right)$$
$$= gM\,\phi(x)\,i\,\mathrm{tr}\left(\gamma_5\int_0^\infty ds\,e^{-isM^2}\langle x|e^{is\varPi^2}|x\rangle\right), \tag{2.197}$$

where M denotes the mass of the fermion. To arrive at (2.197), we have employed the fact that traces of odd numbers of γ^μ's together with a γ_5 vanish. In the last term of (2.197), we encounter the trace of the proper-time evolution operator.

Now Schwinger, identifying the internal fermions with protons, argued that the momentum of the outgoing photons of the pion decay is much smaller than the mass of the loop particle. Therefore, the electromagnetic fields associated with the photons vary slowly compared with the length scale set by the Compton wavelength of the proton. As a consequence, the constant-field approximation for the proper-time transition amplitude appears to be appropriate in the present situation.

However, it is not the proton whose fluctuations constitute the loop; rather, the loop particles have to be identified with the quarks.[7] But the current masses of the light quarks are much smaller than the momenta of the outgoing photons of the decaying pion, which completely spoils the slowly-varying-field assumption. The fact that pion decay is nevertheless describable with our final results can be attributed to the general form of the anomaly, which is already revealed by the constant-field approximation. This will be elucidated in more detail later on.

Hence, let us forget about pion decay and simply proceed with the constant-field/low-photon-energy approximation assuming that a heavy fermion runs in the loop. With the aid of (2.39), we insert the representation of the evolution operator into (2.197) and find[8]

$$
\mathcal{L}_{\text{eff}}^{\text{PS}} = \frac{gM}{(4\pi)^2} \, \phi(x) \int_0^\infty \frac{ds}{s^2} \, e^{-isM^2} \underbrace{\exp\left(-\frac{1}{2}\text{tr} \ln \frac{\sinh eFs}{eFs}\right)}_{= \frac{eas \, ebs}{\sin eas \, \sinh ebs}}
$$

$$
\times \underbrace{\text{tr}\left\{\gamma_5 \, e^{i(e/2)\sigma Fs}\right\}}_{\overset{(C.20)}{=} -4i\sin eas \, \sinh ebs}
$$

$$
= -4i\frac{gM}{(4\pi)^2} \, \phi(x) \, e^2 \int_0^\infty ds \, e^{-isM^2} \, ab
$$

$$
= -\frac{\alpha}{\pi} \frac{g}{M} \, \phi(x) \, \mathcal{G} = \frac{\alpha}{\pi} \frac{g}{M} \, \phi(x) \, \boldsymbol{E} \cdot \boldsymbol{B} \tag{2.198}
$$

$$
= -\frac{1}{4}\frac{\alpha}{\pi} \frac{g}{M} \, \phi(x) \, F_{\mu\nu} \, {}^\star F^{\mu\nu}.
$$

This equation represents the low-energy effective Lagrangian of a pseudoscalar interaction between a spinless meson and a heavy fermion in an external field. Although we have included the coupling of the loop fermion to the external field to all orders, the final result is only of second order in the electromagnetic field strength. Hence, if we had expanded the loop perturbatively in α, then only the graph with two external photons would have contributed to the final result. Note also that we have encountered no

---

[7] In fact, in order to obtain a reasonable value for the pion lifetime, one has to take the number of colors into account.
[8] Remember $\mathcal{G} = (1/4)F_{\mu\nu} \, {}^\star F^{\mu\nu} = -\boldsymbol{E} \cdot \boldsymbol{B} = ab$.

singular terms while calculating $G(x, x|A)$; the dangerous terms vanished by Dirac algebraic arguments.

Pseudovector Interaction

Let us consider a pseudovector interaction of the following particular form:

$$\mathcal{L}^{\mathrm{PV}} = -\mathrm{i}\frac{g}{2M}\, \partial_\mu \phi(x)\, \frac{1}{2\mathrm{i}}\left[\bar{\psi}(x), \gamma_5 \gamma^\mu \psi(x)\right]. \tag{2.199}$$

Classically, this pseudovector interaction Lagrangian is formally equivalent to the pseudoscalar counterpart as defined in (2.194), since

$$\mathcal{L}^{\mathrm{PV}} \stackrel{\mathrm{i.b.p.}}{=} \mathrm{i}\frac{g}{2M}\, \phi\, \frac{1}{2\mathrm{i}}\left\{\left[\partial_\mu\bar{\psi}, \gamma_5 \gamma^\mu \psi\right] + \left[\bar{\psi}, \gamma_5 \partial\!\!\!/\psi\right]\right\} + \text{s.t.},$$

$$\stackrel{\mathrm{e.o.m.}}{=} -\mathrm{i}g\, \phi(x)\, \frac{1}{2}\left[\bar{\psi}(x), \gamma_5 \psi(x)\right] + \text{s.t.},$$

where "s.t." means "surface terms". Here, we have first integrated by parts (i.b.p.); then we have employed the equations of motion (e.o.m.). However, at the quantum level, things become more complicated. Proceeding in the same way as in the pseudoscalar case, we naively arrive at

$$
\begin{aligned}
\mathcal{L}_{\mathrm{eff}}^{\mathrm{PV}} &= -\mathrm{i}\frac{g}{2M}\, \partial_\mu \phi(x)\, \frac{1}{2\mathrm{i}}\langle\left[\bar{\psi}(x), \gamma_5 \gamma^\mu \psi(x)\right]\rangle^A \\
&= -\mathrm{i}\frac{g}{2M}\, \partial_\mu \phi(x)\, \mathrm{tr}\, \gamma_5 \gamma^\mu\, G(x, x|A) \\
&\stackrel{\mathrm{i.b.p.}}{=} \mathrm{i}\frac{g}{2M}\, \phi(x)\, \partial_\mu \mathrm{tr}\, \gamma_5 \gamma^\mu\, G(x, x|A) + \text{s.t.}
\end{aligned}
\tag{2.200}
$$

Now we are in trouble! Not only do we have to face the problem of singularities in $G(x, x|A)$, but we also have to give a meaning to the derivative at this singular coincidence point. The first guess is to introduce an appropriate point splitting regarding the two arguments of the Green's function. But then, we have to keep in mind that $G(x, x'|A)$ is a gauge *variant* quantity while $G(x, x|A)$ is not. In order to ensure gauge invariance, we have to replace the ordinary derivative with covariant derivatives. Following Schwinger, we reinterpret the critical term in the last line of (2.200) as

$$\partial_\mu \mathrm{tr}\, \gamma_5 \gamma^\mu\, G(x, x|A) \tag{2.201}$$

$$\to \lim_{x'', x' \to x}\left\{\left[\partial'_\mu - \mathrm{i}eA_\mu(x')\right] + \left[\partial''_\mu + \mathrm{i}eA_\mu(x'')\right]\right\}\mathrm{tr}\, \gamma_5 \gamma^\mu\, G(x', x''|A).$$

One can easily check that the right-hand side reduces to the left-hand side after naively taking the limit in a formal sense. Now we could follow Schwinger and evaluate the right-hand side of (2.201) in the weak-field limit, i.e. up to second order in the field strength. This would again correspond to a triangle graph, which is known to contribute solely to the axial-vector anomaly to any finite order of perturbation theory.

Instead, we shall try to maintain the coupling to the external field to all orders in order to pursue the question of possible nonperturbative contributions to the meson–photon interaction. Of course, the price we have to pay is that we are strictly tied to the slowly-varying-field approximation.

So, let us employ the representation of the fermionic Green's function in an arbitrary constant electromagnetic field given in (2.46):

$$
G(x, x'|A)
$$

$$
= \Phi(x, x'|A) \frac{1}{(4\pi)^2} \int\limits_0^\infty \frac{ds}{s^2} \left[m - \frac{1}{2}\gamma^\mu [f(s) + eF]_{\mu\nu}(x-x')^\nu \right]
$$

$$
\times \exp\left[-iM^2 s - L(s) + \frac{i}{4}(x-x')f(s)(x-x') \right] \exp\left(i\frac{e}{2}\sigma F s \right), (2.202)
$$

where

$$
f(s) = eF \coth(eFs),
$$

$$
L(s) = \frac{1}{2}\mathrm{tr}\ln\frac{\sinh(eFs)}{eFs} \quad \Rightarrow \quad e^{-L(s)} = \frac{eas\,ebs}{\sin eas\,\sinh ebs},
$$

and

$$
\Phi(x, x'|A) = \exp\left[ie \int\limits_{x'}^{x} d\xi_\mu \left(A^\mu(\xi) + \frac{1}{2}F^{\mu\nu}(\xi - x')_\nu \right) \right] \tag{2.203}
$$

completely carries the gauge dependence of the propagator. Having separated the gauge dependence in this way, we may also write

$$
G(x', x''|A) = \Phi(x', x''|A)\, G(x', x''|A_{\mathrm{SF}}), \tag{2.204}
$$

where $G(x', x''|A_{\mathrm{SF}})$ is the Green's function evaluated in the Schwinger–Fock gauge and depends only on the field strength.

Regarding the required Dirac traces, we find

$$
\mathrm{tr}\,\gamma_5\gamma^\mu\, e^{i(e/2)\sigma Fs} = 0 \qquad \text{(odd number of } \gamma^\mu\text{'s)},
$$

$$
\mathrm{tr}\,\gamma_5\gamma^\mu\gamma^\alpha\, e^{i(e/2)\sigma Fs} = 4i\left(eT\,{}^\star F^{\mu\alpha} + P\,g^{\mu\alpha} - eT^*\,F^{\mu\alpha} \right), \tag{2.205}
$$

where we have extensively employed the results of Appendix C. Furthermore, it is useful to find an explicit expression for the function $f(s)_{\alpha\beta}$; for this, we make use of the spectral representation of $F^{\mu\nu}$ as described in Appendix B. We obtain

$$
f(s)_{\alpha\beta} = \frac{1}{a^2 + b^2}\left(a^2\,g_{\alpha\beta} + F^2_{\alpha\beta} \right)eb\coth ebs
$$

$$
+ \frac{1}{a^2 + b^2}\left(b^2\,g_{\alpha\beta} - F^2_{\alpha\beta} \right)ea\cot eas. \tag{2.206}
$$

Note that $f(s)_{\alpha\beta}$ is symmetric. In view of (2.201), we also need the result of the following derivative construction:

$$\left\{[\partial'_\mu - ieA_\mu(x')] + [\partial''_\mu + ieA_\mu(x'')]\right\}\Phi(x',x''|A)$$

$$\overset{(2.203)}{=}\left[ie\left(A_\mu(x') + \frac{1}{2}F_{\mu\lambda}(x'-x'')^\lambda\right) - ieA_\mu(x')\right.$$

$$\left. -ie\left(A_\mu(x'') + \frac{1}{2}\int_{x''}^{x'}d\xi_\kappa F^\kappa{}_\mu\right) + ieA_\mu(x'')\right]\Phi(x',x''|A)$$

$$= ieF_{\mu\lambda}(x'-x'')^\lambda\,\Phi(x',x''|A). \tag{2.207}$$

It is this term, in particular, which we would not have discovered in the naive calculation of the left-hand side of (2.201).

Let us now combine our findings from (2.201)–(2.207), leading first to

$$\left\{[\partial'_\mu - ieA_\mu(x')] + [\partial''_\mu + ieA_\mu(x'')]\right\}G(x',x''|A)$$

$$\overset{(2.204)}{=}\left(\left\{[\partial'_\mu - ieA_\mu(x')] + [\partial''_\mu + ieA_\mu(x'')]\right\}\Phi(x',x''|A)\right)G(x',x''|A_{\rm SF})$$

$$+\Phi(x',x''|A)\left(\partial'_\mu + \partial''_\mu\right)G(x',x''|A_{\rm SF}). \tag{2.208}$$

Since $G(x',x''|A_{\rm SF}) \equiv G(x'-x''|A_{\rm SF})$ owing to translational invariance, the second term vanishes completely in the constant-field case (we shall come back to this point later on). Upon insertion of (2.205) and (2.207) into (2.208), we arrive at

$$\mathrm{tr}\,\gamma_5\gamma^\mu\left\{[\partial'_\mu - ieA_\mu(x')] + [\partial''_\mu + ieA_\mu(x'')]\right\}G(x',x''|A)$$

$$= -\frac{ie}{2(4\pi)^2}F_{\mu\lambda}(x'-x'')^\lambda\,\Phi(x',x''|A) \tag{2.209}$$

$$\times\int_0^\infty \frac{ds}{s^2}e^{-iM^2 s}(f+eF)_{\alpha\beta}(x'-x'')^\beta\frac{eas\,ebs}{\sin eas\,\sinh ebs}$$

$$\times\exp\left[\frac{i}{4}(x'-x'')f(x'-x'')\right]4i\left(eT\,{}^\star F^{\mu\alpha} + P\,g^{\mu\alpha} - eT^\star\,F^{\mu\alpha}\right).$$

Here, we encounter the Lorentz product,

$$-(x'-x'')^\lambda F_{\lambda\mu}\left(eT\,{}^\star F^{\mu\alpha} + P\,g^{\mu\alpha} - eT^\star\,F^{\mu\alpha}\right)(f+eF)_{\alpha\beta}(x'-x'')^\beta.$$

Since this product is symmetrically contracted with $(x'-x'')$, only symmetric terms with respect to the field strength indices can contribute. With the aid of the explicit representation of the function f (2.206), we find, after a tedious but straightforward calculation,

$$-(x'-x'')^\lambda F_{\lambda\mu}\left(eT\,{}^\star F^{\mu\alpha} + P\,g^{\mu\alpha} - eT^\star\,F^{\mu\alpha}\right)(f+eF)_{\alpha\beta}(x'-x'')^\beta$$

$$= \frac{e}{a^2+b^2}(x'-x'')^\lambda\left[a^2b^2\left(\frac{\sin eas}{\sinh ebs} + \frac{\sinh ebs}{\sin eas}\right)g_{\lambda\beta}\right. \tag{2.210}$$

$$+ \left(b^2 \frac{\sin eas}{\sinh ebs} - a^2 \frac{\sinh ebs}{\sin eas} \right) F_{\lambda\beta}^2 \right] (x' - x'')^\beta .$$

Now, the crucial observation is that the result of (2.210), read together with the last factor of the second line of (2.209), can be related to the total derivative of the function $f(s)_{\alpha\beta}$ with respect to the proper time:

$$\frac{\mathrm{d}f(s)_{\alpha\beta}}{\mathrm{d}s} \overset{(2.206)}{=} -\frac{e^2}{a^2 + b^2} \left[a^2 b^2 \left(\frac{1}{\sinh^2 ebs} + \frac{1}{\sin^2 eas} \right) g_{\lambda\beta} \right.$$
$$\left. + \left(b^2 \frac{1}{\sinh^2 ebs} - a^2 \frac{1}{\sin^2 eas} \right) F_{\lambda\beta}^2 \right]. \quad (2.211)$$

Inserting (2.210) and (2.211) into (2.209), we obtain a comparably simple representation of the desired expression on the right-hand side of (2.201):

$$\operatorname{tr} \gamma_5 \gamma^\mu \Big\{ [\partial_\mu' - ieA_\mu(x')] + [\partial_\mu'' + ieA_\mu(x'')] \Big\} G(x', x''|A)$$

$$= \frac{e^2}{8\pi^2} \Phi(x', x''|A)\, ab \int_0^\infty \mathrm{d}s\, e^{-iM^2 s}\, \exp\left[\frac{i}{4}(x' - x'')f(x' - x'') \right]$$

$$\times (x' - x'')^\alpha \left(-\frac{\mathrm{d}f(s)_{\alpha\beta}}{\mathrm{d}s} \right) (x' - x'')^\beta \quad (2.212)$$

$$= i\frac{\alpha}{2\pi} {}^*F_{\mu\nu} F^{\mu\nu}\, \Phi(x', x''|A) \int_0^\infty \mathrm{d}s\, e^{-iM^2 s}\, \frac{\mathrm{d}}{\mathrm{d}s} \exp\left[\frac{i}{4}(x' - x'')f(x' - x'') \right].$$

Substituting this result back into the starting point, i.e. into the effective Lagrangian in (2.200), yields

$$\mathcal{L}_{\mathrm{eff}}^{\mathrm{PV}} = -\frac{1}{4}\frac{\alpha}{\pi}\frac{g}{M}\, \phi(x)\, {}^*F_{\mu\nu} F^{\mu\nu} \quad (2.213)$$

$$\times \lim_{x', x'' \to x} \left\{ \Phi(x', x''|A) \int_0^\infty \mathrm{d}s\, e^{-iM^2 s}\, \frac{\mathrm{d}}{\mathrm{d}s} \exp\left[\frac{i}{4}(x' - x'')f(x' - x'') \right] \right\}.$$

Comparing this with our result for the pseudoscalar interaction in (2.198), it is obvious that an equivalence exists between the two different interactions on the quantum level if the limit expression in (2.213) finally reduces to 1 for any kind of constant electromagnetic field. By construction, the proper-time integration has to be performed before the limit $x', x'' \to x$ can be taken. For example, if we interchanged this processes in (2.213), then we would find a zero result, since $(\mathrm{d}/\mathrm{d}s)e^0 = 0$.

Employing the fact that $\Phi(x', x''|A) \to 1$ for $x', x'' \to x$, we are left with the question of whether

$$\lim_{x', x'' \to x} \int_0^\infty \mathrm{d}s\, e^{-iM^2 s}\, \frac{\mathrm{d}}{\mathrm{d}s} \exp\left[\frac{i}{4}(x' - x'')f(x' - x'') \right]$$

$$\stackrel{\text{i.b.p.}}{=} \lim_{x',x'' \to x} \left\{ e^{-iM^2 s} \exp\left[\frac{i}{4}(x' - x'')f(x' - x'')\right] \right\}_0^\infty$$

$$+ iM^2 \lim_{x',x'' \to x} \int_0^\infty ds\, e^{-iM^2 s} \exp\left[\frac{i}{4}(x' - x'')f(x' - x'')\right] \quad (2.214)$$

equals 1. First, we shall demonstrate that the boundary term vanishes. For this, note that convergence at the upper bound $s \to \infty$ is ensured by the prescription $M^2 \to M^2 - i\epsilon$. This prescription can also be implemented by rotating the contour of s slightly below the real axis. Hence, the boundary term at $s \to \infty$ vanishes because the mass term approaches zero exponentially. For the lower bound, we need an expansion of $f(s)_{\alpha\beta}$ for small values of s; referring to (2.206), we obtain

$$f(s)_{\alpha\beta} = \frac{1}{s} g_{\alpha\beta} + \frac{e^2}{3} s\, F^2_{\alpha\beta} + \mathcal{O}(s^3). \quad (2.215)$$

In the limit $s \to 0$, only the first term is relevant, and we find that $\exp\left[(i/4)(x' - x'')^2/s\right] \to 0$ for s slightly below the real axis and $(x' - x'')^2 > 0$ space-like. This completes the proof that the boundary terms in (2.214) vanish.

Hence, we are left with the following integral:[9]

$$I(a,b) = iM^2 \lim_{x',x'' \to x} \int_0^\infty ds\, e^{-iM^2 s} \exp\left[\frac{i}{4}(x' - x'')f(x' - x'')\right]. \quad (2.216)$$

To get a better feeling for this integral, let us first make contact with Schwinger's original work and consider the weak-field limit. Since the resulting effective Lagrangian is already of second order in the fields (cf. (2.213)), it suffices to consider the integral (2.216) in the zero-field limit; for this we employ the expansion (2.215), because the weak-field expansion of $f(s)_{\alpha\beta}$ coincides with the small-s expansion:

$$I(a = 0, b = 0) = M^2 \lim_{x',x'' \to x} \int_0^\infty ds\, e^{-iM^2 s} \exp\left[\frac{i}{4}\frac{(x' - x'')^2}{s}\right]$$

$$\stackrel{s \to -is}{=} M^2 \lim_{x',x'' \to x} \int_0^\infty ds\, e^{-M^2 s} \exp\left[-\frac{1}{4}\frac{(x' - x'')^2}{s}\right]$$

$$\stackrel{(3.166)}{=} \lim_{x',x'' \to x} M\,(x' - x'')\, K_1\left[M(x' - x'')\right]$$

$$= 1, \quad (2.217)$$

---

[9] It is amusing to see that a (forbidden) interchange of the limit and the s integration would lead to the correct result: (2.216) $\to$ 1. But this is accidental, as we shall soon demonstrate.

where we have encountered the modified Bessel function $K_1(x)$, which has a simple pole at $x \to 0$ with residue 1. Inserting this into (2.213), we obtain the effective Lagrangian of a pseudovector interaction between a spinless meson and a heavy-fermion field in the presence of a slowly varying *and* weak external electromagnetic field:

$$\mathcal{L}_{\text{eff}}^{\text{PV}} = -\frac{1}{4}\frac{\alpha}{\pi}\frac{g}{M}\,\phi(x)\,F_{\mu\nu}{}^{\star}F^{\mu\nu}. \tag{2.218}$$

This is identical to the outcome for the pseudoscalar interaction and shows the essence of Schwinger's equivalence theorem for the low-energy regime. In this sense, the terminology "low energy" refers to the energy of the outgoing photons (variation of the field strength) as well as the strength of the field.

But we want to go one step further and prove the validity of the equivalence theorem without the weak-field assumption. For this, we have to show that $I(a,b) = 1$ for all values of a and b. As stated above, interchanging the limit and the s integration in (2.216) is not a valid operation. To put this in mathematical language, let $h(x)$ be the limiting value of a sequence of functions $h_n(x)$; then

$$\int\limits_a^b h(x)\,\mathrm{d}x = \lim_{n\to\infty}\int\limits_a^b h_n(x)\,\mathrm{d}x$$

if the $h_n(x)$ are integrable on the interval $M = [a,b]$, and

$$\lim_{n\to\infty}\|h - h_n\|_M = 0, \tag{2.219}$$

where

$$\|h\|_M = \sup\left\{|h(x)| \mid x \in M\right\}$$

is the supremum norm of $h(x)$ for the complete interval $M = [a,b]$. As an example, let us study the zero-field limit, as treated above; here, $h(s) \mathrel{\hat{=}} e^{-iM^2 s}$ and $h_{(x'-x'')}(s) \mathrel{\hat{=}} e^{-iM^2 s}e^{(i/4)(x'-x'')^2/s}$. Obviously, we find for the supremum norm:

$$\lim_{x',x''\to x}\left\|e^{-iM^2 s} - e^{-iM^2 s}\exp\left[\frac{i}{4}\frac{(x'-x'')^2}{s}\right]\right\|_M = 1, \tag{2.220}$$

where M denotes the s integration interval from zero to infinity slightly below the real axis. This does not satisfy the necessary criterion (2.219) for an interchange of limit and s integration. But since a direct evaluation of the integral in $I(a,b)$ in (2.216) is not in sight, we have to find an indirect way to prove that $I(a,b) = 1$.

For this, we choose to work in the special Lorentz frame where the electric and magnetic field are antiparallel, say $\boldsymbol{B} = a\,\hat{\boldsymbol{e}}_z$, $\boldsymbol{E} = -b\,\hat{\boldsymbol{e}}_z$. In this case, the square of the field strength tensor appearing in (2.206) has the form

$$F_{\alpha\beta}^2 = \begin{pmatrix} -b^2 & 0 & 0 & 0 \\ 0 & -a^2 & 0 & 0 \\ 0 & 0 & -a^2 & 0 \\ 0 & 0 & 0 & b^2 \end{pmatrix}.$$

In this special system, the exponent $(x' - x'')f(x' - x'')$ takes the particularly simple form

$$(x' - x'')^\alpha f(s)_{\alpha\beta}(x' - x'')^\beta = \Delta x_\parallel^2 \, eb \coth ebs + \Delta x_\perp^2 \, ea \cot eas, \quad (2.221)$$

where

$$\Delta x_\parallel^\mu = \left((x' - x'')^0, 0, 0, (x' - x'')^3\right),$$
$$\Delta x_\perp^\mu = \left(0, (x' - x'')^1, (x' - x'')^2, 0\right). \quad (2.222)$$

As an exercise, let us briefly investigate the case of purely magnetic fields, i.e. $b \to 0$. Then, we may write

$$(x' - x'')^\alpha f(s)_{\alpha\beta}(x' - x'')^\beta \Big|_{b=0} = \Delta x_\parallel^2 \frac{1}{s} + \Delta x_\perp^2 \, ea \cot eas \quad (2.223)$$

$$= \Delta x^2 \frac{1}{s} + \Delta x_\perp^2 \left(ea \cot eas - \frac{1}{s} \right),$$

where we have introduced the short form $\Delta x = x' - x''$. For $I(a, b)$, we obtain in this limit

$$I(a, 0) \overset{s \to -is}{=} \lim_{\Delta x_\perp^2, \Delta x^2 \to 0} M^2 \int_0^\infty ds \, e^{-M^2 s} \quad (2.224)$$

$$\times \exp\left[-\frac{1}{4}\frac{\Delta x^2}{s} - \frac{1}{4}\Delta x_\perp^2 \left(ea \coth eas - \frac{1}{s} \right) \right].$$

Here, we have replaced the limiting process $x', x'' \to x$, which is essentially four limiting processes, one for each component, by two limiting processes $\Delta x_\perp^2, \Delta x^2 \to 0$. Now, we make the crucial observation that the $\Delta x_\perp^2 \to 0$ limit can be interchanged with the integral, since

$$\lim_{\Delta x_\perp^2 \to 0} \left\| 1 - \exp\left[-\frac{1}{4}\Delta x_\perp^2 \left(ea \coth eas - \frac{1}{s} \right) \right] \right\|_M = 0. \quad (2.225)$$

Equation (2.225) is a consequence of the fact that $ea \coth eas - 1/s$ is bounded. Therefore, we obtain, following (2.217),

$$I(a, 0) = \lim_{\Delta x^2 \to 0} M^2 \int_0^\infty ds \, e^{-M^2 s} \exp\left(-\frac{1}{4}\Delta x^2 \frac{1}{s} \right) = 1. \quad (2.226)$$

Hence, for arbitrarily strong external magnetic fields, the equivalence theorem remains valid. But (2.226) tells us more; since a function which is constant $(= 1)$ on the whole positive real axis is constant on the whole complex plane, we obtain

$$1 = I(a,0)\big|_{a=-ib} = I(-ib,0) \equiv I(0,b), \tag{2.227}$$

which can be easily checked with the aid of the definition of $I(a,b)$ in (2.216). Thus we do not even find a further contribution to the low-energy effective Lagrangian of the pseudovector theory for arbitrarily strong *electric* fields. The latter might have been expected, since the present situation resembles the familiar situation of calculating the effective Heisenberg–Euler Lagrangian for constant electric fields, which is known to reveal a nonperturbative imaginary part.[10]

The final proof that $I(a,b)$ equals 1 is a simple generalization of the preceding considerations. For this, we write for (2.216), with the aid of (2.221),

$$I(a,b)$$

$$= \lim_{\Delta x^2, \Delta x_\perp^2, \Delta x_\parallel^2 \to 0} iM^2 \int_0^\infty ds\, e^{-iM^2 s} \exp\left(\frac{i}{4}\frac{\Delta x^2}{s}\right) \tag{2.228}$$

$$\times \exp\left[\frac{i}{4}\Delta x_\parallel^2\left(eb\coth ebs - \frac{1}{s}\right)\right] \exp\left[\frac{i}{4}\Delta x_\perp^2\left(ea\cot eas - \frac{1}{s}\right)\right].$$

Let us now treat the Δx^2, $\Delta x_\perp^2$ and $\Delta x_\parallel^2$ limits independently of each other, i.e. let us forget about the fact that $\Delta x^2 = \Delta x_\perp^2 + \Delta x_\parallel^2$. Note once again that the convergence of (2.228) is ensured by the prescription $M^2 \to M^2 - i\epsilon$. Furthermore, assuming that $\Delta x^2, \Delta x_\perp^2, \Delta x_\parallel^2 > 0$, we are allowed to rotate the s integral contour by an angle of $-\pi/4$ below the real axis:

$$s \to -e^{i\pi/4}\, s = e^{-i\pi/4} s = \frac{1}{\sqrt{2}}(1-i)\,s. \tag{2.229}$$

The convergence properties are maintained, since

$$e^{-iM^2 s} \to \exp\left(-\frac{1}{\sqrt{2}}(1+i)M^2 s\right),$$

$$\exp\left(\frac{i}{4}\frac{\Delta x^2}{s}\right) \to \exp\left(-\frac{1}{\sqrt{2}}\frac{(1-i)}{4}\frac{\Delta x^2}{s}\right). \tag{2.230}$$

Now the crucial observation is that both combinations, $eb\coth ebs - 1/s$ and $ea\cot eas - 1/s$, are bounded along the new s path. Hence, we are allowed to interchange the $\Delta x_\parallel$ and $\Delta x_\perp$ limits with the s integration, leading to

$$I(a,b) = \lim_{\Delta x^2 \to 0} ie^{i\pi/4} M^2 \int_0^\infty ds\, e^{-ie^{-i\pi/4}M^2 s} \exp\left(\frac{i}{4}e^{i\pi/4}\frac{\Delta x^2}{s}\right)$$

$$= \lim_{\Delta x^2 \to 0} M^2 \int_0^\infty ds\, e^{-M^2 s} \exp\left(-\frac{1}{4}\frac{\Delta x^2}{s}\right)$$

---

[10] It can in fact be checked directly that $I(0,b)$ has a vanishing imaginary part by summing all the residues of the cot poles on the lower imaginary s axis. One finds that the residues are all of the order of $\Delta x_\parallel^2$ and thus vanish for $\Delta x_\parallel^2 \to 0$.

$$= 1. \tag{2.231}$$

This proves Schwinger's equivalence theorem for arbitrary constant external field strengths. Besides, it proves that there are no nonperturbative contributions to the effective pseudovector Lagrangian in an external field. This latter remark will be important for the discussion of the axial-vector anomaly below; it demonstrates that there are no nonperturbative corrections to the anomaly at the one-loop level in the low-energy domain. To summarize the equivalence theorem, the main result can be written in the form

$$\mathcal{L}_{\text{eff}}^{\text{PS}} = -\frac{1}{4}\frac{\alpha}{\pi}\frac{g}{M}\,\phi(x)\,F_{\mu\nu}\,{}^{\star}F^{\mu\nu} = \mathcal{L}_{\text{eff}}^{\text{PV}}, \tag{2.232}$$

where the tree-level interaction Lagrangians for the pseudoscalar and pseudovector interactions are defined in (2.194) and (2.199). Equation(2.232) holds for arbitrary electromagnetic field strengths as long as the fields vary slowly compared with the Compton wavelength of the fermionic loop particle.

2.4.2 Axial-Vector Anomaly

It has often been emphasized in the original literature [2, 84, 107] that the discovery of the axial-vector anomaly (Adler–Bell–Jackiw anomaly) has its roots in Schwinger's work, which we have outlined above. Hence, we want to investigate the connection between the equivalence theorem and the axial-vector anomaly more closely. The anomaly can be summarized in the statement that the axial vector current is not conserved, not only because of an explicit breaking of axial symmetry by mass terms, but also because of the appearance of the $F_{\mu\nu}\,{}^{\star}F^{\mu\nu}$ term induced by quantum effects.

In order to derive this statement, we have to give a proper definition of the axial-vector current, as well as its divergence. On the classical (or even on the operator) level, the following definitions are reasonable:

$$j_5 = \frac{1}{2}[\bar{\psi}, \gamma_5 \psi], \tag{2.233}$$

$$j_5^\mu = \frac{1}{2}[\bar{\psi}, \gamma_5 \gamma^\mu \psi]. \tag{2.234}$$

$$\tag{2.235}$$

The investigation of the pseudoscalar interaction leads us to the conclusion that for the expectation value of the axial (scalar) current, we obtain

$$\langle j_5 \rangle^A = \mathrm{i}\,\mathrm{tr}\,\gamma_5 G(x, x|A) = -\frac{\mathrm{i}}{4}\frac{\alpha}{\pi}\frac{1}{M}\,F_{\mu\nu}\,{}^{\star}F^{\mu\nu},$$

$$\Leftrightarrow \quad \mathcal{L}_{\text{eff}}^{\text{PS}} = -\mathrm{i}g\,\phi(x)\,\langle j_5 \rangle^A. \tag{2.236}$$

Note that $\langle j_5 \rangle^A$ is well defined, since the singularities of $G(x, x|A)$ are removed from $\langle j_5 \rangle^A$ by the Dirac algebra.

Applying the same ideas to the case of the pseudovector interaction, we are tempted to write

$$\text{``}\langle j_5^\mu \rangle^A = i \operatorname{tr} \gamma_5 \gamma^\mu \, G(x,x|A) \text{''}. \tag{2.237}$$

However, this time, the singularities of $G(x,x|A)$ are not protected by the Dirac trace, which causes $\langle j_5^\mu \rangle^A$ of (2.237) to be an ill-defined quantity.

Since we are not interested in $\langle j_5^\mu \rangle^A$ but in its divergence, we shall not pursue the question of how (2.237) can be corrected; instead, let us concentrate on a proper definition of $\partial_\mu \langle j_5^\mu \rangle^A$.

During the study of the pseudovector interaction, we were forced to give a proper meaning to $\partial_\mu \operatorname{tr} \gamma_5 \gamma^\mu \, G(x,x|A)$, which then was achieved by a point-splitting procedure in (2.201). Separating the Green's function into its gauge-dependent part $\Phi(x',x''|A)$ and a part which depends only on the field strength $G(x',x''|A_{\mathrm{SF}})$ (cf. (2.204)), we arrived at (2.208):

$$
\lim_{x',x'' \to x} \left\{ [\partial'_\mu - ieA_\mu(x')] + [\partial''_\mu + ieA_\mu(x'')] \right\} G(x',x''|A)
$$

$$
\overset{(2.208)}{=} \lim_{x',x'' \to x} \left[ie F_{\mu\lambda}(x'-x'')^\lambda \, \Phi(x',x''|A) \right] G(x',x''|A_{\mathrm{SF}})
$$

$$
+ \lim_{x',x'' \to x} \Phi(x',x''|A) \left(\partial'_\mu + \partial''_\mu \right) G(x',x''|A_{\mathrm{SF}}). \tag{2.238}
$$

We remark that the second term is singularity free, since the singularity in $G(x',x''|A_{\mathrm{SF}})$ must be of the form $\lim_{x',x'' \to x} G(x',x''|A_{\mathrm{SF}}) \sim \lim_{x',x'' \to x} h(x'-x'')$, with some function h, where $h(0) \to \infty$. Hence, the singularity vanishes in the second term because of the symmetry in $\partial'_\mu \leftrightarrow \partial''_\mu$. Since $\Phi(x',x''|A) \to 1$ for $x',x'' \to x$, we are led to interpret the second term in (2.238) as a building block for an appropriate definition of the divergence of the axial-vector current:

$$
\partial_\mu \langle j_5^\mu(x) \rangle = \lim_{x',x'' \to x} \Phi(x',x''|A) \, (\partial'_\mu + \partial''_\mu) \, i \operatorname{tr} \gamma_5 \gamma^\mu \, G(x',x''|A_{\mathrm{SF}}). \tag{2.239}
$$

We want to stress that this definition is gauge-invariant, although it is not obvious, since, after taking the limit $x',x'' \to x$, the left-hand side and the first term on the right-hand side of (2.238) are separately gauge-invariant. As shown above, the trace of $\gamma_5 \gamma^\mu$ times the first term on the right-hand side of (2.238) yields

$$
\lim_{x',x'' \to x} ie F_{\mu\lambda}(x'-x'')^\lambda \, \Phi(x',x''|A) \operatorname{tr} \gamma_5 \gamma^\mu \, G(x',x''|A_{\mathrm{SF}})
$$

$$
= \frac{i\alpha}{2\pi} F_{\mu\nu} {}^\star F^{\mu\nu}. \tag{2.240}
$$

Substituting (2.239) and (2.240) into the right-hand side of (2.238) and applying the equivalence theorem to the left-hand side, i.e. $\operatorname{tr}[\gamma_5 \gamma^\mu \text{LHS of (2.238)}] = -2M \langle j_5 \rangle^A$, we finally arrive at

$$
\partial_\mu \langle j_5^\mu \rangle = -2iM \langle j_5 \rangle + \frac{\alpha}{2\pi} F_{\mu\nu} {}^\star F^{\mu\nu}. \tag{2.241}
$$

This is the well-known equation for the divergence of the axial-vector current and represents the Adler–Bell–Jackiw anomaly. However, by employing the equivalence theorem, we have proved (2.241) only for the special case of constant external fields. Hence, Schwinger's work on the constant-field case is only capable of deriving the anomaly in a certain energetic regime, namely the low-energy domain.

Within the usual diagrammatic approach [2], one also arrives at (2.241), but this time without assuming that the electromagnetic field has to be constant. For example, starting with the left-hand side of (2.240), one may expand the Green's function in an arbitrary external field as a power series in the coupling constant:

$$
\begin{aligned}
G &= S_{\mathrm{F}} + \mathrm{i}e S_{\mathrm{F}} \, \gamma A \, G \\
&= S_{\mathrm{F}} + \mathrm{i}e \, S_{\mathrm{F}} \, \gamma A \, S_{\mathrm{F}} + (\mathrm{i}e)^2 \, S_{\mathrm{F}} \, \gamma A \, S_{\mathrm{F}} \, \gamma A \, S_{\mathrm{F}} + \ldots .
\end{aligned}
\tag{2.242}
$$

Then one observes that the zeroth-order term vanishes owing to the Dirac algebra, the second term leads to the correct anomaly and the e^2 term, as well as all higher-order terms, vanishes in the limit $x', x'' \to x$, since these terms do not contribute to the singular behavior of G. (Note that the latter statement holds only perturbatively.) It is exactly this second term in (2.242) which is diagrammatically represented by the famous triangle graph.

To summarize our considerations, we want to stress that the equivalence theorem corresponds to the axial-vector anomaly in the domain of slowly varying fields only. This is the only regime where the equivalence theorem is applicable. Moreover, by giving up the weak-field assumption, we were able to prove the equivalence theorem for arbitrary field strength, i.e. to all orders in the external field. Translating this into the language of the anomaly, we have proved that there are no nonperturbative corrections to the anomaly in the case of slowly varying external fields.

Without the constant-field approximation, one can show perturbatively that (2.241) for the divergence of the axial-vector current holds for arbitrary fields to any finite order of perturbation theory. The price that one has to pay for this stronger statement is that one loses control over the nonperturbative domain. Within these techniques, the question still remains open as to whether the anomaly receives nonperturbative corrections for rapidly varying fields. However, Fujikawa's path integral procedure, which is truly nonperturbative for arbitrary A fields [81], demonstrates that (2.241) is not modified.

Let us remark in this context that the validity of (2.241) for arbitrary electromagnetic fields does not tell us anything about the question of whether the equivalence theorem holds for arbitrary photon energies.

Let us conclude this chapter with an interesting observation for the constant-field case. Inserting our findings for $\langle j_5 \rangle^A$ for constant fields (cf. (2.236)) into (2.241), we find that the divergence of the axial-vector current vanishes:

$$
\partial_\mu \langle j_5^\mu \rangle = 0 .
\tag{2.243}
$$

This is already clear from its definition (2.239), since $G(x', x''|A_{SF})$ for constant fields depends only on $x' - x''$. Hence (2.239) vanishes simply by antisymmetry. Nevertheless, the result appears rather unfamiliar, because it signals the conservation of the axial-vector current at the quantum level, although this current is not conserved at the classical level owing to the explicit breaking of the axial symmetry by mass terms. Therefore, the constant-field case is an exceptional situation which creates an "inverse anomaly": a classically and explicitly broken symmetry is restored by quantum effects.

Since the fields are considered to vary slowly compared with the Compton wavelength of the fermionic loop particle, one can reinterpret the constant-field limit as a heavy-fermion limit. In this sense, very heavy fermions, although breaking the axial symmetry very violently, do not violate the axial-vector current conservation after quantum fluctuations are taken into account.

Finally, let us stress once more that our considerations remain strictly at the one-loop level. Similarly to the Fujikawa method, we do not take photonic fluctuations into account; hence our statements about nonperturbative contributions to the anomaly do not touch questions about higher-loop corrections. In particular, we cannot comment on the correctness of the Adler–Bardeen theorem [3].

3. Nonlinear Electrodynamics: Effective-Action Approach

In the preceding chapter, we studied examples of quantum electrodynamic effects induced by external fields by means of the electron Green's function, the polarization tensor and the induced current. The "operator" language we employed made explicit reference to the quantum nature of the underlying physics. In contrast, the physical effects that we extracted from this formalism, such as optical birefringence, resemble closely the features of classical physics.

An important tool that interpolates between the full quantum theory and classical field theory is provided by the concept of the effective action: once the high-energy degrees of freedom which are invisible in the classical low-energy domain are integrated out, the resulting effective action for the remaining degrees of freedom can be employed to define a new quasi-classical theory.[1] In particular, classical equations of motion can be derived whose solution provides for the description of quantum effects in the language of classical physics. In this sense, electrodynamics represents an excellent laboratory for investigating and applying this concept.

While classical electrodynamics is distinguished by Lorentz invariance, gauge invariance and linearity, quantum electrodynamics features only the first two principles. Nonlinear effects clearly are of quantum origin and arise as a result of the polarization of the vacuum. Self-interactions of the electromagnetic field are a by-product of the creation and subsequent annihilation of charged pairs. Hence, these self-interactions are nonlocal in nature on a microscopic scale, but are, of course, causal, since the interaction with the charged pairs is causal. Nevertheless, this nonlocal nature can be disregarded on a scale where the virtual creation and annihilation processes can no longer be resolved. This scale is set by the Compton wavelength of the electron, $\lambda = 1/m$.

Taking these considerations into account, the one-loop effective Lagrangian of QED for constant fields as derived in Sect. 2.1, (2.49), read together with the classical Maxwell Lagrangian, can be employed to define a local classical field theory that enriches classical electrodynamics by nonlinear self-

---

[1] In this work, we do not elucidate the question of whether the effective action might be used as a defining action for a new quantum theory of the residual degrees of freedom. This type of approach is reviewed in [111] and [136]. A subtle example demonstrating the limits of this approach can be found in [32].

interactions of the fields. The validity of this theory is, on the one hand, limited by the validity of the loop expansion; on the other hand, it is restricted to scales larger than the Compton wavelength of the electron; in particular, the fields must vary only slowly on that scale.

3.1 A First Look at Light Propagation

It is proposed that the study of light propagation in a perturbed QED vacuum is a useful tool for investigating the manifold properties of such a quantum ground state. As an introductory example, we solve the equations of motion for low-frequency light propagating in an external field for the simplest nonlinear extension of classical electrodynamics: the Heisenberg–Euler Lagrangian. We obtain a modified light cone condition, and discuss polarization dependences as well as velocity shifts.

The main purpose of this section is to illustrate on a simple level the various techniques and necessary approximations that will become important in the sections to follow.

Since we are interested in the borderline at which classical and quantum physics meet, i.e. where nonlinear effects begin to play a significant role, the study of the propagation of light, which is trivial in the classical vacuum theory, serves as an important example. In this introductory section, we briefly present our strategy by means of the simplest example: the lowest-order correction to the classical Lagrangian. The more sophisticated investigations in the later sections are based on the same philosophy as described in the following.

In (2.50), we found the lowest-order correction to the Maxwell Lagrangian; in combination, the "next-to-leading-order" classical electrodynamics can be defined by the lowest-order Heisenberg–Euler Lagrangian

$$\mathcal{L} = -\mathcal{F} + c_1\,\mathcal{F}^2 + c_2\,\mathcal{G}^2, \quad c_1 = \frac{8}{45}\frac{\alpha^2}{m^4}, \quad c_2 = \frac{14}{45}\frac{\alpha^2}{m^4}. \tag{3.1}$$

The equations of motion derived à la Euler–Lagrange reduce to

$$0 = \frac{\partial\mathcal{L}}{\partial A_\nu} - \partial_\mu\frac{\partial\mathcal{L}}{\partial(\partial_\mu A_\nu)} = -\partial_\mu\frac{\partial\mathcal{L}}{\partial F_{\alpha\beta}}\frac{\partial F_{\alpha\beta}}{\partial(\partial_\mu A_\nu)} = -2\partial_\mu\frac{\partial\mathcal{L}}{\partial F_{\mu\nu}}, \tag{3.2}$$

where the differentiation with respect to $F_{\mu\nu}$ must be performed under the constraint of antisymmetry of the indices. With the aid of the identities

$$\frac{\partial\mathcal{F}}{\partial F_{\alpha\beta}} = \frac{1}{2}F^{\alpha\beta}, \qquad \frac{\partial\mathcal{G}}{\partial F_{\alpha\beta}} = \frac{1}{2}{}^\star F^{\alpha\beta}, \tag{3.3}$$

we end up with

$$0 = \partial_\mu\big(F^{\mu\nu} - 2c_1\,\mathcal{F}\,F^{\mu\nu} - 2c_2\,\mathcal{G}\,{}^\star F^{\mu\nu}\big), \tag{3.4}$$

which represents the vacuum field equation for the lowest-order nonlinear electrodynamics. This equation is accompanied by the usual Bianchi identity $\partial_\mu{}^\star F^{\mu\nu} = 0$.

Applying these equations to the case of propagating light in a background field, we decompose the field strength tensor into the background part $F^{\mu\nu}$ and the plane wave field $f^{\mu\nu}$, $F^{\mu\nu} \to F^{\mu\nu} + f^{\mu\nu}$. We assume the background field to be constant or, at most, slowly varying and disregard any derivative of $F^{\mu\nu}$. Furthermore, we linearize the field equations with respect to the plane wave field; this is equivalent to neglecting self-interactions of the propagating light itself. In a certain sense, the plane wave is thus reminiscent of a test charge in classical electrodynamics. It is important to note that the linearization does not necessarily imply that the background field is much stronger than the plane wave field. The zero-field limit for the background field will be well defined.

With these assumptions, we obtain for the field equations, with the aid of the Bianchi identity,

$$0 = \partial_\mu f^{\mu\nu} - c_1 F_{\alpha\beta} F^{\mu\nu} \partial_\mu f^{\alpha\beta} - c_2 {}^\star F_{\alpha\beta} {}^\star F^{\mu\nu} \partial_\mu f^{\alpha\beta}. \tag{3.5}$$

Since $f^{\mu\nu}$ is assumed to be a plane wave field, we perform a Fourier transformation to momentum space, where $f^{\mu\nu}$ can be written as $f^{\mu\nu} = k^\mu a^\nu - k^\nu a^\mu$. Here, $k^\mu = (\omega, \mathbf{k})$ denotes the wave vector of the propagating light; consequently, the wave vector replaces the derivative in (3.5); $\partial^\mu \to ik^\mu$. Finally, we impose the Lorentz gauge constraint on the gauge potential a^μ of the plane wave, $k_\mu a^\mu = 0$, and arrive at the light cone condition

$$0 = k^2 a^\nu - 2c_1 F_{\alpha\beta} F^{\mu\nu} k_\mu k^\alpha a^\beta - 2c_2 {}^\star F_{\alpha\beta} {}^\star F^{\mu\nu} k_\mu k^\alpha a^\beta. \tag{3.6}$$

Obviously, we recover the trivial light cone condition $k^2 = 0$ in the classical limit where $c_{1,2} = 0$. For the two different polarization states that solve this field equation, we may try the ansatz

$$a_1^\mu \sim {}^\star F k^\mu \equiv {}^\star F^{\mu\lambda} k_\lambda, \qquad a_2^\mu \sim F k^\mu \equiv F^{\mu\lambda} k_\lambda, \tag{3.7}$$

which is motivated by our findings in Sect. 2.2, (2.128). Employing the fundamental algebraic identity (B.4), we find upon insertion of this ansatz into (3.6)

$$\begin{aligned} a_1^\mu : \quad & 0 = \left(k^2 - 2c_2 z_k + 2c_2 \mathcal{F} k^2\right) {}^\star F k^\nu - \left(2c_1 \mathcal{G} k^2\right) F k^\nu, \\ a_2^\mu : \quad & 0 = \left(k^2 - 2c_1 z_k\right) F k^\nu - \left(2c_2 \mathcal{G} k^2\right) {}^\star F k^\nu, \end{aligned} \tag{3.8}$$

where $z_k = (F^{\mu\nu} k_\nu)^2$ denotes the Lorentz- and gauge-invariant quantity that we have already encountered in (2.63). Owing to the last terms on the right-hand sides of (3.8), it is obvious that the polarization states a_1^μ and a_2^μ do not solve the field equation *exactly*; however, since we are in the weak-field domain, we expect k^2 to deviate only slightly from the trivial light cone condition: $k^2 = 0 + \mathcal{O}(c_{1,2})$. Hence, the terms proportional to $c_{1,2}\mathcal{G} k^2$ or $c_2 \mathcal{F} k^2$ will be of higher order in $c_{1,2}$ and should be omitted for reasons of consistency. We find

$$a_1^\mu : \quad k^2 = 2c_2 z_k, \qquad a_2^\mu : \quad k^2 = 2c_1 z_k. \tag{3.9}$$

Introducing the phase velocity v via $k^\mu = (k^0, \mathbf{k}) = k(v, \hat{\mathbf{k}})$, where $\mathrm{k} \equiv |\mathbf{k}|$, we obtain

$$a_1^\mu : \quad v_1 = 1 - c_2 \frac{z_k}{\mathrm{k}^2} = 1 - \frac{14}{45} \frac{\alpha^2}{m^4} \frac{z_k}{\mathrm{k}^2},$$

$$a_2^\mu : \quad v_2 = 1 - c_1 \frac{z_k}{\mathrm{k}^2} = 1 - \frac{8}{45} \frac{\alpha^2}{m^4} \frac{z_k}{\mathrm{k}^2}, \tag{3.10}$$

which exactly corresponds to our findings in (2.125). An appropriate representation of the invariant z_k in terms of $\boldsymbol{E}$ and $\boldsymbol{B}$ fields is given by (2.130):
$z_k = \mathrm{k}^2 \left[B^2 \sin^2 \theta_B + E^2 \sin^2 \theta_E + 2\hat{\boldsymbol{k}} \cdot (\boldsymbol{B} \times \boldsymbol{E}) \right]$.

Within the Lagrangian formalism developed so far, this result was first obtained by Białynicka-Birula and Białynicki-Birula [28] (correcting and generalizing a result of Klein and Nigam [115]). For purely magnetic fields, a similar calculation was performed by Adler [4]. With a noncovariant notation, the weak-field computation is nicely reviewed in [125].

The following remarks should be made:

(1) The light cone condition derived from the field equations is of second order in k^μ by construction, since the effective Lagrangian (3.1) contains no derivatives of the field strength. The latter property, in turn, follows from disregarding the nonlocalities of the effective action. As a consequence, the phase and group velocities of the propagating plane wave coincide, because the phase velocity does not depend on the frequency. This agrees with the fact that the results are only valid for soft photons, i.e. for low frequencies, $\omega \ll m$, which cannot resolve virtual processes.

(2) The phase velocities do not exceed the vacuum velocity $c\ (=1)$, because the coefficients $c_{1,2}$ are strictly positive. However, even if they were not, we could not draw the conclusion that causality is violated, because the signal velocity is equivalent to the front velocity; but the latter is related to the phase velocity in the infinite-frequency limit $\omega \to \infty$ (cf. the discussion of causality in Sect. 3.3.4).

(3) From (3.8), we could read off the linear combinations of Fk^μ and $^*Fk^\mu$ which, as the correct polarization states, solve the field equations exactly [28]. In the following, however, we shall be particularly interested in light cone conditions averaged over polarization (and propagation direction), which might be called "sum rules". In this case, the calculation following (3.6) can be simplified by multiplying (3.6) by a_ν and summing over the two polarization states according to the rule

$$\sum_{\text{pol.}} a^\nu a^\beta \sim g^{\nu\beta}. \tag{3.11}$$

This yields

$$0 = 2\,k^2 \left(1 + 2c_2\,\mathcal{F}\right) - 2(c_1 + c_2)z_k. \tag{3.12}$$

Neglecting the higher-order terms in $c_{1,2}$ again, we finally find

$$k^2 = (c_1 + c_2)z_k \quad \Rightarrow \quad v = 1 - \frac{1}{2}(c_1 + c_2)\frac{z_k}{k^2} = 1 - \frac{11}{45}\frac{\alpha^2}{m^4}\frac{z_k}{k^2}. \quad (3.13)$$

In fact, this is the arithmetic average over the polarization states of (3.10).

3.2 Light Cone Condition

We derive the light cone condition for low-frequency light propagating in an arbitrary constant external field. The dynamics of the field is determined by a Lagrangian which is assumed to depend on the two invariants of the electromagnetic field but is otherwise left arbitrary. We show that the deformation of the light cone is determined by the (vacuum expectation value of the) energy–momentum tensor of the field, and a field-dependent factor which we call the *effective-action charge*. By inserting the one-loop QED effective action, the light cone condition is applied to the cases of weak fields and strong and superstrong magnetic fields, yielding refractive indices for the various electromagnetically perturbed vacua. We observe that a vacuum modified by a superstrong magnetic field shows striking similarities to a magnetized plasma. This domain is also studied with the aid of a renormalization-group-improved effective action.

The comparably simple procedure of the previous section will be generalized in the present section in order to obtain a light cone condition for a plane wave propagating in an electromagnetic field of arbitrary strength. Again, the basic quantity will be the effective action of QED in an arbitrary constant background field. The success of this strategy will further support the picture of considering the modified vacuum as a medium.

In the following, we mimic the steps of the preceding calculation, but this time drop the weak-field assumption for the background field. To be concrete, we first demand that the following condition is satisfied:

(I) *Any length scale which we can construct from the parameters of the system (except from the field strength) is large compared with the Compton wavelength of the lightest massive particle (the electron) $\lambda_c = 1/m$. This ensures that the effective Lagrangian is local, implies that the field is "slowly" varying and restricts the plane wave to be of "low" frequency (compared with λ_c).*

This property is sufficient for considering the effective Lagrangian to be a function only of the gauge and Lorentz invariants of the electromagnetic field; a convenient choice for the two linearly independent invariants is given by the standard (pseudo)scalars

$$\mathcal{F} = \frac{1}{4}F_{\mu\nu}F^{\mu\nu} = \frac{1}{2}(\boldsymbol{B}^2 - \boldsymbol{E}^2), \quad \mathcal{G} = \frac{1}{4}F_{\mu\nu}{}^{\star}F^{\mu\nu} = -\boldsymbol{E}\cdot\boldsymbol{B}. \quad (3.14)$$

For example, the Maxwell Lagrangian can simply be written as $\mathcal{L}_M = -\mathcal{F}$. The general Lagrangian therefore reads

$$\mathcal{L} = \mathcal{L}(\mathcal{F}, \mathcal{G}). \quad (3.15)$$

Note that a parity-conserving fundamental theory (e.g. standard QED) demands that $\mathcal{L}$ is an even function of $\mathcal{G}$.

Following (3.2) and (3.3), we obtain the equations of motion from $\mathcal{L}$ by variation:

$$0 = \partial_\mu \frac{\partial \mathcal{L}}{\partial(\partial_\mu A_\nu)} - \frac{\partial \mathcal{L}}{\partial A_\mu} = \partial_\mu \left(\partial_\mathcal{F}\mathcal{L}\, F^{\mu\nu} + \partial_\mathcal{G}\mathcal{L}\,{}^\star F^{\mu\nu}\right), \qquad (3.16)$$

where $\partial_\mathcal{F}, \partial_\mathcal{G}$ denote the partial derivatives with respect to the field strength invariants (3.14) (and should not be confused with space–time derivatives ∂_μ). If we take advantage of the Bianchi identity and move ∂_μ to the right, we arrive at

$$0 = (\partial_\mathcal{F}\mathcal{L})\,\partial_\mu F^{\mu\nu} + \frac{1}{2} M^{\mu\nu}_{\alpha\beta}\,\partial_\mu F^{\alpha\beta}\,, \qquad (3.17)$$

where $M^{\mu\nu}_{\alpha\beta}$ is given by

$$\begin{aligned} M^{\mu\nu}_{\alpha\beta} &= F^{\mu\nu} F_{\alpha\beta}\,(\partial^2_\mathcal{F}\mathcal{L}) + {}^\star F^{\mu\nu}\,{}^\star F_{\alpha\beta}\,(\partial^2_\mathcal{G}\mathcal{L}) \\ &\quad + \partial_\mathcal{FG}\mathcal{L}\left(F^{\mu\nu}\,{}^\star F_{\alpha\beta} + {}^\star F^{\mu\nu} F_{\alpha\beta}\right). \end{aligned} \qquad (3.18)$$

Note that M is antisymmetric in the upper, as well as the lower, indices: $M^{\mu\nu}_{\alpha\beta} = -M^{\nu\mu}_{\alpha\beta} = M^{\nu\mu}_{\beta\alpha}$.

Next, we apply this field equation to the case of a plane wave field $f^{\mu\nu}$ propagating in a background field $F^{\mu\nu}$, i.e. we decompose the field strength according to $F^{\mu\nu} \to F^{\mu\nu} + f^{\mu\nu}$. In doing so, we specify the second basic condition as follows:

(II) *Vacuum modifications (= self-interactions) of the plane wave field are negligible; this is formally achieved by a linearization of the field equation with respect to $f^{\mu\nu}$.*

Furthermore, neglecting derivatives of the background field (owing to condition I), we can replace the partial derivatives by the wave vector of the plane wave field in momentum space; this leads us to

$$0 = (\partial_\mathcal{F}\mathcal{L})\,k_\mu f^{\mu\nu} + \frac{1}{2} M^{\mu\nu}_{\alpha\beta}\,k_\mu f^{\alpha\beta}\,. \qquad (3.19)$$

As mentioned above, we are primarily interested in an average over polarization states. Representing the plane wave field strength by $f^{\mu\nu} = k^\mu a^\nu - k^\nu a^\mu$ (imposing the Lorentz gauge $k^\mu a_\mu = 0$), multiplying (3.19) by a_ν and summing over polarization states according to (3.11), therefore yields

$$0 = 2(\partial_\mathcal{F}\mathcal{L})\,k^2 + M^{\mu\nu}_{\alpha\nu}\,k_\mu k^\alpha\,. \qquad (3.20)$$

Equation (3.20) already represents the desired light cone condition and actually indicates that the familiar $k^2 = 0$ will, in general, not hold for arbitrary Lagrangians. Our final task is to put $M^{\mu\nu}_{\alpha\nu}$ into a convenient shape. Using the powerful fundamental algebraic relations of the field strength tensors given in Appendix B, (B.3) and (B.4), we obtain

$$M^{\mu\nu}_{\alpha\nu} = 2\left[(1/2)F^{\mu\nu} F_{\alpha\nu}(\partial^2_\mathcal{F} + \partial^2_\mathcal{G})\mathcal{L} + \delta^\mu_\alpha(\mathcal{G}\partial_\mathcal{FG}\mathcal{L} - \mathcal{F}\partial^2_\mathcal{G}\mathcal{L})\right]\,. \qquad (3.21)$$

Since the remaining indices μ and α will be contracted with k_μ and k^α upon insertion into (3.20), we note that the first term of $M^{\mu\nu}_{\alpha\nu}$ will create the invariant z_k, while the second term results in a term proportional to k^2. This proves that the deformation of the light cone caused by electromagnetic background fields will always be of the form $k^2 \sim z_k$. However, it is extremely useful to choose a different way of representing the light cone condition, by introducing the Maxwell energy–momentum tensor,

$$T^\mu{}_\alpha = F^{\mu\nu} F_{\alpha\nu} - \mathcal{F} \delta^\mu_\alpha , \tag{3.22}$$

In principle, the Maxwell energy–momentum tensor is devoid of any physical meaning, since we are simply not dealing with the Maxwell Lagrangian. The right quantity to deal with is therefore the vacuum expectation value (VEV) of the energy–momentum tensor, defined by[2]

$$\langle T^{\mu\nu} =\rangle \frac{2}{\sqrt{-g}} \frac{\delta \Gamma}{\delta g_{\mu\nu}}, \quad \Gamma = \int d^4x \sqrt{-g}\, \mathcal{L}, \tag{3.23}$$

where Γ denotes the effective action. Performing the calculation, we arrive at

$$\langle T^{\mu\nu} \rangle = -T^{\mu\nu}(\partial_\mathcal{F}\mathcal{L}) + g^{\mu\nu}\left(\mathcal{L} - \mathcal{F}\partial_\mathcal{F}\mathcal{L} - \mathcal{G}\partial_\mathcal{G}\mathcal{L}\right). \tag{3.24}$$

On the quantum level, the conformal symmetry is broken, since the fermionic measure of the functional integral is not invariant under space–time scale transformations. As a consequence, the trace of the vacuum expectation value of the energy–momentum tensor does not vanish any more, but develops an anomaly given by

$$\langle T^\alpha{}_\alpha \rangle = 4(\mathcal{L} - \mathcal{F}\partial_\mathcal{F}\mathcal{L} - \mathcal{G}\partial_\mathcal{G}\mathcal{L}). \tag{3.25}$$

By differentiation, we find

$$\partial_\mathcal{F}\langle T^\alpha{}_\alpha \rangle = -4(\mathcal{F}\partial^2_\mathcal{F}\mathcal{L} + \mathcal{G}\partial_{\mathcal{F}\mathcal{G}}\mathcal{L}),$$
$$\partial_\mathcal{G}\langle T^\alpha{}_\alpha \rangle = -4(\mathcal{G}\partial^2_\mathcal{G}\mathcal{L} + \mathcal{F}\partial_{\mathcal{F}\mathcal{G}}\mathcal{L}). \tag{3.26}$$

Inserting the various formulas related to the energy–momentum tensor into (3.21), we obtain two equivalent representations of $M^{\mu\nu}_{\alpha\nu}$:

$$M^{\mu\nu}_{\alpha\nu} = 2\left[-\frac{1}{2}\frac{(\partial^2_\mathcal{F}+\partial^2_\mathcal{G})\mathcal{L}}{\partial_\mathcal{F}\mathcal{L}}\langle T^\mu{}_\alpha\rangle - \left(\frac{1}{2}\mathcal{F}(\partial^2_\mathcal{F}+\partial^2_\mathcal{G})\mathcal{L} \right.\right.$$
$$\left.\left. +\frac{1}{4}\partial_\mathcal{F}\langle T^\beta{}_\beta\rangle - \frac{(1/8)(\partial^2_\mathcal{F}+\partial^2_\mathcal{G})\mathcal{L}}{\partial_\mathcal{F}\mathcal{L}}\langle T^\beta{}_\beta\rangle\right)\delta^\mu_\alpha\right], \tag{3.27}$$

$$M^{\mu\nu}_{\alpha\nu} = 2\left[\frac{1}{2}(\partial^2_\mathcal{F}+\partial^2_\mathcal{G})\mathcal{L}\; T^\mu{}_\alpha - \left(\frac{1}{2}\mathcal{F}(\partial^2_\mathcal{F}+\partial^2_\mathcal{G})\mathcal{L} + \frac{1}{4}\partial_\mathcal{F}\langle T^\beta{}_\beta\rangle\right)\delta^\mu_\alpha\right].$$
$$\tag{3.28}$$

---

[2] Note that the variation with respect to the metric tensor is just a trick to calculate the symmetric energy–momentum tensor. With some care, the same result can be obtained by canonical methods.

Substituting these representations of $M^{\mu\nu}_{\alpha\nu}$ into the field equation (3.20) and solving for k^2, we obtain the final versions of the light cone condition:

$$k^2 = Q \langle T^{\mu\nu} \rangle k_\mu k_\nu, \qquad\qquad k^2 = Q' \, T^{\mu\nu} k_\mu k_\nu. \qquad (3.29)$$

The functions Q and Q' are given by

$$Q = \frac{1}{2}(\partial^2_{\mathcal{F}} + \partial^2_{\mathcal{G}})\mathcal{L}\left\{(\partial_{\mathcal{F}}\mathcal{L})^2 - (\partial_{\mathcal{F}}\mathcal{L})\left[\frac{\mathcal{F}}{2}(\partial^2_{\mathcal{F}}+\partial^2_{\mathcal{G}})\mathcal{L}+\frac{1}{4}\partial_{\mathcal{F}}\langle T^\alpha{}_\alpha\rangle\right]\right.$$
$$\left. +\frac{1}{8}\left[(\partial^2_{\mathcal{F}}+\partial^2_{\mathcal{G}})\mathcal{L}\right]\langle T^\alpha{}_\alpha\rangle\right\}^{-1}, \qquad (3.30)$$

$$Q' = \frac{(1/2)(\partial^2_{\mathcal{F}} + \partial^2_{\mathcal{G}})\mathcal{L}}{-(\partial_{\mathcal{F}}\mathcal{L}) + \left[(\mathcal{F}/2)(\partial^2_{\mathcal{F}} + \partial^2_{\mathcal{G}})\mathcal{L} + (1/4)\partial_{\mathcal{F}}\langle T^\alpha{}_\alpha\rangle\right]}. \qquad (3.31)$$

The classical limit $k^2 = 0$ is immediately recovered, since the Q factors simply vanish for $\mathcal{L} = \mathcal{L}_\mathrm{M} = -\mathcal{F}$. We regard the light cone condition involving the vacuum expectation value of the energy–momentum tensor as the more fundamental one, since only the information contained in $\langle T^{\mu\nu} \rangle$ provides for a proper definition of the energy, momentum, etc. of the electromagnetic background field. Nevertheless, the representation involving the Maxwell energy–momentum tensor is the simpler one and can therefore be evaluated more easily. Of course, both representations are equivalent by virtue of (3.24).

Since an average over the different polarization states entered the derivation of the light cone condition, we may call it a "sum rule". While the factor of $\langle T^{\mu\nu} \rangle k_\mu k_\nu$ (or $T^{\mu\nu} k_\mu k_\nu$) essentially carries the information about the geometry of the system determined by $\boldsymbol{E}$, $\boldsymbol{B}$ and $\boldsymbol{k}$, the value of the Q factor dictates the strength of the light cone deformation. In the following, we shall call the Q factor the *effective-action charge* for reasons that will be elucidated below.

Note that the validity of the light cone conditions in (3.29) is not restricted to results of perturbation theory or to only small modifications of the Maxwell Lagrangian. As long as we do not leave the parameter space that is limited by conditions (I) and (II), the light cone condition is an exact statement in the sense of an effective theory. The better we determine $\mathcal{L}(\mathcal{F},\mathcal{G})$, the larger the domain of validity will be.

For the special case of weak electromagnetic background fields, this representation of the sum rule was first found by Shore [153]; the present discussion of the light cone condition is as given by [60, 86].

In the remainder of this section, we calculate further representations of the sum rule by choosing a certain reference frame and introducing the modulus $\mathrm{k} \equiv |\boldsymbol{k}|$ of the spatial components of the wave vector in that frame:

$$k^\mu = |\boldsymbol{k}|\frac{k^\mu}{|\boldsymbol{k}|} = \mathrm{k}\left(\frac{k^0}{\mathrm{k}},\hat{\boldsymbol{k}}\right) = \mathrm{k}(v,\hat{\boldsymbol{k}}), \qquad (3.32)$$

where we have again introduced the phase velocity by means of $v = k^0/\mathrm{k}$. For (3.29), we obtain

$$v^2 = 1 - Q \langle T^{\mu\nu}\rangle \bar{k}_\mu \bar{k}_\nu, \quad \bar{k}^\mu = \frac{k^\mu}{k}. \tag{3.33}$$

Here and in the following, we may replace Q by Q' and $\langle T^{\mu\nu}\rangle$ by $T^{\mu\nu}$. Another representation of the light cone condition is found by averaging over propagation directions, i.e. integrating over $\hat{k} \in S^2$. We furthermore assume the resulting velocity shift to be small in order to set $\bar{k}^0 = v \simeq 1$ on the right-hand side of (3.33):

$$v^2 = \frac{1 - (Q/3)\big(\langle T^{00}\rangle + \langle T^\alpha{}_\alpha\rangle\big)}{1 + Q\,\langle T^{00}\rangle}. \tag{3.34}$$

For $Q\langle T^{00}\rangle \ll 1$ and $\langle T^\alpha{}_\alpha\rangle$ of even lower order, this reduces to

$$v^2 = 1 - \frac{4}{3}\,Q\,\langle T^{00}\rangle = 1 - \frac{4}{3}\,Q\,u, \tag{3.35}$$

where u denotes the (renormalized) energy density of the modified vacuum. Let us point out again that the last two equations denote approximations to the light cone condition, while (3.29) and (3.33) are exact representations.

3.2.1 Effective-Action Charge

To illustrate the various versions of the light cone condition given above, let us assume for the moment that the effective Lagrangian under consideration is dominated by the Maxwell term $\mathcal{L}_M$ and yields only a small quantum correction $\mathcal{L}_c$, $\mathcal{L}_c \ll \mathcal{L}_M$. Since the numerator of the Q factor (3.30) is determined only by the quantum correction $\mathcal{L}_c$ anyway, we can approximate the denominator by 1 (from the first term), because additional terms will be of order $\mathcal{L}_c$ and therefore negligible. The Q factor thus reduces to

$$Q \simeq \frac{1}{2}(\partial_{\mathcal{F}}^2 + \partial_{\mathcal{G}}^2)\mathcal{L} \implies \nabla^2 \mathcal{L} \simeq 2Q, \quad \text{for} \quad \frac{\mathcal{L}_c}{\mathcal{L}_M} \ll 1. \tag{3.36}$$

Here, we have introduced the two-dimensional Laplace operator ∇^2 acting in the space of the field invariants $\mathcal{F}$ and $\mathcal{G}$. It is the similarity of this equation to the two-dimensional Poisson equation of classical electrodynamics which justifies calling the Q factor the *effective-action charge*. In fact, we shall demonstrate that the shape of Q as a function of the field invariants (and later of further parameters) in many cases resembles a localized, charge-like distribution. Hence, it will support our intuition to think of Q as a charge which can be associated with the effective Lagrangian in the role of an *electrostatic potential*; for example, the classical vacuum $\mathcal{L}_M = -\mathcal{F}$ is *uncharged*. This agrees with the structure of, for example, (3.35). There, the velocity shift increases linearly with the energy density of the modified vacuum and, therefore, similarly with respect to the field invariants. If the effective-action charge Q indeed obeys a charge-like distribution, i.e. it decreases for increasing invariants, the increase of the velocity shift will slow down (and may halt) for increasing field invariants.

Nevertheless, let us stress that the concept of Q as a charge is nothing but a picture and not an interpretation. There may certainly exist various counterexamples for which the picture of a localized charge distribution will fail; but this will be without consequences for the correctness of our formalism.

There have been earlier attempts to interpret the role of the Q factor; in particular, Shore [153] suggested a deeper connection between the velocity shift and the scale anomaly. He observed that the coefficients of the field invariants in the trace of the vacuum expectation value of the energy–momentum tensor in the weak-field limit,

$$\langle T^\alpha{}_\alpha \rangle = -4 \left(\frac{8}{45} \frac{\alpha^2}{m^4} \mathcal{F}^2 + \frac{14}{45} \frac{\alpha^2}{m^4} \mathcal{G}^2 \right), \tag{3.37}$$

coincide in modulus and sign with the coefficients of the velocity shift (see, e.g. (3.10)) for the different polarization states. He therefore conjectured that the sign of the conformal anomaly may be linked to causality in photon propagation.

Within the framework developed so far, we are in a position to clarify whether such a deeper connection indeed exists in the general case or whether the coincidence arises accidentally. Our task is to investigate the possibility of expressing the effective-action charge Q in terms of the anomaly. It is sufficient to consider the simpler case of a small quantum correction to the Maxwell Lagrangian $\mathcal{L}_c \ll \mathcal{L}_M$. Then, the effective action charge Q can be rewritten by virtue of (3.25) and (3.26) as

$$Q \simeq \frac{1}{2}(\partial_\mathcal{F}^2 + \partial_\mathcal{G}^2)\mathcal{L} = -\frac{1}{2}\left(\frac{\mathcal{G}}{\mathcal{F}} + \frac{\mathcal{F}}{\mathcal{G}} \right) \partial_{\mathcal{F}\mathcal{G}}\mathcal{L} - \frac{1}{8}\left(\frac{1}{\mathcal{F}}\partial_\mathcal{F} + \frac{1}{\mathcal{G}}\partial_\mathcal{G} \right) \langle T^\alpha{}_\alpha \rangle. \tag{3.38}$$

It is obvious from (3.38) that there is no immediate connection between $\langle T^\alpha{}_\alpha \rangle$ and the velocity shift (3.10). The findings of Shore arise from the special structure of the lowest-order Heisenberg–Euler Lagrangian (3.1), where $\partial_{\mathcal{F}\mathcal{G}}\mathcal{L} = 0$ and $(1/\mathcal{F})\partial_\mathcal{F} = 2/\mathcal{F}^2$ (similarly for $\mathcal{G}$). In general, higher order mixed terms exist, since they are not forbidden by gauge, Lorentz or parity invariance. With regard to (3.38), the introduction of the scale anomaly into the numerator appears regrettably to be artificial rather than interpretable.

3.2.2 Weak Electromagnetic Fields

As a first application and a cross-check, we may consider the case of light propagating in a weak electromagnetic background field again; incidentally, the weak-field condition here reads $E, B \ll m^2/e$. A simple improvement of our earlier results can be achieved by employing the two-loop-corrected version of the weak-field-approximated Heisenberg–Euler Lagrangian this time. According to Ritus [140, 141], the two-loop version is given by

$$\mathcal{L} = -\mathcal{F} + c_1' \, \mathcal{F}^2 + c_2' \, \mathcal{G}^2 \,, \tag{3.39}$$

$$c_1' = \frac{8\alpha^2}{45m^4} \left(1 + \frac{40}{9} \frac{\alpha}{\pi} \right), \quad c_2' = \frac{14\alpha^2}{45m^4} \left(1 + \frac{1315}{252} \frac{\alpha}{\pi} \right). \tag{3.40}$$

In the following, we denote the two-loop-corrected coefficients by $c'_{1,2}$, while the one-loop coefficients remain unprimed: $c_1 = 8\alpha^2/45m^4$, $c_2 = 14\alpha^2/45m^4$. Obviously, the quantum correction is small compared with the classical Maxwell term. Hence, we may simply employ the light cone condition as stated in (3.35); we immediately obtain for the polarization and the propagation-direction-averaged velocity

$$Q = c'_1 + c'_2 = \frac{2}{45}\frac{\alpha^2}{m^4}\left(11 + \frac{1955}{36}\frac{\alpha}{\pi}\right),\tag{3.41}$$

$$v = 1 - \frac{4\alpha^2}{135m^4}\left(11 + \frac{1955}{36}\frac{\alpha}{\pi}\right)\left[\frac{1}{2}\left(\boldsymbol{E}^2 + \boldsymbol{B}^2\right)\right].\tag{3.42}$$

The term proportional to 11 clearly corresponds to the one-loop part and is equivalent to our preceding findings (note that the average of the quantity z_k/k^2, e.g. in (3.13), over propagation directions yields $z_k/\mathrm{k}^2 = (2/3)(\boldsymbol{E}^2 + \boldsymbol{B}^2)$). The numerical value of the two-loop correction amounts to $(1955/36) \times (\alpha/\pi) \simeq 0.12629\ldots$, which is a 1% effect (compared with 11). At this point, we should stress that (3.41) contains the complete α^3 two-loop correction to the effective-action charge Q; modifications arising from an exact treatment of the denominator of Q in (3.30) contribute to the order of α^4.

In the spirit of the previous section, it should be pointed out that (3.41) represents the value of the effective-action charge at the origin in field space (at $\mathcal{F}, \mathcal{G} = 0$). As we shall soon demonstrate, this value plays a significant role in various other cases of modified light propagation.

3.2.3 Strong Magnetic Fields

While (3.39) represents an expansion of the effective Lagrangian about the origin in field space, we shall now extract information about the behavior of $\mathcal{L}$ and Q along the complete positive $\mathcal{F}$ axis. The positive $\mathcal{F}$ axis implies the existence of a reference frame in which the field is purely magnetic. Any other point in field space is additionally related to electric-field components and therefore represents a nonstable vacuum state.

As our starting point, we use Schwinger's well-known formula for the one-loop effective QED Lagrangian [148],

$$\mathcal{L} = -\mathcal{F} - \frac{1}{8\pi^2}\int\limits_0^{i\infty}\frac{ds}{s^3}e^{-m^2 s}\left[(es)^2 ab\,\coth eas\,\cot ebs - \frac{2}{3}(es)^2\mathcal{F} - 1\right],\tag{3.43}$$

where we have employed the following representation of the secular invariants a and b (see Appendix B, (B.5)):

$$a = \left(\sqrt{\mathcal{F}^2 + \mathcal{G}^2} + \mathcal{F}\right)^{1/2}, \quad b = \left(\sqrt{\mathcal{F}^2 + \mathcal{G}^2} - \mathcal{F}\right)^{1/2}.\tag{3.44}$$

It is understood that the convergence of the integral in (3.43) is implicitly ensured by the prescription $m^2 \to m^2 - i\epsilon$. (Note that we have not performed a proper-time Wick rotation yet.) To obtain the effective-action charge, we have to evaluate the Laplacian ∇^2 of the effective action in field space; for this, a transition to the secular invariants is convenient, which has been achieved in (B.12) of Appendix B:

$$\nabla^2 \equiv \frac{1}{a^2 + b^2}\left(\partial_a^2 + \partial_b^2\right). \tag{3.45}$$

For the term in the square brackets in (3.43), we can straightforwardly perform the differentiation. In the limit of purely magnetic fields, $b \to 0$ and $a \to |\boldsymbol{B}| \equiv B$, we find

$$\nabla^2\left[\cdots\right]_{b=0} = \frac{2(es)^2}{B^2}\left(\frac{eBs\coth eBs - 1}{\sinh^2 eBs} - \frac{1}{3}eBs\coth eBs\right). \tag{3.46}$$

The complete formula for the numerator of the effective-action charge might be written (with the substitution $z = eBs$) as

$$\frac{1}{2}\nabla^2\mathcal{L} = -\frac{1}{2B^2}\frac{\alpha}{\pi}\int_0^{i\infty}\frac{dz}{z}e^{-(B_{\mathrm{cr}}/B)z}\left(\frac{z\coth z - 1}{\sinh^2 z} - \frac{1}{3}z\coth z\right). \tag{3.47}$$

With some effort, the evaluation of the integral can be performed analytically by standard means which resemble dimensional regularization. Details are given in Appendix D (cf. (D.47)). The result is

$$\frac{1}{2}\nabla^2\mathcal{L}$$
$$= \frac{1}{2B^2}\frac{\alpha}{\pi}\left[\left(\frac{B_{\mathrm{cr}}^2}{2B^2} - \frac{1}{3}\right)\psi\left(1 + \frac{B_{\mathrm{cr}}}{2B}\right) - \frac{2B_{\mathrm{cr}}}{B}\ln\Gamma\left(\frac{B_{\mathrm{cr}}}{2B}\right) - \frac{B_{\mathrm{cr}}}{2B}\right.$$
$$\left. - \frac{3B_{\mathrm{cr}}^2}{4B^2} + \frac{B_{\mathrm{cr}}}{B}\ln 2\pi + \frac{1}{6} + 4\zeta'\left(-1, \frac{2B_{\mathrm{cr}}}{B}\right) + \frac{B}{3B_{\mathrm{cr}}}\right], \tag{3.48}$$

where ψ denotes the logarithmic derivative of the Γ function and ζ' is the first derivative of the Hurwitz zeta function with respect to the first argument [91].

Remembering that we are dealing with a perturbative approximation of the effective action, we can conclude that the domain of validity of our calculations is marked by the condition $\mathcal{L}_{\mathrm{c}} \ll \mathcal{L}_{\mathrm{M}}$. To be precise, as long as the quantum corrections are small compared with the classical behavior of the system, we can completely trust our results. Nevertheless, the one-loop Lagrangian contains, in principle, information over a wider range of field strength; in particular, the infinite-B-field limit will be discussed in the following subsection. Here, we restrict ourselves to the "severe" region in which $\mathcal{L}_{\mathrm{c}} \ll \mathcal{L}_{\mathrm{M}}$.

Thus, we neglect higher-order contributions to the effective-action charge Q stemming from the denominator of (3.30), which results in $Q \simeq (1/2)\nabla^2\mathcal{L} \hat{=}(3.48)$. For strong fields, the last term of (3.48) $\propto (B/(3B_{\mathrm{cr}})$ dominates the expression in the square brackets. We find, obviously, that the effective-action charge decreases with

$$Q(B) \simeq \frac{1}{2}\nabla^2 \mathcal{L} = \frac{1}{6}\frac{\alpha}{\pi}\frac{1}{B_{cr}}\frac{1}{B}\left[1 + \mathcal{O}(1)\right] \quad \text{for} \quad B > B_{cr}, \tag{3.49}$$

as we move along the positive $\mathcal{F}$ axis in field space (Fig. 3.1). Indeed, the effective-action charge exhibits a localized charge-like distribution, which serves as an example of how this picture of the Q factor should be viewed. To lowest order in $\mathcal{L}_c$, the contraction of the energy–momentum tensor VEV may be cast into the form

$$\langle T^{\mu\nu}\rangle \bar{k}_\mu \bar{k}_\nu = \boldsymbol{B}^2 - (\boldsymbol{B}\cdot\hat{\boldsymbol{k}})^2 + \mathcal{O}(\mathcal{L}_c) = \boldsymbol{B}^2 \sin^2\theta + \mathcal{O}(\mathcal{L}_c), \tag{3.50}$$

where θ measures the angle between the $\boldsymbol{B}$ field and the propagation direction.

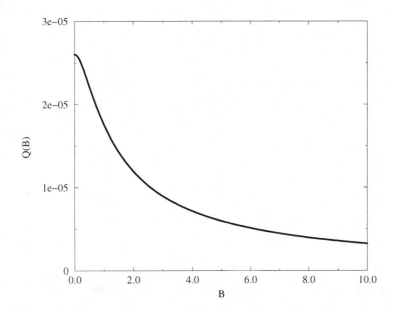

Fig. 3.1. Effective-action charge $Q(B)$ in units of $1/m^4$ versus magnetic field B in units of the critical field strength $B_{cr} = m^2/e$. At the origin, we find the value $Q = c_1 + c_2$, while for strong fields the decrease is proportional to B^{-1}

Finally, the light cone condition (3.33) yields, for arbitrary background fields consistent with the one-loop approximation,

$$v^2 = 1 - \frac{\alpha}{\pi}\frac{\sin^2\theta}{2}\left[\left(\frac{B_{cr}^2}{2B^2} - \frac{1}{3}\right)\psi\left(1 + \frac{B_{cr}}{2B}\right) - \frac{2B_{cr}}{B}\ln\Gamma\left(\frac{B_{cr}}{2B}\right) - \frac{3B_{cr}^2}{4B^2}\right.$$
$$\left. - \frac{B_{cr}}{2B} + \frac{B_{cr}}{B}\ln 2\pi + \frac{1}{6} + 4\zeta'\left(-1, \frac{B_{cr}}{2B}\right) + \frac{B}{3B_{cr}}\right]. \tag{3.51}$$

Equation (3.51) corresponds to the polarization-averaged velocity shift for strong magnetic fields that we found in Sect. 2.2, (2.93)–(2.95). Hence, we have proved the coincidence of the original results of Tsai and Erber, obtained via the polarization tensor, and the outcome of the quasi-classical effective-action approach. This has also independently been shown by Heyl and Hernquist [99, 100], who additionally treated the field invariant $\mathcal{G}$ perturbatively.

Although the velocity shift increases proportionally to the magnetic field for large B, the total amount of the velocity shift remains relatively small for values of B of the order of the critical field strength B_{cr} (Fig. 3.2). To be precise, the terms in the neglected denominator of the effective-action charge Q are of the order of α^2/π^2 and multiply the field invariant $\mathcal{F} \sim B^2$. We therefore expect our approximation to be appropriate as long as $B/B_{\mathrm{cr}} < \pi/\alpha \simeq 430$.

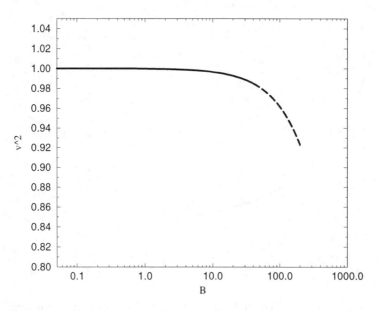

Fig. 3.2. Squared velocity v^2 versus magnetic field B in units of the critical field strength $B_{\mathrm{cr}} = m^2/e$. The *dashed curve* indicates the region in which higher-order contributions become important

3.2.4 Superstrong Magnetic Fields

The following considerations might be judged to be of purely academic interest, since we investigate here the propagation of light in magnetic fields whose strength lies beyond the limit of $B/B_{\mathrm{cr}} \sim \pi/\alpha \simeq 430$. While the results for strong magnetic fields given in the previous section may be relevant to "usual" astrophysical objects such as neutron stars, only the recently discovered, mysterious "magnetars" [67] might produce magnetic field strengths of the order considered in the present section.

Formally, we are interested in the infinite-B-field limit of the light cone condition (3.29) obtained by insertion of the one-loop effective Lagrangian (3.43). For this, we have to improve the previous calculation by no longer neglecting terms of higher order in the quantum corrections $\mathcal{L}_c$. In particular, we have to take the denominator of the effective-action charge, as well as the higher-order terms of (3.50), completely into account. For reasons of convenience, we switch to the representation of the light cone condition involving the Maxwell energy–momentum tensor and the modified effective-action charge Q' given in (3.31). The denominator of Q' can be read off from (3.31):

$$\text{denom}(Q') = -\partial_{\mathcal{F}}\mathcal{L} + \frac{\mathcal{F}}{2}\nabla^2\mathcal{L} + \frac{1}{4}\partial_{\mathcal{F}}\langle T^\alpha{}_\alpha\rangle. \tag{3.52}$$

The Laplacian of $\mathcal{L}$ has been calculated in (3.48), and the strong-field behavior is exhibited in (3.49). We need to know the $\mathcal{F}$ derivatives of $\mathcal{L}$ and $\langle T^\alpha{}_\alpha\rangle$. Since no $\mathcal{G}$ derivative is involved, it is sufficient to consider these functions in the limit of $\mathcal{G} \to 0$. Regarding the effective Lagrangian, we can make use of Weisskopf's [164] well-known formula ($\mathcal{F} \to (1/2)B^2$):

$$\mathcal{L}(B \gg B_{\text{cr}}, \mathcal{G} = 0) = -\mathcal{F} + \frac{\alpha}{6\pi}\,\mathcal{F}\,\ln\frac{2\mathcal{F}}{B_{\text{cr}}^2}$$

$$\Rightarrow \quad \partial_{\mathcal{F}}\mathcal{L} = -1 + \frac{\alpha}{6\pi}\ln\frac{2\mathcal{F}}{B_{\text{cr}}^2} + \frac{\alpha}{6\pi} = \frac{\mathcal{L}}{\mathcal{F}} + \frac{\alpha}{6\pi}. \tag{3.53}$$

Employing (3.25), we can deduce the correct form of $\langle T^\alpha{}_\alpha\rangle$ in a simple manner from (3.53):

$$\langle T^\alpha{}_\alpha\rangle = -\frac{2\alpha}{3\pi}\,\mathcal{F} \quad \Rightarrow \quad \partial_{\mathcal{F}}\langle T^\alpha{}_\alpha\rangle = -\frac{2\alpha}{3\pi}. \tag{3.54}$$

Upon insertion of (3.49), (3.53) and (3.54) into (3.52) and (3.31), we arrive at

$$Q' = \frac{\alpha/(6\pi B_{\text{cr}}B)\left[1 + \mathcal{O}(1)\right]}{1 - (\alpha/3\pi)\left[1 + \ln(B/B_{\text{cr}})\right] + (\alpha/12\pi)(B/B_{\text{cr}}) + \mathcal{O}(1)}$$

$$= \frac{2}{B^2}\left[1 + \mathcal{O}(B_{\text{cr}}/B)\right]. \tag{3.55}$$

Remembering that Q' represents an average over polarization states, we can reconstruct the information about the contribution from each mode by employing the findings of Sect. 2.2. In (2.97), we observed that the dominant contribution leading to a linear increase of the velocity shift with respect to the magnetic field appeared only in the $\|$ mode (where the polarization vector is in the plane spanned by the magnetic field and the polarization direction). In contrast, the velocity shift of the $\perp$ mode approached a constant value in the limit $B \to \infty$. Thus, we have to conclude that the dominant contribution to $\nabla^2 \mathcal{L}$ proportional to $1/B$ in (3.49) is produced by the $\|$ mode, while the $\perp$ mode contributes only to the terms of order $\mathcal{O}(1)$. Hence, we obtain

$$Q'_\| = \frac{2}{B^2}[1 + \mathcal{O}(B_{cr}/B)], \quad Q'_\perp = \mathcal{O}(B_{cr}/B^3). \tag{3.56}$$

The averaged form in (3.55) is recovered by using $Q' = Q'_\| + Q'_\perp$. The second improvement required for evaluating the light cone condition (3.33) in the superstrong-B-field regime concerns the exact treatment of the bilinear form $T^{\mu\nu}\bar{k}_\mu\bar{k}_\nu$, where $\bar{k}^\mu = (v, \hat{k})$. With the aid of (3.22) and the representation of the field strength tensor in terms of E and B fields (Appendix A, (A.8)–(A.12)), we find

$$T^{\mu\nu}\bar{k}_\mu\bar{k}_\nu = B^2 \sin^2 \theta_B + E^2 \sin^2 \theta_E - 2v\,\hat{k} \cdot (E \times B)$$
$$- \frac{1}{2}(E^2 + B^2)(1 - v^2), \tag{3.57}$$

where θ_B denotes the angle between $\hat{k}$ and B, and θ_E the angle between $\hat{k}$ and E. Inserting (3.57) into the light cone condition (3.33), we obtain a quadratic equation for the phase velocity of low-frequency photons,

$$\left(1 + \frac{Q'}{2}(E^2 + B^2)\right) v^2 - 2Q'\,\hat{k} \cdot (E \times B)\,v$$
$$+ Q'\left(B^2 \sin \theta_B + E^2 \sin \theta_E - \frac{1}{2}(E^2 + B^2)\right) - 1 = 0, \tag{3.58}$$

where Q' can be replaced by $Q'_\|$ or $Q'_\perp$ for a particular mode. For vanishing electric fields, we obtain

$$v^2 = \frac{1 - Q'\,B^2 \sin^2 \theta_B + Q'(B^2/2)}{1 + Q'(B^2/2)} = 1 - \frac{\sin^2 \theta_B}{(1/2) + (1/Q'B^2)}. \tag{3.59}$$

Substituting the Q' factors for the different modes into (3.59) finally leads us to

$$v^2_\| = 1 - \sin^2 \theta_B + \mathcal{O}(B_{cr}/B), \quad v^2_\perp = 1 + \mathcal{O}(B_{cr}/B). \tag{3.60}$$

First of all, observe that the phase velocities remain bounded even in the limit of an infinite magnetic field, $0 \le v \le 1$, i.e. the results are still interpretable. In fact, the $\perp$ mode that is characterized by a polarization vector perpendicular to the magnetic field and to the propagation direction ceases to be influenced by the magnetic field; it propagates as in the trivial vacuum, without regard to the direction of propagation. In contrast, the phase velocity of the $\|$ mode is directly related to the direction of the wave vector; propagation along the direction of the magnetic field ($\sin \theta_B = 0$) is preferred, without restrictions, while perpendicular propagation is strictly forbidden, since $v_\|(k \perp B) = 0$. As a consequence, waves in the $\|$ mode will eventually propagate along the magnetic field lines, irrespective of their incidence angle.

It is highly remarkable that this outcome can be immediately interpreted in terms of wave propagation in a magnetized plasma. The $\parallel$ mode is reminiscent of the Alfvén mode; the latter is electrodynamically transverse like the $\parallel$ mode and exhibits an identical angle dependence $\sim 1 - \sin^2 \theta_B$. Since the $\parallel$ mode describes a propagation along the magnetic field lines, it is illustrative to interpret it as an oscillation of the magnetic field lines. The properties of the $\perp$ mode resemble those of the fast magnetoacoustic mode with an Alfvén speed much greater than the speed of light, which are also not influenced by the direction of the magnetic field in this limit. This analogy between physics in superstrong magnetic fields and magnetohydrodynamics was first discovered by Melrose and Stoneham [123] by studying the polarization tensor in such intense fields to one-loop order.

In this sense, the results of (3.60) seem to fit nicely into the picture which is evoked by the concept of vacuum polarization. The polarization of virtual particles caused by such superstrong magnetic fields seems to induce plasma properties in the vacuum, resembling the behavior of real particles.

However, there is obviously a blemish in this appealing interpretation: we should not forget that we are dealing with a one-loop approximation to the effective Lagrangian, which is known to fail for an unbounded increase of the field strength. In fact, to study the infinite-field-strength limit is a truly nonperturbative task, inasmuch as the electrodynamics of strong fields is very reminiscent of the electrodynamics of short distances [141], including, for example, the Landau pole problem.

Nevertheless, despite being tied to perturbation theory, we may be able to draw conclusions with the aid of the renormalization group which go beyond a low-order perturbative approximation. First, note that, regarding (3.59), an analogy to waves in a plasma can immediately be drawn if $Q'B^2 \to 2$ for $B \to \infty$, or, equivalently, $Q'\mathcal{F} \to 1$ for $\mathcal{F} \to \infty$ ($\mathcal{F}\hat{=}(1/2)B^2$). Then, the results of (3.60) hold. Employing the explicit representation (3.31) of Q',

$$\mathcal{F}Q' = \frac{(\mathcal{F}/2)\nabla^2 \mathcal{L}}{-(\partial_{\mathcal{F}}\mathcal{L}) + (\mathcal{F}/2)\nabla^2 \mathcal{L} + (1/4)\partial_{\mathcal{F}}\langle T^{\alpha}{}_{\alpha}\rangle}, \tag{3.61}$$

this will generally be the case if $-(\partial_{\mathcal{F}}\mathcal{L}) + (1/4)\partial_{\mathcal{F}}\langle T^{\alpha}{}_{\alpha}\rangle$ can be neglected compared with $(\mathcal{F}/2)\nabla^2 \mathcal{L}$ for $\mathcal{F} \to \infty$. As recognized above, the one-loop approximation exactly satisfies this criterion; there, $(\mathcal{F}/2)\nabla^2 \mathcal{L}$ increases linearly with B, while $\partial_{\mathcal{F}}\mathcal{L}$ increases only logarithmically and $\partial_{\mathcal{F}}\langle T^{\alpha}{}_{\alpha}\rangle$ even approaches a constant.

As far as $\nabla^2 \mathcal{L}$ is concerned, little is known beyond the one-loop calculation, since its computation first requires the knowledge of $\mathcal{L}$ with its complete dependence on $\mathcal{F}$ and $\mathcal{G}$, which has only been computed up to two-loop order. However, it is not unreasonable to assume that $(\mathcal{F}/2)\nabla^2 \mathcal{L}$ remains an increasing function of $\mathcal{F}$ of the type $\mathcal{F}^{\rho}$ with $\rho > 0$ ($\rho = 1/2$ in the one-loop case). If this does not hold at higher-loop order or in the nonperturbative regime, the function $\mathcal{F}Q'$ might increase logarithmically (then, we cannot conclude anything and further discussion is pointless) or tend to zero; in the latter case, we would find, according to (3.59), that $v = 1$ for $B \to \infty$. This is at least a reasonable result that satisfies the a priori assumptions on v, $0 \leq v \leq 1$, although it seems to be counterintuitive, since then the plane wave would propagate as in the trivial vacuum. Therefore, we assume a power-like increase of $(\mathcal{F}/2)\nabla^2 \mathcal{L}$ for the remainder of this section.

As far as the other terms in the denominator of (3.61), $-(\partial_{\mathcal{F}}\mathcal{L}) + (1/4)\partial_{\mathcal{F}}\langle T^\alpha{}_\alpha \rangle$, are concerned, we need to know $\mathcal{L}$ only in the limit $\mathcal{G} \to 0$ and $\mathcal{F} = (1/2)B^2 \to \infty$. Indeed, as is known from higher-order perturbative calculations [140], this renormalized effective Lagrangian in the perturbative regime can be expanded according to

$$
\mathcal{L}_\mathrm{R}(\mathcal{F})\big|_{\mathcal{F} \to \infty} = -\mathcal{F}\left[1 + \frac{\alpha}{\pi}\left(a_{10} + a_{11}\ln\frac{\sqrt{2\mathcal{F}}}{\gamma\pi B_\mathrm{cr}}\right)\right.
$$
$$
\left. + \sum_{n=2}^{\infty}\left(\frac{\alpha}{\pi}\right)^n \sum_{k=0}^{n-1} a_{nk}\left(\ln\frac{\sqrt{2\mathcal{F}}}{\gamma\pi B_\mathrm{cr}}\right)^k\right], \tag{3.62}
$$

where the coefficients a_{ij} are constants determined partly by fixing the physical parameters (e and m) and partly by the coefficients of the Callan–Symanzik β function. The constant γ is related to Euler's constant: $C = 0.57721 = \ln\gamma$. The validity of this expansion is limited to $\ln(\sqrt{2\mathcal{F}})/(\gamma\pi B_\mathrm{cr}) \ll 1$. From the intrinsic logarithmic nature of this function, it is obvious that $\partial_{\mathcal{F}}\mathcal{L}$ and $\langle T^\alpha{}_\alpha\rangle$ diverge at most logarithmically for increasing $\mathcal{F}$. Under these mild assumptions, we are therefore allowed to conclude that our previous findings, particularly $\mathcal{F}Q' \to 1$ for $\mathcal{F} \gg B_\mathrm{cr}^2$, hold to any finite order in perturbation theory, i.e. in the complete perturbative domain which is characterized by $\ln(\sqrt{2\mathcal{F}})/(\gamma\pi B_\mathrm{cr}) \ll 1$.

This is, of course, a stronger statement than the ones derived from first-order perturbation theory. In particular, the plasma analogy can be regarded as a felicitous interpretation of the physics of the vacuum in superstrong magnetic fields far beyond the critical field strength.

Employing the renormalization group improvement [55], one can even compute the sum over the leading logarithms of the perturbative expansion exactly, which yields

$$
\mathcal{L}_\mathrm{R}(\mathcal{F})\big|_{\mathcal{F}\to\infty} = -\mathcal{F} + \frac{\alpha}{6\pi}\mathcal{F}\left[\ln\frac{2\mathcal{F}}{\gamma^2\pi^2 B_\mathrm{cr}^2} + \frac{6}{\pi^2}\zeta'(2)\right]
$$
$$
- \frac{3}{4}\frac{\alpha}{\pi}\mathcal{F}\ln\left(1 - \frac{\alpha}{6\pi}\ln\frac{2\mathcal{F}}{\gamma^2\pi^2 B_\mathrm{cr}^2}\right). \tag{3.63}
$$

Now, (3.63) is valid for an extended field strength interval, subject to the condition

$$
\frac{\alpha}{\pi}\frac{\ln[1/(1-z)]}{1-z} \ll 1, \qquad z = \frac{\alpha}{6\pi}\ln\frac{2\mathcal{F}}{\gamma^2\pi^2 B_\mathrm{cr}^2}. \tag{3.64}
$$

Again, we discover at most a logarithmically diverging behavior of $\partial_{\mathcal{F}}\mathcal{L}$ and of $\langle T^\alpha{}_\alpha\rangle$. Hence, the increasing powers of logarithms in higher-order perturbation theory appear to be harmless. The physical response of the system to a propagating plane wave seems to be describable by a simple one-loop approximation even for superstrong magnetic fields.

Of course, we should remind the reader of the fact that truly nonperturbative effects which might exist have not been taken into account; we have also imposed an assumption on the behavior of $\nabla^2 \mathcal{L}$ for large magnetic fields which could not be confirmed but was at least made plausible. It is, furthermore, a deficit that our theory describes only a world built out of photons and electrons. For increasing field strengths, quantum effects of further charged particles which couple to the photon take part in the plasma, and might slightly modify the results.

Finally, we would like to stress that it is not the formal $B \to \infty$ limit which is physically interesting. This limit is only an auxiliary device to obtain information about the system at a field strength larger than the critical field strength. Hence, we believe that the physics of a plasma is a good first-order approximation to the physics of superstrong magnetic fields.

3.3 Light Cone Condition for Various Vacua

We investigate the influence of various perturbations of the QED vacuum on the deformation of the light cone; in particular, we study thermalized and Casimir vacua as well as combined perturbations. For weak perturbations (low-energy domain), we *derive* the so-called "unified formula" [120] for the light velocity shift/refractive index, which covers all of the well-known cases, for example the Scharnhorst effect, describing a Casimir-induced increase of the speed of soft photons.

Finally, we address the question of the measurability of the various effects. For this, it is essential to distinguish accurately between phase, group and front velocities on the one hand, and, on the other hand, between velocity and interference measurements. In toto, there is no evidence for a violation of causality. We estimate the order of magnitude of the effects for various experimental conditions.

3.3.1 "Unified Formula"

In the preceding sections, we discussed extensively the propagation of a plane wave in a vacuum which is modified by an intense electromagnetic background field. While an experimental verification of the theoretical predictions for such a system might be expected in the near future [167], this seems not to be the case for the systems which we are now going to consider. However, we intend to demonstrate that light can be a useful tool to test the peculiar properties of the quantum vacuum subject to various external perturbations. Studying the propagation of light in different modified vacua is therefore a subject of fundamental theoretical interest, the aim being to gain more understanding of the vacuum, rather than a search for experimentally tiny effects. It should, however, be noted that reliable results for the velocity shift of photons in modified vacua can only be obtained in the low-frequency domain, which most of the results cited below refer to.

In addition to the investigation of electromagnetic modifications, which started in the early 1950s [158], Drummond and Hathrell [66] initiated the study of light propagating in various gravitational backgrounds. They discovered that vacuum polarization, e.g. the electron–positron loop, in a gravitational background can be understood as giving a "size" to the photon on which the gravitational field can act via tidal forces. On a formal level, the electromagnetic field characterizing the plane wave field couples to the curvature tensors and/or covariant derivatives; hence, the equations of motion for the photon are altered in a highly nontrivial manner. In addition, the strong principle of equivalence is violated, which only allows for a photon kinetic term of the form $F^{\mu\nu}F_{\mu\nu}$.

In [66] the surprising phenomenon was revealed that some parts of the modified light cone may lie "outside" the vacuum light cone. In particular, for Schwarzschild, Robertson–Walker and gravitational-wave space–times, directions and polarizations were identified for which the photon velocity is faster than c. As an example, we cite the phase velocity for low-frequency photons for a homogeneous, isotropic Robertson–Walker background with Friedman cosmology,

$$v = 1 + \frac{11}{45}\alpha\, G_{\mathrm{N}}\frac{\rho + p}{m^2}, \tag{3.65}$$

where G_{N} is Newton's constant and ρ and p denote the energy density and pressure of the perfect fluid which approximates the matter distribution of the universe. Assuming that the matter satisfies the positive-energy condition $\rho + p > 0$, the velocity indeed exceeds the vacuum velocity c (=1).

In subsequent papers, Daniels and Shore showed that "faster than light" velocities also appeared in the Reissner–Nordström [49] and the Kerr [50] background. The variety of effects of these different space–times can be ordered with the help of two important theorems established by Shore [153].

The first theorem, the polarization sum rule, relates the average velocity over the two physical polarizations to the matter energy–momentum tensor. As a consequence, in Ricci-flat space–times, the velocity shifts of the two polarization modes must be equal but opposite in sign in order to maintain an average velocity equal to c.

The second theorem, the horizon theorem, states that the light cone for photons traveling normal to an event horizon of a general but stationary space–time remains $k^2 = 0$. As an application, this explains in simple terms why, for example, photon propagation is not modified in the de Sitter space–time; here, every point in space–time belongs to an event horizon of some observer. Hence, the velocity of light must be c everywhere.

Further interesting studies on this subject of photon propagation in gravitational backgrounds have been performed [42] for dilaton black-hole space–times and [37] for topological black-hole solutions. In particular, dilatonic effects become important or even dominant for orbits close to the event horizon; topological structures have been shown to have an influence on light propagation in nonstationary black-hole space–times such as the radiating topological black hole, while the light cone condition is not sensitive to the asymptotic behavior and topological structures in the stationary case.

Substituting photons by massless neutrinos, Ohkuwa [128] discovered similar effects regarding their propagation in gravitational backgrounds. The major difference from photons is marked by the fact that superluminal velocities are not possible for Ricci-flat space–times.

To obtain a superluminal velocity shift induced by vacuum polarization, it is not necessary to combine the theories of quantum electrodynamics and general relativity (a combination which one may feel suspicious of). As was demonstrated by Scharnhorst [143], superluminal phase velocities also appear in simple QED in the standard parallel-plate Casimir configuration for propagation normal to the plates. On a formal level, the Scharnhorst effect arises from imposing periodic boundary conditions on the photonic fluctuations of the QED vacuum polarization processes; the effect was originally calculated from the two-loop polarization tensor after insertion of an internal photon line modified by the periodic boundary conditions (Fig. 3.3). The result is

$$v = 1 + \frac{11\pi^2}{8100} \frac{\alpha^2}{m^4} \frac{1}{a^4} \cos^2 \theta, \tag{3.66}$$

and depends on the plate separation a and the angle θ between the direction of propagation and the normal to the plates. A different derivation of the Scharnhorst effect was given by Barton [20] with the aid of the Heisenberg–Euler Lagrangian. Our treatment of the light cone condition given below is in the spirit of Barton's work.

Since a finite temperature can also be introduced by imposing boundary conditions on the fluctuating fields, it is plausible that light propagation will also be modified in a thermalized vacuum. Similarly to the Scharnhorst effect, a two-loop calculation of the polarization tensor with a modified internal photon line leads to such a result. For a homogeneous, isotropic thermal vacuum, the velocity shift for low-frequency photons was first calculated by Tarrach [157] (the numerical coefficient was later corrected by Barton [20]) and yields

$$v = 1 - \frac{44\pi^2}{2025} \frac{\alpha^2}{m^4} T^4. \tag{3.67}$$

The phase velocity obviously remains smaller than c; this is formally related to the fact that the boundary conditions on the fluctuating internal photons which are considered as thermalized have been imposed in the direction of imaginary time.

It should be stressed that the energy scale of the vacuum modifications in (3.65)–(3.67) is assumed to be much smaller than the energy scale set by the electron mass, i.e. $T \ll m$, $1/a \ll m$. In this sense, these findings can be viewed as low-energy results.

A remarkable observation was made by Latorre, Pascual and Tarrach [120], who pointed out that striking similarities exist between (3.65)–(3.67)

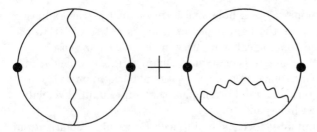

Fig. 3.3. Feynman diagrams for the two-loop polarization tensor. In order to obtain the low-energy velocity shift for a Casimir vacuum or a thermalized vacuum, the internal photon is modified by the appropriate boundary conditions

and also the electromagnetically modified phase velocity, e.g. (3.13). Taking the average over polarization states (and propagation directions, if necessary), each formula can be written identically (in a "unified" way) as

$$v = 1 - \frac{44}{135} \frac{\alpha^2}{m^4} u, \tag{3.68}$$

where we have to replace one α by $G_N m^2$ in the case of a gravitational background. In (3.68), $u = \langle T^{00} \rangle$ denotes the renormalized energy density of the modified vacuum, i.e.

- $u = (1/2)(E^2 + B^2)$ for electromagnetic background fields
- the Planck energy density $u = (\pi^2/15)T^4$ for the thermal background
- $u = -(\pi^2/720)(1/a^4)$ for the Casimir vacuum
- $u = -\rho = -(3/4)(\rho + p)$ for a radiation-dominated Friedman–Robertson–Walker vacuum, in which $p = \rho/3$.

The so-called "unified formula" attributes the appearance of superluminal phase velocities to negative energy densities. At this stage, we would like to stress that this formula has only been *identified* and not at all *proved*. This raises several questions. Does a proof exist that this "unified formula" also holds for further modified vacua or are the similarities purely accidental? Does the numerical prefactor common to all these examples hint at any deeper physical realm (as has sometimes been speculated in the literature)? How can the fact that (3.65) and (3.13) derive from one-loop calculations while (3.66) and (3.67) are based on the two-loop level be understood?

The appropriate tool for investigating the nature of the "unified formula"[3] obviously is the effective-Lagrangian approach; in this case, it is superior to the approach via the polarization tensor, since it provides a way to establish certain approximations more conveniently. However, the problem of light propagation in gravitational backgrounds has to be excluded from the following considerations, because the combination of QED and gravity has to be established on a more fundamental level than that of effective theories.

---

[3] We continue to write this phrase in quotes, because we doubt its unifying power.

3.3.2 Light Cone Condition for Low-Energy Phenomena

Our starting point is the light cone condition for purely EM-field-modified vacua and low-frequency photons as given in (3.29):

$$k^2 = Q \langle T^{\mu\nu} \rangle_{\mathcal{F}\mathcal{G}} \, k_\mu k_\nu. \tag{3.69}$$

Here, we have introduced the invariants $\mathcal{F}$ and $\mathcal{G}$ (cf. Appendix A) as subscripts to remind the reader that we are dealing with the expectation value of the energy–momentum tensor with respect to the electromagnetically modified vacuum state. Since the "unified formula", by construction, refers to the domain in which the energy scales of the modified vacua are smaller than the electron mass (the energy scale of the fundamental theory, QED), it is natural to assume the quantum corrections to the effective action to be small compared with the classical Maxwell action ($\mathcal{L}_c \ll \mathcal{L}_M$). In this approximation, we found that the effective-action charge Q is a pure number (cf. (3.40) and (3.41)):

$$Q \simeq \frac{1}{2}(\partial_{\mathcal{F}}^2 + \partial_{\mathcal{G}}^2)\mathcal{L} = \frac{22}{45}\frac{\alpha^2}{m^4} \quad \left(Q^{2-\text{loop}} = \frac{2}{45}\frac{\alpha^2}{m^4}\left(11 + \frac{1955}{36}\frac{\alpha}{\pi} \right) \right), \tag{3.70}$$

where we have cited the results obtained from the two-loop as well as the one-loop Heisenberg–Euler Lagrangian.

In the language of path integrals, the one-loop approximation has been obtained by performing a Gaussian integration over the fermion fields. Here, the background field has been assumed as constant, and the nonconstant contributions of the fluctuating photon fields which are also contained in the Dirac operator have been neglected. If, instead, we had taken the fluctuating photon field into account, we would have obtained an exact effective action (depending, of course, on infinitely many field operators); after insertion of this action into the functional integral over the fluctuating photon field, we would still be dealing with the exact theory of QED. In this sense, the Heisenberg–Euler Lagrangian can be viewed as the lowest-order approximation to this particular effective action. Insertion of the Heisenberg–Euler Lagrangian into the functional integral is the philosophy of the modern approach of effective theories [111]. (In fact, for the present purpose, it is even sufficient to consider the Maxwell action inserted into the functional integral.)

In view of these considerations, we regard the field quantities on the right-hand side of the light cone condition (3.69) as being subject to quantization with the aid of the functional integral over the fluctuating photons. Modifying the vacuum by either Casimir plates or finite temperature (or both) is reflected by imposing boundary conditions on fluctuating photons.[4] Hence,

---

[4] In the case of finite temperature, it is sufficient to impose the boundary conditions on the photon fields only, i.e. to thermalize the photons only, since we are at present studying only the low-temperature domain, $T \ll m$. In the case of the Casimir vacuum, we consider the ideal situation of infinitely thin (and perfectly conducting) plates which do not influence the electrons. So there also is no need for boundary conditions on the fermion field.

the functions of the field quantities on the right-hand side of (3.69) have to be replaced by their vacuum expectation values for the appropriate vacuum, and we may write

$$k^2 = \langle Q \langle T^{\mu\nu} \rangle_{\mathcal{F}\mathcal{G}} \rangle_z k_\mu k_\nu, \tag{3.71}$$

where the subscript z labels the vacuum modification under consideration. Since Q is a pure number in the low-energy domain, we obtain

$$k^2 = Q \langle T^{\mu\nu} \rangle_{\mathcal{F}\mathcal{G}z} k_\mu k_\nu. \tag{3.72}$$

Introducing the phase velocity v and averaging over propagation directions leads us to the formula (cf. (3.35))

$$v = 1 - \frac{2}{3} Q \langle T^{00} \rangle_{\mathcal{F}\mathcal{G}z} = 1 - \frac{2}{3} Q u = 1 - \frac{44}{135} \frac{\alpha^2}{m^4} u, \tag{3.73}$$

where u denotes the (renormalized) energy density of the modified vacuum. Equation (3.73) is identical to the "unified formula" for the propagation of low-frequency photons in modified vacua given by Latorre, Pascual and Tarrach [120]. Our derivation represents a proof for vacua which are modified by electromagnetic fields, Casimir configurations, finite temperature or any further external influences which enter the theory via the functional integral over the photon fields. Examples will be given below.

Let us first of all mention that the "proof" appears to be almost trivial and should serve to demystify the "unified formula". Above all, there is nothing mysterious about the numerical prefactor containing the number 11. We would like to stress that this factor is common to all the different vacua simply because it has the same origin: it is nothing but the effective-action charge of the Heisenberg–Euler Lagrangian in the weak-field limit. Even in the case of a gravitational background, which is not covered here, it should not be too surprising that we encounter this factor: it is always the fermion loop, with its spin-$(1/2)$ character, that leads to the appearance of typical numerical coefficients. For example, in scalar electrodynamics, the Scharnhorst effect has been calculated by Cougo-Pinto et al. [46], and the result can be written as

$$v = 1 - \frac{8}{135} \frac{\alpha^2}{m_0^4} u, \tag{3.74}$$

where m_0 denotes the mass of the scalar particle, and the Casimir energy density u to lowest order in scalar QED is, of course, the same as in spinor QED.

Casimir Vacuum (Scharnhorst Effect)

Inserting the energy density of the idealized parallel-plate Casimir configuration, $u = -(\pi^2/720a^4)$, into our formula (3.73), we arrive at the direction-averaged version of Scharnhorst's result:

$$v = 1 + \frac{1}{3} \frac{11\pi^2}{(90)^2} \frac{\alpha^2}{m^4} \frac{1}{a^4}. \tag{3.75}$$

Since propagation parallel to the plates is not modified, the averaged velocity shift yields one-third of the perpendicular velocity shift (3.66). The low-energy approximation, $1/a \ll m$, can be inverted to give $a \gg \lambda_c$ ($\lambda_c = 1/m$, the Compton wavelength). In fact, we have to impose an even stronger condition on the plate separation a: on the one hand, our formalism is based on the soft-photon approximation, i.e. the wavelength of the propagating light should be much larger than the Compton wavelength, $\lambda \gg \lambda_c$; on the other hand, the plate separation has to exceed the wavelength of the light probe by far in order to justify considering the Casimir region as a dielectric medium, $a \gg \lambda$. Therefore, we have to treat the plate separation a as a macroscopic parameter. In terms of what is experimentally practicable, a reasonable value is of the order of a few microns. In combination with the smallness of the prefactors, an experimental verification is obviously out of the question. Whether measurability of the Scharnhorst effect is in principle possible at all is discussed in Sect. 3.3.4.

It is worthwhile to point out that our procedure for deriving the Scharnhorst effect works very economically; remember that the Scharnhorst effect is, diagrammatically, produced at the two-loop level, since the polarization tensor has to be evaluated with an internal modified-photon line (see Fig. 3.3). In the present derivation, we employed only the one-loop Heisenberg–Euler Lagrangian and additionally inserted information about the vacuum expectation value of field correlators, from which the energy–momentum tensor can be constructed. Therefore, our formalism mimics a two-loop calculation on the one-loop level.

Whether this can successfully be extended to higher-loop orders is not clear at present. We may try to estimate higher-order corrections to the Scharnhorst effect in the following way: first, we employ our knowledge about the two-loop value of the effective-action charge Q as given in (3.70); secondly, we make use of the radiative corrections to the Casimir energy density as obtained by Bordag, Robaschik and Wieczorek [31].[5] We obtain [61]:

$$v = 1 + \frac{1}{3} \frac{4}{45} \frac{\alpha^2}{m^4} \left(11 + \frac{1955}{36} \frac{\alpha}{\pi} \right) \left(\frac{\pi^2}{720} \frac{1}{a^4} - \frac{\alpha}{\pi} \frac{\pi^3}{2560} \frac{1}{ma^5} \right). \tag{3.76}$$

However, we have to stress that (3.76) does not represent an exact result of quantum field theory, but has to be viewed as an estimate of the possible higher-order corrections. In particular, the functional integration over the photonic fluctuations has been treated on two different footings: on the one

---

[5] Employing the method of effective theories, the next-to-leading-order contribution to the Casimir energy is found to be of the order of $1/m^4a^8$. For the discussion of the Scharnhorst effect, this result was used in [60]. However, as stressed in [32], the effective-theory approach naturally misses the radiative correction proportional to $1/ma^5$, which has been proved to be the dominant correction even for a more realistic treatment of the plates.

hand, in the derivation of the two-loop Heisenberg–Euler Lagrangian, and on the other hand, in the determination of the vacuum expectation value of the energy–momentum tensor.

Nevertheless, in an exact calculation, the appearance of these terms can be expected; but it is not clear whether the estimate (3.76) really includes all of the important terms. Translated into a diagrammatic language, the above formula does not, for example, contain the contribution of a three-loop diagram with two internal photon lines which are both modified by the Casimir boundary conditions. It is plausible that this diagram contributes to higher order in $1/a$, but exact answers to this and similar questions concerning higher-order calculations are presently out of reach.

Finite Temperature ($T \ll m$)

Inserting the Planck energy density, $u = (\pi^2/15)T^4$, into the generalized light cone condition (3.73), the correct result for the lowest-order velocity shift as given in (3.67) is obtained:

$$v = 1 - \frac{44}{135}\frac{\alpha^2}{m^4}\frac{\pi^2}{15}T^4. \tag{3.77}$$

Since the boundary conditions are imposed on the photon fields in *imaginary* time, there is no additional condition that the temperature must satisfy in relation to the wavelength of the light probe, as in the case of the Scharn-horst effect. Only $T \ll m$ must hold in order to justify the approximation of nonthermalized fermions.

With the same reservations as cited in the preceding paragraph, higher-order corrections to (3.77) could be estimated.

Scharnhorst Effect at Finite Temperature

The combination of thermal and Casimir phenomena is in itself worth study-ing, because both effects enter the formalism via boundary conditions but lead to opposite results for the velocity shift. Here, we want to investigate where and why each effect dominates the velocity shift. The determining or-der parameter is the dimensionless combination Ta. Nevertheless, the plate separation a has to be considered as a macroscopic quantity ($a \simeq \mathcal{O}(\mu m)$).

Even in the low-energy regime, the hierarchy of scale parameters can in principle be ordered in two different ways: (i) low-temperature domain, $Ta \ll 1 \ll ma$; (ii) intermediate-temperature domain, $1 \ll Ta \ll ma$.

For both cases, we consider light propagation perpendicular to the plates in the 3-direction, for which we employ the light cone condition in the form of (3.72):

$$v = 1 - \frac{1}{2}Q\left(\langle T^{00}\rangle + \langle T^{33}\rangle\right). \tag{3.78}$$

Here, we have tacitly assumed the velocity shift to be small. For the remainder, we confine ourselves to the one-loop approximation, considering, as usual, a low frequency for the propagating light; hence, the effective-action charge Q is given by $Q = (22\alpha^2)/(45m^4)$ (see (3.70)).

The various contributions to the vacuum expectation value of $T^{\mu\nu}$ for the appropriate temperature regime can be taken from the work of Brown and Maclay [35]. For low temperature, $Ta \ll 1 \ll ma$, the required components read

$$\langle T^{00}\rangle_T^a = -\frac{\pi^2}{720}\frac{1}{a^4} + \frac{\zeta(3)}{\pi^2}\frac{T^3}{a}, \quad \langle T^{33}\rangle_T^a = -\frac{\pi^2}{240}\frac{1}{a^4},$$

$$\text{for} \quad Ta \to 0, \qquad (3.79)$$

for the parallel-plate configuration ($\zeta(3) = 1.202056\ldots$). According to (3.78), the light cone condition for propagation perpendicular to the plates yields

$$v = 1 + \frac{11}{(90)^2}\frac{\alpha^2}{m^4}\frac{\pi^2}{a^4}\left[1 - \frac{180\zeta(3)}{\pi^4}(Ta)^3\right]. \qquad (3.80)$$

In the low-T limit, the $(Ta)^3$ term can be neglected and we only rediscover Scharnhorst's result. But, astonishingly, we do not find an additional velocity shift proportional to T^4, which could have been expected from (3.77). This can be understood from the fact that in the Casimir vacuum only those fluctuations of the photon field exist that respect the boundary conditions. In the direction perpendicular to the plates, the fluctuation modes are therefore quantized in a discrete fashion. At low temperature, none of these perpendicular modes can be excited, because the energy gap is too large. The $(Ta)^3$ term in (3.80) will become important for $Ta = \mathcal{O}(1)$, i.e. $T > 2000$ K for $a \simeq 1\mu$m. This shows that the Scharnhorst effect is not sensitive to temperature perturbations.

For increasing temperature, we encounter an intermediate-temperature region characterized by the condition $1 \ll Ta \ll ma$, which corresponds to 0.2 eV$\ll T \ll 0.5$ MeV. This implies that taking the effective-action charge at its constant, low-energy value is still a good approximation.

Using further results of Brown and Maclay [35],

$$\langle T^{00}\rangle_T^a = \frac{\pi^2}{15}T^4, \quad \langle T^{33}\rangle_T^a = \frac{\pi^2}{45}T^4 - \frac{\zeta(3)}{4\pi}\frac{T}{a^3}, \quad \text{for} \quad Ta \gg 1, \quad (3.81)$$

we find for the phase velocity in the low-frequency limit

$$v = 1 - \frac{44\pi^2}{(45)^2}\frac{\alpha^2}{m^4}T^4\left[1 - \frac{45\zeta(3)}{16\pi^3}\frac{1}{(Ta)^3}\right]. \qquad (3.82)$$

In this limit, only the modifications caused by the thermalized photon gas become important. A term proportional to $1/a^4$ as a consequence of some missing zero-point fluctuations does not occur, since higher (perpendicular) modes have been thermally excited.

This clearly demonstrates that a combination of Casimir and thermal vacuum modifications does not simply result in a sum of both effects. Finite-size effects can reveal a nontrivial interplay, which, in this case, can be interpreted in terms of the peculiar features of the quantum vacuum.

Two-loop effects can again be estimated by taking the two-loop corrections to the effective-action charge and energy densities into account. As in the preceding investigations, the next-to-leading-order correction is simply obtained by replacing $Q^{1-\text{loop}}$ by $Q^{2-\text{loop}}$ (cf. (3.70)).

3.3.3 Light Cone Conditions for High-Energy Phenomena

As an example of light propagation in a modified vacuum whose energy scale lies beyond the electron mass scale, we discussed light propagation in arbitrarily strong magnetic fields in Sect. 3.2. There, an exact treatment of the problem was possible, since the effective Lagrangian for the problem is known with respect to its dependence on the complete set of invariants of the physical system.

The same procedure should, in principle, be applicable to further modified vacua. But, owing to the difficulties of computing such a Lagrangian, this simple plan breaks down. Either the number of invariants exceeds the computational power, as is the case for an extension to a light probe of higher frequency, or a one-loop Lagrangian is not sufficient and the two-loop version for the full parameter range is out of reach, as observed for the finite-temperature and Scharnhorst effects; or finally, the required theory simply does not exist, as is the case for light propagation in curved spaces, for which quantum gravity would have to be employed.

A one-loop calculation of the effective Lagrangian of QED at finite temperature in which the fermions are considered to be thermalized can in fact be performed, retaining the dependence on the complete set of invariants. Furthermore, the two-loop order is obtainable, at least for low temperature. Owing to the conceptual and technical difficulties of this program, we dedicate two whole sections to this enterprise (Sects. 3.5 and 3.6). The implications for light propagation at high temperature will be discussed afterwards. There, we find that the further invariant structure beyond the purely electromagnetic part becomes highly important and dominates the subject of light propagation.

In view of these results, we come to the conclusion that the simple structure of the "unified formula" is closely connected to a simple set of invariants. As long as we are dealing with only a small number of invariants (say, one or two), the right-hand side of the light cone condition should be expressible in terms of the energy–momentum tensor for purely algebraic reasons. When the number of invariants increases, it becomes more and more improbable that the algebraic structure coincides with that of the corresponding energy–momentum tensor; then, the right-hand side must necessarily be more complicated.

To conclude, the "unified formula" is a product of the low-energy domain; for more complex vacuum modifications and in the high-energy domain, a unifying generalization is not expected to exist.

3.3.4 Measurability and Causality

Although the success of QED is overwhelming as far as scattering experiments and atomic physics are concerned, the experimental proof of nonlinear interactions between classical electromagnetic fields as predicted by the theory is still missing. But there is justified hope that this outstanding verification of the theory will be achieved soon. In fact, the aim of the current experiments (e.g. [17, 18, 104, 167]) is to measure the velocity shifts for a low-frequency light probe (laser) propagating in a magnetic background field in an indirect way: since the magnetically modified vacuum closely resembles an anisotropic medium, the effect of birefringence arises. Suppose a linearly polarized laser beam passes perpendicularly through a magnetic field B with polarization at an angle to the direction of B. As was found, for example, in (3.10), the component of the laser electric field parallel to B will travel more slowly than the component of the laser electric field perpendicular to B. This results in a phase shift between the components of the laser electric field. Hence, a linearly polarized laser beam becomes elliptically polarized. The phase shift is given by

$$\Phi^{\mathrm{QED}} = 2\pi \left(n_{\parallel} - n_{\perp}\right)\frac{L}{\lambda}, \tag{3.83}$$

where λ denotes the wavelength of the laser beam. The refractive indices $n_{\parallel, \perp}$ of the two polarization modes are the inverses of the velocities in (3.10). Inserting these into (3.83), we find

$$\Phi^{\mathrm{QED}} = \frac{1}{15}\alpha \left(\frac{B}{B_{\mathrm{cr}}}\right)^2 \frac{L}{\lambda} \left(1 + \frac{25}{4}\frac{\alpha}{\pi}\right), \tag{3.84}$$

where we have also added the two-loop correction by employing the constants c_1' and c_2' as given in (3.40).

The numerical value of the critical field strength is $B_{\mathrm{cr}} = 4.4 \times 10^9$ T, which is huge compared with a typical laboratory value of $B = 8$ T. With a wavelength of $\lambda = 1064$ nm (Nd–YAG laser), any realistic value of L that is comparable to the size of a laboratory leads to a vanishingly small value for Φ. The solution to this problem is to increase the optical path by a high-finesse Fabry–Pérot optical resonator, which is expected to increase the phase shift up to a (hopefully) measurable value of $\Phi \simeq 5 \times 10^{-11}$ rad.

While we have concentrated on constant fields as external perturbations, a background of intense laser fields similarly provides for a variety of vacuum effects [23]; for example, the phenomenon of birefringence has been discussed in [8] (for a comprehensive review, see [24]).

There should also be no obstacles in principle against measuring the temperature-induced velocity shift given, for example, in (3.77). Although the thermalized vacuum resembles an isotropic medium, one can imagine an interference experiment in which a split laser beam partially passes through a "hot" region and partially propagates in a vacuum. By interfering the two beams, a phase shift should be detectable. However, being tied to laboratory dimensions diminishes the hope of experimental verification. Even the cosmic laboratory does not seem to provide an appropriate experimental setting; for example, the velocity shift induced by the cosmic microwave background (CMB), $T_{CMB} = 2.726$ K, amounts to

$$\delta v_{CMB} = -5.101 \times 10^{-43}, \tag{3.85}$$

which would even imply vanishing effects if the optical path was of the size of the universe.

The situation is similar for the cases of a gravitational background and the Scharnhorst effect. The shifts of the phase velocities for low-frequency photons could in principle be made visible by interference experiments; needless to say, in practice measurability is impossible.

For the case of a gravitational background, two assumptions have been made, namely, that the light probe is of low frequency, $\lambda \gg \lambda_c = 1/m$, and that the wavelength is small compared with the characteristic length l_R associated with the curvature ($R \sim l_R^{-2}$). As a conclusion, we find $\lambda_c \lll l_R$, so the velocity shift $\delta v = \mathcal{O}(\alpha \lambda_c^2 / l_R^2)$ is necessarily tiny.

Similarly disappointing is the result of the phase velocity shift for the Scharnhorst effect. The experimental possibility of a plate separation of 1 μm implies a velocity shift of $\delta v = 1.55 \times 10^{-32}$.

Even more interesting than the question of the measurability of the phase velocities is whether the superluminal velocities of the gravitational case and the Scharnhorst effect can be exploited for sending faster-than-c signals. At least as far as the particular result of Scharnhorst is concerned, the answer is no, as was shown by Ben-Menahem [27]. The argument goes as follows: since high-frequency photons were excluded from the calculation and therefore their effect cannot be estimated, the wave packet of the light probe has to be constructed from frequencies smaller than the electron mass. Hence, the wavefront will be smeared out over a distance of $1/m$. To measure a faster-than-c velocity for propagation perpendicular to the plates, the wavefront must be found at a distance δx, larger than $1/m$, beyond the light cone after propagation for some time t. That is,

$$\delta x = \delta v \, t \overset{(3.66)}{=} \frac{11\pi^2\alpha^2}{(90)^2} \frac{1}{(ma)^4} \, t \overset{!}{>} \frac{1}{m}. \tag{3.86}$$

Since t can maximally be as large as the plate separation a, we find

$$\frac{11\pi^2\alpha^2}{(90)^2} \frac{1}{(ma)^3} \overset{!}{>} 1. \tag{3.87}$$

Obviously, this inequality cannot hold since the prefactor and $1/(ma)$ are much smaller than 1. To circumvent this restriction, the wavefront must be sharpened by inclusion of higher frequencies. For this, (3.66) is, of course, an insufficient approximation.

To decide whether superluminal signal propagation is possible for a Casimir configuration or a gravitational background, it is necessary to study the velocity shift or, alternatively, the refractive index $n = 1/v$ in the infinite-frequency limit. This is because the signal velocity can only be uniquely defined by the wavefront velocity, and the construction of a sharp wavefront requires the inclusion of infinite frequencies. Unfortunately, nothing is known about the refractive index of the Casimir vacuum or of a vacuum in a curved background apart from $n(\omega \to 0)$ at present.[6] In principle, the two-loop polarization tensor for a photon propagating in the vacuum contains information for any value of four-momentum of the propagating light. But it is not clear at present whether such a low-order calculation suffices to draw conclusions about the infinite-frequency limit or whether this is a truly non-perturbative task. In particular, it has not yet been investigated whether the logarithms (depending on the external four-momentum) which will appear with increasing powers in higher-order computations spoil the perturbation series expansion or whether they are completely harmless, as is the case for superstrong magnetic fields (Sect. 3.2).

As an alternative approach to the infinite-frequency limit, there is a non-perturbative relation between the different frequency scales which is based on the principle of local causality: the Kramers–Kronig dispersion relation. In the following, we elucidate the line of argument given in [21,66,144]. Suppose $f(\omega)$ is an analytic function in the upper half plane vanishing in the limit $|\omega| \to \infty$. According to Kramers and Kronig, the real and imaginary parts are related to each other according to

$$\mathrm{Re}\, f(\omega) = \frac{1}{\pi}\, \mathrm{P} \int\limits_{-\infty}^{\infty} d\omega' \frac{\mathrm{Im}\, f(\omega')}{\omega' - \omega},$$

$$\mathrm{Im}\, f(\omega) = -\frac{1}{\pi}\, \mathrm{P} \int\limits_{-\infty}^{\infty} d\omega' \frac{\mathrm{Re}\, f(\omega')}{\omega' - \omega}. \tag{3.88}$$

It is reasonable to assume that these conditions are satisfied by the combination

$$f(\omega) = \frac{n(\omega) - n(\omega_0)}{\omega - \omega_0}, \qquad \omega_0 \quad \text{arbitrary}. \tag{3.89}$$

Substituting the equation for the imaginary part into the one for the real part and inserting (3.89), we find

---

[6] A complete two-loop calculation for the next-to-leading-order dispersive terms for the Casimir vacuum [120] turned out to be incorrect.

$$\operatorname{Re} n(\omega) - \operatorname{Re} n(\omega_0) = \frac{1}{\pi} \operatorname{P} \int\limits_{-\infty}^{\infty} d\omega' \, \frac{\operatorname{Im}\left[n(\omega') - n(\omega)\right](\omega - \omega_0)}{(\omega' - \omega)(\omega' - \omega_0)}. \tag{3.90}$$

In the limit $\omega_0 \to \infty$, this expression reads

$$\operatorname{Re} n(\omega) = \operatorname{Re} n(\infty) + \frac{1}{\pi} \operatorname{P} \int\limits_{-\infty}^{\infty} d\omega' \, \frac{\operatorname{Im} n(\omega')}{\omega' - \omega}. \tag{3.91}$$

Using the *reality condition*, $n(-\omega) = n^*(\omega)$, which implies that the field quantities are purely real, we deduce that $\operatorname{Re} n(\omega)$ is an even function and $\operatorname{Im} n(\omega)$ is an odd function. For $\omega \to 0$, we therefore obtain

$$\operatorname{Re} n(\infty) = \operatorname{Re} n(0) - \frac{2}{\pi} \operatorname{P} \int\limits_{0}^{\infty} d\omega' \, \frac{\operatorname{Im} n(\omega')}{\omega'}. \tag{3.92}$$

Assuming that the amplitude of the light probe is not amplified by the modified vacuum, i.e. that the medium can at most absorb the light probe, the imaginary part of the complex index of refraction has to be nonnegative: $\operatorname{Im} n(\omega) \geq 0$. Then, we obtain the inequality

$$1 > n(0) \geq n(\infty), \tag{3.93}$$

for both the Casimir configuration and the gravitational background. Hence, the low-frequency refractive index serves as an upper bound for the signal velocity which turns out to be superluminal!

To circumvent this result, one possibility is that the assumptions leading to the Kramers–Kronig dispersion relation are not satisfied. However, no idea has been suggested up to now that gives a hint in this direction.

As a second possibility, one can, of course, assume that the medium is not passive, i.e. the imaginary part of the refractive index becomes negative for certain values of the frequency. In this region, the amplitude of the light probe is amplified. This, of course, implies energy transfer from the vacuum to the medium and therefore appears to be in conflict with the principle of energy conservation.

Neither possibility is furnished by conventional physical concepts; hence, we have to face the question of whether superluminal velocities in a curved background or in the Casimir vacuum shake our physical fundaments. One standard argument against superluminal signal propagation is the violation of causality. Such a violation is proved by constructing a causal paradox. For example, in special relativity, if a signal is sent with a space-like velocity from an observer A to an observer B, the propagation is certainly causal with respect to the reference frame of A. But there exist reference frames in which this signal corresponds to a motion backwards in time. Now, if B's rest frame is such a reference frame, he/she is able to send the signal back to a point C that lies on the past world line of A. The result of this paradox demonstrates the violation of causality. But it also shows that the construction of noncausal signals requires not only superluminal velocities but also the invariance of physics under Poincaré transformations; only if the physics in the rest frame of B is equivalent to the physics in A's rest frame can the signal be sent back with a noncausal result. (Un)fortunately, this is not the case here, since the boundary conditions of the Casimir vacuum explicitly break Poincaré symmetry from the very beginning. Hence, there is no reason to conclude that causality is violated (although this still is under discussion; for a controversial viewpoint, see [63]).

In the context of gravitational backgrounds, global Poincaré invariance is lost; but the principle of equivalence, as the nearest analogue, imposes the equivalence of the laws of physics in all local inertial frames. However, this is explicitly violated by the quantum interplay between the fermion loop and the curvature; the resulting effective theory of electrodynamics in curved spaces explicitly depends on the curvature at a given space–time point. Therefore, violation of causality cannot be inferred for this modified vacuum either.

To conclude, there is no reason why the above-mentioned results, which hint at the possibility of superluminal signal velocities, should be wrong or excluded from the theory by additional assumptions. In the standard frame of conventional quantum field theory, there seems to be place enough to embody such superluminal effects.

Nevertheless, whether our quantum theories, which are based on the principle of local causality, in fact respect causality in a final sense has to be investigated further; this will be a worthwhile challenge.

3.4 Photon Splitting

Since Adler's famous paper on photon splitting and photon dispersion in magnetic fields [4], this subject has drawn much attention owing to its phenomenological impact as well as its theoretical multifariousness. As an astrophysical application, photon splitting can provide a mechanism for the production of linearly polarized gamma rays in the vicinity of compact objects with strong magnetic fields. For details about photon splitting in the context of astrophysics, see e.g. [19]. From the theoretical viewpoint, the results of Adler were confirmed by various very different methods (e.g. [5, 16]).

In the present section, we present a pedestrian approach to the basics of photon splitting required for deducing information about this process from the effective Lagrangian of QED. This basic point of view will later be useful for the discussion of finite-temperature effects on the photon-splitting process. The latter have to be taken into account, for example, in order to arrive at a more realistic description of the splitting process near pulsars.

Kinematics

The approximations for this concise investigation are the following: we assume the photons involved to be soft, i.e. all frequencies should be much smaller than the pair production threshold $2m$; this justifies the use of the Heisenberg–Euler Lagrangian. For reasons of simplicity, we consider field strengths much smaller than the critical field strength, $B \ll B_{\mathrm{cr}}$. Furthermore, we neglect dispersion effects during our calculations, but, of course, take their influence into account in the final discussion. The external field is assumed to be constant so that it is not capable of transferring momentum to the photons. Finally, we shall not go beyond the one-loop calculation; hence, no fluctuating photons are involved in the calculation.

Taking all these assumptions into account, the momentum k of the initial photon and the momenta k_1 and k_2 of the final photons satisfy a very simple energy–momentum relation:

$$k = k_1 + k_2 \qquad \Longleftrightarrow \qquad \omega(1, \hat{\boldsymbol{k}}) = \omega_1(1, \hat{\boldsymbol{k}}_1) + \omega_2(1, \hat{\boldsymbol{k}}_2), \qquad (3.94)$$

where ω_i denote the corresponding frequencies. This energy–momentum conservation implies that the propagation directions are collinear:

$$\hat{\boldsymbol{k}} = \hat{\boldsymbol{k}}_1 + \hat{\boldsymbol{k}}_2. \qquad (3.95)$$

Another obvious consequence is the vanishing of all of their scalar products:

$$k^2 = k_1^2 = k_2^2 = k \cdot k_1 = k \cdot k_2 = k_1 \cdot k_2 = 0. \qquad (3.96)$$

After these kinematical considerations, we now turn to the algebraic structure of the field strength of photon-splitting matrix elements. The complete field strength tensor can be decomposed into contributions from the propagating photons and from the external field:

$$F_{\mu\nu} = F_{\mu\nu}^{\mathrm{ext}} + f_{\mu\nu} + f_{\mu\nu}^1 + f_{\mu\nu}^2, \qquad (3.97)$$

with a self-explanatory notation. Below, when there is no danger of confusion, we shall drop the superscript $(^{\mathrm{ext}})$ characterizing the external constant background field. In perturbation theory to one-loop order, all the various field strength contributions couple to the fermion loop. Besides the single couplings of f, f^1 and f^2 to that loop, we can have infinitely many vertices involving the external field. However, according to Furry's theorem, we must have an even total number of vertices attached to the loop, which implies that there is an odd number of external field vertices. A trivial consequence is the vanishing of the third-order diagram.

Astonishingly, the fourth-order diagram (box graph) also vanishes (if dispersion effects are neglected), as we shall now demonstrate. The matrix element of the splitting process, as a gauge and Lorentz scalar, should be built out of a sum of terms of the form

$$f, f^1, f^2, \underbrace{F, \dots, F}_{\mathrm{odd}}, \{1; \, kk; \, kkkk; \dots\}, \qquad (3.98)$$

with the Lorentz indices contracted. Note that the above considerations show that there is essentially only one momentum vector (up to scalar multiples). An odd number of k's is forbidden in (3.98), since one index would then be left open. Moreover, a higher number of k's than of F's is not allowed, since then, two k's would have to be contracted with each other with a zero result. Owing to the uniqueness of the momentum vector, the field strength tensors of the propagating photons can be decomposed as

$$f_{\mu\nu} = k_\mu \epsilon_\nu - k_\nu \epsilon_\mu, \quad f_{\mu\nu}^i = k_\mu \epsilon_\nu^i - k_\nu \epsilon_\mu^i, \quad i = 1, 2, \tag{3.99}$$

where we have introduced polarization vectors ϵ which obey the usual on-shell orthogonality relation with k_μ. As a consequence, we find

$$k_\mu f^{\mu\nu} = k_\mu f^{i\mu\nu} = 0, \quad \text{and} \quad k_\mu F^{\mu\nu} k_\nu = 0. \tag{3.100}$$

Not only must each of the k's be contracted with different field strength tensors owing to antisymmetry, but also one index of each f or f^i must be contracted with an external field strength index, since, for example,

$$f_{\mu\nu} f^{1\nu}{}_\lambda = -k_\mu k_\lambda (\epsilon_\nu \epsilon^{i\nu}), \quad \Longrightarrow \quad f^{\mu\nu} f_{\mu\nu}^1 = 0. \tag{3.101}$$

Applying all these arguments to the fourth-order diagram, it must necessarily be of the form $f \, f^1 \, f^2 \, F$, with all indices contracted and without any k. Since there is only one external field strength tensor, the indices of one propagating photon tensor are completely contracted with indices of other propagating photon tensors only. According to (3.100) and (3.101), this fourth-order term vanishes.

The lowest possible order is given by the sixth-order diagram (hexagon graph) involving three external field vertices. For this, there exist two different combinations: $f \, f^1 \, f^2 \, F F F$ and $f \, f^1 \, f^2 \, F F F k k$. Indeed, the first term is nonvanishing in various combinations of index contractions; the second term is again zero: as stated in (3.100), each of the k's must be contracted with an external field F. Then, the part $F F F k k$ has four open indices. The remaining part $f \, f^1 \, f^2$ has six open indices, so that at least two of them have to be contracted with each other. According to (3.101), this creates two indices associated with k's, which vanish upon contraction with the $F F F k k$ part.

In fact, following Adler, one can prove a general theorem for photon-splitting matrix elements which states that matrix elements involving $2n + 1$ external field vertices and $2n$ factors of k vanish identically; this implies that only $2n - 2$ or fewer factors of k are allowed in such matrix elements. For the lowest-order matrix elements discussed above, we can draw the conclusion that the sixth-order term is the dominant one in the weak-field limit; moreover, to this order of calculation, no external momenta have to be taken into account, and the constant-field limit is exact! It is this constant-field limit which is perfectly described by the Heisenberg–Euler Lagrangian, exploited in the following.

Regarding dispersive effects, it has to be remarked that dispersion will modify the light cone condition $k^2 = 0$ so that the kinematics discussed above no longer holds in a strict sense. As a consequence, the fourth-order diagram is nonvanishing; but, since the deviation from the trivial light cone is an order-α effect, the fourth-order diagram represents only an order-α correction to the sixth-order diagram. Nevertheless, dispersive effects are important for establishing superselection rules (see below).

Calculation of the Matrix Element for the Constant-Field Limit

Starting from the Heisenberg–Euler effective action representing the sum of all ring diagrams coupled to an external field, we can generate three external photons by differentiating the action three times with respect to the field. Denoting the complete field by $F = f + f^1 + f^2 + F^{\text{ext}}$, the photon splitting matrix element in an external magnetic field is given by

$$\mathcal{M} \sim f_{\mu\nu} f^1_{\rho\sigma} f^2_{\kappa\lambda} \frac{\partial^3 \mathcal{L}}{\partial F_{\mu\nu} \partial F_{\rho\sigma} \partial F_{\kappa\lambda}} \bigg|_{\mathcal{G}=0, \mathcal{F}=(1/2)B^2}. \tag{3.102}$$

In the present investigations, the effective action depends only on the two invariants of the electromagnetic field: $\mathcal{L} = \mathcal{L}(\mathcal{F}, \mathcal{G})$. In order to obtain the derivatives, it is useful to introduce the three-vectors

$$\boldsymbol{F}^{\pm} = \boldsymbol{B} \pm \mathrm{i} \boldsymbol{E}, \tag{3.103}$$

$$\Longrightarrow \mathcal{F} = \frac{1}{2}(\boldsymbol{B}^2 - \boldsymbol{E}^2) = \frac{1}{4}(\boldsymbol{F}^{+2} + \boldsymbol{F}^{-2}),$$

$$\mathcal{G} = -\boldsymbol{E} \cdot \boldsymbol{B} = \frac{\mathrm{i}}{4}(\boldsymbol{F}^{+2} - \boldsymbol{F}^{-2}), \tag{3.104}$$

for the total field strength tensor as well as for its components f, f^i, F^{ext}. Then, we need the partial derivatives:

$$\frac{\partial \mathcal{F}}{\partial F_r^{\pm}} = \frac{1}{2} F_r^{\pm}, \qquad \frac{\partial^2 \mathcal{F}}{\partial F_r^{\pm} \partial F_s^{\mp}} = 0, \qquad \frac{\partial^2 \mathcal{F}}{\partial F_r^{\pm} \partial F_s^{\pm}} = \frac{1}{2} \delta_{rs}, \tag{3.105}$$

where r, s denote spatial indices. A similar set of equations holds for $\mathcal{G}$. The important observation is that a term proportional to δ_{rs} cannot contribute to the photon-splitting matrix element, since it contracts two field strength quantities of the propagating photons which obey

$$\boldsymbol{f}^{\pm} \cdot \boldsymbol{f}^{1\pm} = \boldsymbol{f}^{\pm} \cdot \boldsymbol{f}^{2\pm} = \boldsymbol{f}^{1\pm} \cdot \boldsymbol{f}^{2\pm} = 0, \tag{3.106}$$

since the photons propagate along the same direction. Therefore, only single derivatives of $\mathcal{F}$ (and of $\mathcal{G}$) with respect to the field strength quantities need be retained. This translates into the fact that only those third derivatives of $\mathcal{L}$ that contain three derivatives with respect to $\mathcal{F}$ and $\mathcal{G}$ will contribute. Owing to parity invariance, an odd number of derivatives with respect to $\mathcal{G}$ acting on $\mathcal{L}$ will give a vanishing result in the limit $\mathcal{G} \to 0$ (e.g. a purely magnetic field).

The relevant nonvanishing derivatives can be worked out for an arbitrary $\mathcal{L}$:

$$\left.\frac{\partial^3 \mathcal{L}}{\partial F_r^\pm \partial F_s^\pm \partial F_t^\pm}\right|_B = \left(\frac{\partial^3 \mathcal{L}}{\partial \mathcal{F}^3} + 3\frac{\partial^3 \mathcal{L}}{\partial \mathcal{F} \partial \mathcal{G}^2}\right)_B \frac{1}{8} B_r B_s B_t,$$

$$\left.\frac{\partial^3 \mathcal{L}}{\partial F_r^\pm \partial F_s^\pm \partial F_t^\mp}\right|_B = \left(\frac{\partial^3 \mathcal{L}}{\partial \mathcal{F}^3} - \frac{\partial^3 \mathcal{L}}{\partial \mathcal{F} \partial \mathcal{G}^2}\right)_B \frac{1}{8} B_r B_s B_t, \tag{3.107}$$

where the subscript B is an abbreviation for taking the magnetic-field limit, i.e. $\boldsymbol{E} \to 0, |\boldsymbol{B}| \to B^{\mathrm{ext}}$; finally, we have omitted the superscript $(^{\mathrm{ext}})$ of the three B factors which, in fact, represent only the external field.

At this stage, we would like to point out that these considerations have to be modified in the finite-temperature case: then the Lagrangian depends additionally on another invariant, which can be constructed from the heat-bath vector: $\mathcal{E} = (u_\mu F^{\mu\nu})^2$. As a consequence, the box graph (fourth order) is nonvanishing and contributes significantly (see Sect. 3.6.3).

Let us now return to the expression for the photon-splitting matrix element in (3.102), where the sum over all possible index combination now becomes

$$\mathcal{M} = f_r^+ f_s^{1+} f_t^{2+} \left.\frac{\partial^3 \mathcal{L}}{\partial F_r^+ \partial F_s^+ \partial F_t^+}\right|_B + f_r^- f_s^{1-} f_t^{2-} \left.\frac{\partial^3 \mathcal{L}}{\partial F_r^- \partial F_s^- \partial F_t^-}\right|_B$$

$$+ (f_r^{1+} f_s^{2+} f_t^- + f_r^{1+} f_s^+ f_t^{2-} + f_r^{2+} f_s^+ f_t^{1-}) \left.\frac{\partial^3 \mathcal{L}}{\partial F_r^+ \partial F_s^+ \partial F_t^-}\right|_B$$

$$+ (f_r^{1-} f_s^{2-} f_t^+ + f_r^{1-} f_s^- f_t^{2+} + f_r^{2-} f_s^- f_t^{1+}) \left.\frac{\partial^3 \mathcal{L}}{\partial F_r^- \partial F_s^- \partial F_t^+}\right|_B . \tag{3.108}$$

The photon field strengths to be inserted into (3.108) can be represented by[7]

$$\boldsymbol{f}^\pm = \omega(\hat{\boldsymbol{k}} \times \hat{\boldsymbol{\epsilon}} \pm i\hat{\boldsymbol{\epsilon}}), \tag{3.109}$$

$$\boldsymbol{f}^{1\pm} = \omega_1(\hat{\boldsymbol{k}} \times \hat{\boldsymbol{\epsilon}}_1 \pm i\hat{\boldsymbol{\epsilon}}_1), \qquad \boldsymbol{f}^{2\pm} = \omega_2(\hat{\boldsymbol{k}} \times \hat{\boldsymbol{\epsilon}}_2 \pm i\hat{\boldsymbol{\epsilon}}_2),$$

where we have introduced the normalized polarization vectors $\hat{\boldsymbol{\epsilon}}$ and $\hat{\boldsymbol{\epsilon}}_i$ for the electric-field component of the propagating plane waves. In the present case of a background magnetic field, the polarization vectors of interest are the ones which are oriented either perpendicular ($\perp$) or parallel ($\|$) to the plane spanned by $\hat{\boldsymbol{k}}$ and the direction of the magnetic field.[8] The associated eight matrix elements are written as

$$\mathcal{M}\left[\begin{pmatrix} \| \\ \perp \end{pmatrix} \to \begin{pmatrix} \| \\ \perp \end{pmatrix}_1 + \begin{pmatrix} \| \\ \perp \end{pmatrix}_2\right]. \tag{3.110}$$

---

[7] Contrary to Adler's notation, we use rationalized units, so that $F^{\mu\nu}_{\mathrm{Adler}} = \sqrt{4\pi} F^{\mu\nu}$, $e_{\mathrm{Adler}} = e/\sqrt{4\pi}$, and $e^2_{\mathrm{Adler}} = \alpha = e^2/(4\pi)$.

[8] Note that Adler's definition of the $\|, \perp$ modes relies on the direction of the magnetic field vector of the photon and thus is opposite to ours.

Four of these matrix elements vanish immediately because of the algebra of the polarization vectors ($\| \to \|_1 + \|_2$, $\| \to \perp_1 + \perp_2$, $\perp \to \perp_1 + \|_2$, $\perp \to \|_1 + \perp_2$). For the remaining four, we take over a result from Adler's considerations of the dispersive effects: namely, the kinematics of the dispersive case forbids three processes ($\perp \to \perp_1 + \perp_2$, $\| \to \|_1 + \|_2$, $\| \to \perp_1 + \|_2$). This arises from the fact that the phase velocity of both modes is decreased, but the $\perp$ mode is always faster than the $\|$ mode. As a result, the only allowed process is described by the matrix element $\mathcal{M}(\perp \to \|_1 + \|_2)$.

From this matrix element, i.e. the splitting amplitude, we obtain the absorption coefficient κ via the formula

$$\kappa = \frac{1}{32\pi\omega^2} \int\limits_0^\omega d\omega_1 \int\limits_0^\omega d\omega_2 \, \delta(\omega - \omega_1 - \omega_2) \, \mathcal{M}^2. \tag{3.111}$$

Inserting our findings from (3.107)–(3.109) into (3.111) and performing the frequency integration, we finally arrive at the expression for the absorption coefficient for photon splitting in the low-frequency limit derived from an effective Lagrangian:

$$\frac{\kappa}{m} = \frac{\pi^2 \alpha^3}{3 \times 5} \left(\frac{eB}{m^2} \right)^6 \left(\frac{\omega}{m} \right)^5 \sin^6 \theta_B \left(\left. \frac{\partial^3 \mathcal{L}}{\partial \mathcal{F} \partial \mathcal{G}^2} \right|_B \right)^2 \frac{m^{16}}{e^{12}}. \tag{3.112}$$

Here, θ_B denotes the angle between the propagation direction and the magnetic field; $\sin^2 \theta_B = 1 - (\hat{\boldsymbol{k}} \cdot \hat{\boldsymbol{B}})^2$. In the weak-field limit, we may insert the corresponding term $\sim \mathcal{F}\mathcal{G}^2$ from the expansion of the Heisenberg–Euler Lagrangian (2.49),

$$\left. \frac{\partial^3 \mathcal{L}}{\partial \mathcal{F} \partial \mathcal{G}^2} \right|_0 = \frac{13 \, e^6}{3^2 \times 5 \times 7 \, \pi^2 \, m^8}, \tag{3.113}$$

into (3.112), leading to

$$\frac{\kappa}{m} = \frac{13^2}{3^5 \times 5^3 \times 7^2} \frac{\alpha^3}{\pi^2} \left(\frac{eB}{m^2} \right)^6 \sin^6 \theta_B \left(\frac{\omega}{m} \right)^5, \tag{3.114}$$

$$\implies \quad \kappa \simeq 0.116 \, \text{cm}^{-1} \left(\frac{eB}{m^2} \right)^6 \sin^6 \theta_B \left(\frac{\omega}{m} \right)^5, \qquad B \ll B_{\text{cr}}$$

for the photon-splitting absorption coefficient in the weak-field limit $B \ll B_{\text{cr}} = m^2/e$. With the aid of the Heisenberg–Euler Lagrangian, one can also derive the corresponding expression for strong magnetic fields, $B \gg B_{\text{cr}}$, which we cite in passing [16,99]:

$$\kappa = 0.472 \, \text{cm}^{-1} \sin^6 \theta_B \left(\frac{\omega}{m} \right)^5, \qquad B \gg B_{\text{cr}}. \tag{3.115}$$

Note that the strong-field limit becomes independent of the field. The ω^5 dependence of the absorption coefficient stems directly from the properties of the phase space in the low-frequency limit. In particular, it has nothing to do with the number of external field vertices of the calculation (box, hexagon, etc.); thus, this typical frequency dependence will also be encountered in the finite-temperature case.

Let us recall that the absorption coefficients given above refer to the distinct splitting process in which a ⊥-polarized photon splits into two ∥-polarized photons, while the other channels are strongly suppressed. Now, if the magnetic field is sufficiently strong that the absorption length $1/\kappa$ is much shorter than the extension of the field, only ∥-polarized photons can pass through the magnetic field directly, while ⊥-polarized photons will split into ∥-polarized ones. This characterizes a mechanism for the production of linearly polarized radiation. In fact, the requirements for such a scenario are satisfied by the extremal conditions in the vicinity of a pulsar with $B \sim B_{\mathrm{cr}}$ and an extension of $\sim 10^6$ cm.

Of course, it has to be stressed that this scenario applies only to photons with a frequency below the pair-production threshold. In contrast, for $\omega > 2m$, the splitting absorption coefficient cannot compete with the pair-production absorption coefficient, which is several orders of magnitude larger than that for photon splitting. In Sect. 3.6.3, we shall present some other processes which enter into competition with photon splitting at finite temperature.

3.5 QED Effective Action at Finite Temperature to One Loop

The QED effective Lagrangian in the presence of an arbitrary constant electromagnetic background field at finite temperature is derived in the imaginary-time formalism to one-loop order. The calculation is performed on a maximally Lorentz-covariant and gauge-invariant stage. We point out that gauge invariance requires a careful study: the boundary conditions in imaginary time reduce the set of gauge transformations of the background field, which puts restrictions on the choice of gauge and allows for a further gauge invariant. The additional invariant enters the effective action by a topological mechanism and can be identified with a chemical potential; it is furthermore related to Debye screening. In agreement with the real-time formalism, we do not find a thermal correction to Schwinger's pair-production formula to one-loop order.

Following the philosophy of treating the quantum field-theoretic vacuum as a medium, the low-energy effective action for a modified vacuum can be viewed as the defining quantity of a classical theory for the low-energy degrees of freedom. We applied this idea to the case of light propagation in nontrivial vacua in the previous section, where we derived classical equations of motions for a light probe from the quantum effective action. There, we were able to carry out this procedure exactly for the case of a purely electromagnetically modified vacuum only. The generalization to other modified vacua, especially in the high-energy limit, relied on additional assumptions and approximations.

An exact treatment of these modified vacua requires a knowledge of the corresponding effective action, with its full dependence on the complete set of invariants of the physical system. In fact, effective Lagrangians for modified vacua are often only known in a certain parameter range or in various limiting cases. Hence, we dedicate this section to a study of the effective QED Lagrangian at finite temperature, with its complete dependence on the invariants (cf. [87]).

The study of generalizations of the Heisenberg–Euler Lagrangian that include finite-temperature effects was initiated by Dittrich [54], who considered the case of a constant external magnetic field at finite temperature using the imaginary-time formalism. An extension of this work to the case of arbitrary constant electromagnetic fields turned out to be qualitatively more substantial than was naively expected. With the real-time formalism, this situation was investigated by Cox, Hellman and Yildiz [48], and Loewe and Rojas [121]. A more comprehensive study of the problem was performed by Elmfors and Skagerstam [72], who corrected the preceding findings and additionally introduced a chemical potential. An attempt employing the imaginary-time formalism was made by Ganguly, Kaw and Parikh [82] for the case of an external electric field. Recently, the finite-temperature effective action of electromagnetic fields was studied by Shovkovy [154] in the worldline approach, where finite temperature is also introduced via an imaginary-time formalism.

Our intention is to derive the effective action of an arbitrary constant electromagnetic field at finite temperature in the imaginary-time formalism. Similarly to the above-mentioned papers, our approach is based on Schwinger's proper-time formalism [148] and refers to the one-loop level. By assigning a four-velocity vector to the motion of the observer with respect to the heat bath, a manifestly covariant notation is obtained [165] which enables us to formulate the problem in terms of gauge-invariant and covariant quantities. We employ the imaginary-time formalism for two reasons: first, the equivalence of the real-time and imaginary-time formalisms for this particular problem has been questioned in the literature; secondly, the imaginary-time formalism allows a solution to the problem of gauge invariance in terms of topological considerations.

It is, of course, obligatory to point out that the implications of the present calculation may not be immediately interpretable, since the presence of an electric field violates the thermal-equilibrium assumption of the imaginary-time formalism, i.e. an electric field in general is incompatible with maintaining a thermal distribution. In particular, a constant electric field transfers energy to thermally fluctuating charged particles. In turn, these tend to distribute themselves so as to cancel the field, until there is either no field or no particles in the region of interest. On a formal level, it is not clear whether the periodicity in imaginary time can be identified with the physical temperature of the heat bath. However, there are field configurations allowing (almost) thermal equilibrium, e.g. a shallow potential well as suggested in [72], for which the constant-field approximation may be applicable.

The sceptical reader should, however, bear in mind that the knowledge of the effective action given below, depending on the complete set of invariants of an electromagnetic field, including an additional Lorentz vector (temperature times heat-bath velocity), will be useful even in the limit of vanishing electric fields in order to derive generalized Maxwell equations.

While this one-loop study is particularly devoted to an investigation of the fundamental characteristics of thermal effective-action QED, phenomenological applications are postponed to the end of the next section, where we shall have the two-loop contribution at our disposal as well. In the low-temperature domain, the latter will turn out to be the dominant part.

3.5.1 Imaginary-Time Formalism

The one-loop effective action of QED is characterized by the fact that the fluctuating charged fermions which couple to the external field to all orders have been integrated out. Finite temperature is introduced via the imaginary-time formalism by postulating antiperiodic boundary conditions for these fluctuating fermions in the direction of imaginary time, with period $\beta = 1/T$. In this way, a relation between the thermodynamic partition function of the grand canonical ensemble and the generating functional of QED is established.

Regarding the complete generating functional, the external field is treated as a background field [1]. To maintain invariance of the fermionic integral under gauge transformations of the background field, it is important to restrict the gauge functions $\Lambda(x)$ to be β-periodic in imaginary time,[9]

$$\{\Lambda_{\mathrm{p}}\}: \quad \Lambda_{\mathrm{p}}(x^{\mu} + \mathrm{i}\beta u^{\mu}) = \Lambda_{\mathrm{p}}(x^{\mu}), \tag{3.116}$$

where u^{μ} denotes the four-velocity vector of the heat bath. Although the QED action and the integration measure are invariant under arbitrary gauge transformations $\Lambda(x)$ of the background field, the antiperiodic boundary conditions will be modified if $\Lambda(x) \notin \{\Lambda_{\mathrm{p}}\}$; in particular, $\psi(0) = -\psi(\beta) \rightarrow \psi(0) = -\mathrm{e}^{\mathrm{i}e[\Lambda(\beta)-\Lambda(0)]}\psi(\beta)$. At zero temperature, the fermion determinant can depend only on the field strength $F^{\mu\nu}$ that arises from the background field; the explicit form of A_{μ} is subject to arbitrary gauge transformations. In contrast, the restricted class of gauge transformations Λ_{p} at finite temperature allows for further gauge-invariant quantities of the type

---

[9] In principle, one could additionally allow an integer multiple of $2\pi/e$ on the right-hand side of (3.116). But, since such a term does not contribute to the present situation, we shall simply omit it in the following. Moreover, such an additional term can arise only from an imaginary gauge function, which is not desirable. In space–times with an even number of spatial dimensions, however, such terms will contribute to the induced Chern–Simons terms [103].

$$\bar{A}_u(x) = \frac{1}{\beta} \int_0^\beta d\tau \, A_u(x^\mu + i\tau u^\mu), \quad A_u = A^\mu u_\mu, \tag{3.117}$$

where x denotes the components of x^μ orthogonal to u^μ. Already, at this stage, one might suspect that the physical meaning of $\bar{A}_u$ is related to a chemical potential μ, which would enter the QED action by adding $\mu\gamma^\mu u_\mu$ to the Dirac operator: $\slashed{\Pi} = (-i\slashed{\partial} - e\slashed{A}) \rightarrow (-i\slashed{\partial} - e\slashed{A} + \mu\slashed{u})$. In the following, we shall further establish this relation between $\bar{A}_u$ and μ and, in particular, demonstrate that the appearance of $\bar{A}_u$ in the effective action is of topological origin. Instead of employing the functional integral formalism, we shall closely follow Schwinger's proper-time formalism as outlined in Sect. 2.1, which provides for a detailed study of gauge invariance.

In (2.10), we defined the effective Lagrangian via the proper-time transition amplitude:

$$\mathcal{L}^{(1)} = \frac{i}{2} \text{tr}_\gamma \int_0^\infty \frac{ds}{s} \, e^{-ism^2} \langle x| e^{is(\gamma\Pi)^2} |x\rangle. \tag{3.118}$$

However, since we first calculated the transition amplitude for arbitrary x and x' and then took the coincidence limit, the more appropriate definition of the effective Lagrangian is

$$\mathcal{L}^{(1)} = \lim_{x' \to x} \frac{i}{2} \text{tr}_\gamma \int_0^\infty \frac{ds}{s} \, e^{-ism^2} K(x, x'; s|A), \tag{3.119}$$

whereby (2.11)

$$K(x, x'; s|A) = \langle x| e^{is(\gamma\Pi)^2} |x'\rangle. \tag{3.120}$$

We determined the proper-time transition amplitude via the Green's function equation (2.5) for a Dirac particle; for the auxiliary scalar propagator $\Delta(x, x'|A)$, this equation reads (cf. (2.28))

$$D[A]\,\Delta(x, x'|A) = [m^2 - (\gamma\Pi)^2]\,\Delta(x, x'|A) = \delta(x - x'), \tag{3.121}$$

where $D[A]$ is an abbreviation for the differential operator. The formal solution to this equation establishes the connection between $\Delta(x, x'|A)$ and $K(x, x'; s|A)$:

$$\Delta(x, x'|A) = i \int_0^\infty ds \, e^{-im^2 s} \, K(x, x'; s|A). \tag{3.122}$$

Insertion of (3.122) into (3.121) yields the differential equation for the proper-time transition amplitude, which was solved in the Schwinger–Fock gauge,

$$A_{\text{SF}}^\mu = -\frac{1}{2} F^{\mu\nu}(x - x')_\nu, \tag{3.123}$$

leading to

$$K(x, x'; s|A_{\mathrm{SF}}) = \int \frac{\mathrm{d}^4 p}{(2\pi)^4}\, \mathrm{e}^{-\mathrm{i}p(x-x')}\, \mathrm{e}^{\mathrm{i}(e/2)\sigma F s}\, \mathrm{e}^{-Y(\mathrm{i}s)}\, \mathrm{e}^{-pX(\mathrm{i}s)p}, \quad (3.124)$$

where $\sigma F = \sigma_{\mu\nu} F^{\mu\nu}$, $\sigma_{\mu\nu} = (\mathrm{i}/2)[\gamma_\mu, \gamma_\nu]$, and the quantities Y and X depend additionally on the field strength:

$$Y(s) = \frac{1}{2}\mathrm{tr}\,\ln[\cos(e\mathsf{F}s)], \quad X(s) = \frac{\tan(e\mathsf{F}s)}{e\mathsf{F}}. \quad (3.125)$$

We have used matrix notation, e.g. $F_\mu{}^\nu \equiv (\mathsf{F})_\mu{}^\nu$. In Sect. 2.1, the representation of the Green's function and the effective Lagrangian were obtained with the aid of this solution for $K(x, x'; s|A_{\mathrm{SF}})$. This all applies to the case of constant external fields at zero temperature (in the Schwinger–Fock gauge!).

To introduce finite temperature via the imaginary-time formalism, one is tempted to replace the p_0 integration in (3.124) by a sum over Matsubara frequencies.[10] However, this would lead to an incorrect, or at least incomplete, result, since the gauge dependence of the Green's functions has to be taken into account.

As is demonstrated in (2.40)–(2.45), the complete gauge dependence can be treated multiplicatively by a *holonomy* factor. In particular, the transition amplitude in an arbitrary gauge is related to that in the Schwinger–Fock gauge by

$$K(x, x'; s|A) = \Phi(x, x'|A)\, K(x, x'; s|A_{\mathrm{SF}}), \quad (3.126)$$

where the holonomy factor reads

$$\Phi(x, x'|A) = \exp\left[\mathrm{i}e \int_{x'}^{x} \mathrm{d}\xi_\mu \left(A^\mu(\xi) + \frac{1}{2}F^{\mu\nu}(\xi - x')_\nu\right)\right]. \quad (3.127)$$

Note that the integrand is curl-free, and hence the integral in (3.127) is path-independent as long as the configuration space is simply connected. Concerning the effective Lagrangian (3.119) at zero temperature, the holonomy factor plays no role, since $\Phi(x, x'|A) \to 1$ in the coincidence limit $x \to x'$. Consequently, the effective action is gauge-invariant.

The situation changes substantially at finite temperature: since the imaginary time becomes compactified according to the antiperiodic boundary conditions, the configuration space is no longer simply connected. As a consequence, the holonomy factor is invariant under continuous deformations of the integration path but can pick up a winding number by closing the path via the antiperiodic boundary.

The simplest way to establish antiperiodicity in imaginary time is to apply the method of image sources to the Green's function equation. Therefore, let x and x' belong to the same topological sector, i.e. there is a straight path

---

[10] For theories without gauge symmetries, this procedure has been applied successfully in [92].

from x to x' which does not cross the imaginary-time boundaries. We define the reflection points of x' along the imaginary-time axis by (Fig. 3.4)

$$x'_n = x' - i\beta nu. \tag{3.128}$$

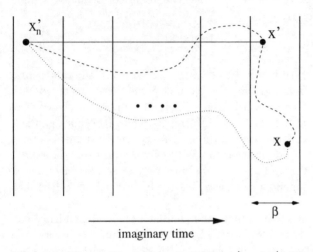

Fig. 3.4. The positions of the points x, x' and x'_n. The *dotted line* represents an arbitrary path from $x'_n = x' - i\beta nu$ to x. As a first step, this path is continuously deformed in such a way that x' becomes an element of the path (*dashed line*). As a second step, the path from x'_n to x' can be deformed to a straight line (*solid line*), which gives rise to (3.136)

Applying the image-source construction to (3.121), for example, we obtain

$$\sum_{n=-\infty}^{\infty} (-1)^n \delta(x, x'_n) = \sum_{n=-\infty}^{\infty} (-1)^n D[A]\, \Delta(x, x'_n|A)$$
$$= D[A]\, \Delta^T(x, x'|A), \tag{3.129}$$

where u^μ denotes the four-velocity of the heat bath, the periodicity scale is set by the inverse temperature β, the factor $(-1)^n$ stems from the *anti*periodic boundary conditions and we have defined the thermal Green's function as

$$\Delta^T(x, x'|A) = \sum_{n=-\infty}^{\infty} (-1)^n \, \Delta(x, x'_n|A). \tag{3.130}$$

Transition to Fourier space and separation of the temperature-dependent parts leads us to

$$\Delta^T(x, x'|A) = \int \frac{\mathrm{d}^4 p}{(2\pi)^4} e^{-ip(x-x')} \Delta(p)\, \Phi(x, x'|A) \tag{3.131}$$

$$\times \sum_{n=-\infty}^{\infty} (-1)^n \, e^{-ip(i\beta n u)} \Phi(x', x'_n|A)\, e^{-in(e/2)\beta[iuF(x-x')]}.$$

Here we have employed the separation

$$\Phi(x, x'_n | A) = \Phi(x, x' | A)\, \Phi(x', x'_n | A)\, e^{-in(e/2)\beta[iuF(x-x')]}, \tag{3.132}$$

which requires special care in its derivation. First, note that the Schwinger–Fock gauge field depends on a base point. Therefore, the holonomy factor does not depend only on its second argument via the boundaries of the line integral, since this argument also acts as the base point of the Schwinger–Fock gauge field in the integrand (see (3.127)). To achieve more transparency, let us rewrite the holonomy factor in the form

$$\Phi(x, x'_n | A) = \exp\left[ie \int_{x'_n}^{x} d\xi_\mu \left[A^\mu(\xi) - A^\mu_{\mathrm{SF}}(\xi, x'_n) \right] \right]. \tag{3.133}$$

Now, we deform the integration path in such a way that, on the one hand, x' becomes an element of the path and, on the other hand, the path of $\Phi(x, x' | A)$ lies entirely in the topological trivial sector:

$$\Phi(x, x'_n | A) = \exp\left[ie \int_{x'_n}^{x'} d\xi_\mu \left[A^\mu(\xi) - A^\mu_{\mathrm{SF}}(\xi, x'_n)) \right] \right]$$

$$\times \exp\left[ie \int_{x'}^{x} d\xi_\mu \left[A^\mu(\xi) - A^\mu_{\mathrm{SF}}(\xi, x'_n) \right] \right]$$

$$= \Phi(x', x'_n | A) \exp\left[ie \int_{x'}^{x} d\xi_\mu \left[A^\mu(\xi) - A^\mu_{\mathrm{SF}}(\xi, x'_n) \right] \right]. \tag{3.134}$$

To write the second exponential factor in terms of a holonomy, we have to insert the correct base point for the Schwinger–Fock gauge potential:

$$\exp\left[ie \int_{x'}^{x} d\xi_\mu \left[A^\mu(\xi) - A^\mu_{\mathrm{SF}}(\xi, x'_n) \right] \right]$$

$$= \exp\left[ie \int_{x'}^{x} d\xi_\mu \left[A^\mu(\xi) - A^\mu_{\mathrm{SF}}(\xi, x') + A^\mu_{\mathrm{SF}}(\xi, x') - A^\mu_{\mathrm{SF}}(\xi, x'_n) \right] \right]$$

$$= \Phi(x, x' | A) \exp\left[ie \int_{x'}^{x} d\xi_\mu \left[A^\mu_{\mathrm{SF}}(\xi, x') - A^\mu_{\mathrm{SF}}(\xi, x'_n) \right] \right]$$

$$= \Phi(x, x' | A) \exp\left[ie \int_{x'}^{x} d\xi_\mu \left(\frac{1}{2} F^{\mu\nu}(x' - x'_n)_\nu \right) \right]$$

$$= \Phi(x, x' | A) \exp\left[-in\frac{e}{2}\beta[iu_\mu F^{\mu\nu}(x - x')_\nu] \right]. \tag{3.135}$$

Inserting (3.135) into (3.134) proves the decomposition given in (3.132).

Concerning $\Phi(x', x'_n|A)$, we can deform the integration path into a straight line (SL) along the imaginary u^μ direction:

$$\Phi(x', x'_n|A) = \exp\left[ie \int_{\substack{x'-i\beta n u \\ \text{SL}}}^{x'} d\xi_\mu\, A^\mu(\xi)\right].\tag{3.136}$$

As mentioned above, the exponent in (3.136) is an invariant quantity under periodic gauge transformations Λ_p but will depend on the explicit form of A^μ in a certain manner. At this stage, it is important to point out that the background field potential is not necessarily subject to periodic boundary conditions, since it does not correspond to thermalized particles; it is not an integration variable even in the complete theory. To specify the form of A^μ, more physical input is required: since the effective Lagrangian for a homogeneous system such as a constant-field configuration has to be independent of x, the coincidence limit $x \to x'$ of the thermal transition amplitude $K^T(x, x'; s|A)$ must also be independent of x; the same requirement holds for $\Delta^T(x, x'|A)$. With regard to (3.131), this is only satisfied if $\Phi(x', x'_n|A)$ is independent of x'. Thereby, we obtain the gauge condition

$$0 = \int_0^1 d\tau\, \partial_{x'}^\mu A_u[x'^\mu - i\beta n u^\mu + \tau(i\beta n u^\mu)].\tag{3.137}$$

Condition (3.137) is satisfied if

$$A_u \equiv \bar{A}_u = \text{const.},\tag{3.138}$$

which is the generic choice. Any other solution is gauge-equivalent to (3.138), $A_u \to A_u + \partial_u \Lambda_\mathrm{p}$. Equation (3.137) also fixes the choice of the spatial components $\boldsymbol{A}$: since A^μ should produce a constant electric field via

$$\text{const.} = \boldsymbol{E} = \boldsymbol{\nabla} A_u - \partial_u \boldsymbol{A} \overset{(3.138)}{=} -\partial_u \boldsymbol{A},$$

the generic choice of $\boldsymbol{A}$ in the heat-bath rest frame $(\partial_u \hat{=} \partial/\partial t)$ reads

$$\boldsymbol{A} = -\boldsymbol{E}t + \boldsymbol{a}(\boldsymbol{x}),\tag{3.139}$$

where the function $\boldsymbol{a}(\boldsymbol{x})$ is defined by $\boldsymbol{B} = \boldsymbol{\nabla} \times \boldsymbol{a}$. Again, other choices of $\boldsymbol{A}$ are given by its gauge transforms with respect to Λ_p. Note that these gauge conditions are different from those found in [72] by employing real-time methods.[11] In the real-time formalism, the A^0 component of the gauge field

----

[11] In [72], it was argued that a gauge condition of the form $dA^\mu/dt = 0$ is required for obtaining a clear separation of fermionic and electromagnetic energies. This implies that the constant electric background field is produced by a spatially nonconstant A^0, $\boldsymbol{E} = -\boldsymbol{\nabla} A^0$, and has to be interpreted as a spatially nonconstant chemical potential (see later). This demonstrates that different gauge choices which are not gauge-equivalent with respect to $\{\Lambda_\mathrm{p}\}$ correspond to different physical situations.

enters the propagators and the effective action via the thermal distribution function, which is given as an external condition. Hence, there is no intrinsic criterion for the correct choice of A^μ. Even worse, without a knowledge about the compactification of space–time, the class of gauge transformations is not restricted, so there is no natural way to ensure gauge invariance of the effective Lagrangian.

Taking these considerations into account, the holonomy factor (3.136) eventually yields

$$\Phi(x', x'_n | A) = \exp\left[ie(i\beta n)\bar{A}_u\right]. \tag{3.140}$$

With the aid of a Poisson resummation, we obtain for the sum in (3.131)

$$\sum_{n=-\infty}^{\infty} (-1)^n \, e^{-ip(i\beta nu)} \Phi(x', x'_n | A) \, e^{-in(e/2)\beta[iuF(x-x')]} \tag{3.141}$$

$$= 2\pi i T \sum_{n=-\infty}^{\infty} \delta\left[p_u - e(\bar{A} - A_{\mathrm{SF}})_u + i\pi T(2n+1)\right],$$

where $p_u = u^\mu p_\mu$ and $A_{\mathrm{SF}\,u} = -\frac{1}{2} u_\mu F^{\mu\nu}(x - x')_\nu$. Inserting (3.141) into (3.131) leads us to the final expression for $\Delta^T(x, x'|A)$. Similarly, the thermal transition amplitude $K^T(x, x'; s|A)$ and the thermal fermion Green's function can be derived. Note that these objects contain temperature-dependent contributions as well as the zero-temperature part.

We observe that the Matsubara prescription finally reads

$$\int \frac{dp_u}{2\pi} f(p_u^2) \rightarrow iT \sum_{n=-\infty}^{\infty} f\left\{-\left[\pi T(2n+1+\frac{i}{\pi} \frac{e(\bar{A}-A_{\mathrm{SF}})_u}{T})\right]^2\right\}. \tag{3.142}$$

The explicit appearance of $A_{\mathrm{SF}\,u}$ hints at the fact that this modified Matsubara prescription will be applied to an object which has been calculated in the Schwinger–Fock gauge. Of course, we could have started with a different choice of gauge, which then would appear in (3.142) instead of A_{SF}^μ.

Equation (3.142) finally states that it is a gauge-field-shifted momentum in the u^μ direction, $[p - e(A - A_{\mathrm{SF}})]_u$, which is replaced by Matsubara frequencies, instead of the canonical momentum. This implies a dependence of the Green's functions and the transition amplitude on the gauge field invariant $\bar{A}_u$ even in the coincidence limit $x' \rightarrow x$ (note that $A_{\mathrm{SF}} \rightarrow 0$ for $x' \rightarrow x$). As a consequence, the effective Lagrangian will be invariant under periodic gauge transformations Λ_p but not under arbitrary gauge transformations Λ. Of course, this was expected from our initial considerations. The physical role of $\bar{A}_u$ will be elucidated in Sect. 3.5.4.

3.5.2 Covariant Formulation

The imaginary-time formalism has often been criticized because it exhibits the explicit noncovariant feature of leading to discrete energies but continuous momenta for the quantized fields. In the present work, we want to demonstrate that it is nevertheless possible to establish covariance at any stage of this calculation, since the above-mentioned discrepancy between energy and momentum appears only in internal propagators, all of which are integrated out. Manifest covariance is achieved by constructing a reference frame that completely relies on the covariant and gauge-invariant building blocks of the problem.

These building blocks in the present problem of a constant electromagnetic field at finite temperature are the field strength tensors, $F^{\mu\nu}$ and $^\star F^{\mu\nu} = (1/2)\epsilon^{\mu\nu\alpha\beta}F_{\alpha\beta}$; furthermore, we encounter the heat-bath vector n^μ [105], which is, on the one hand, characterized by the value of its invariant scalar product,[12] $n^\mu n_\mu = -T^2$, where T denotes the heat-bath temperature, and, on the other hand, is related to the heat-bath four-velocity via the invariant parameter T: $n^\mu = T u^\mu$. There are ten independent components in $F^{\mu\nu}$, $^\star F^{\mu\nu}$ and n^μ. The number of generators of the Lorentz group is six; hence we can transform six components to zero, since there is no little group that leaves $F^{\mu\nu}$, $^\star F^{\mu\nu}$ and n^μ invariant.[13] Therefore, we are left with four Lorentz- and gauge-invariant scalars (or pseudoscalars). For reasons of convenience, we choose the following set (cf. Appendix B):

$$a = \left(\sqrt{\mathcal{F}^2 + \mathcal{G}^2} + \mathcal{F}\right)^{1/2}, \quad b = \left(\sqrt{\mathcal{F}^2 + \mathcal{G}^2} - \mathcal{F}\right)^{1/2}, \quad T = \sqrt{-n^\mu n_\mu},$$

$$\mathcal{E} = \frac{1}{T^2}\left(n_\alpha F^{\alpha\mu}\right)\left(n_\beta F^\beta{}_\mu\right) \equiv \left(u_\alpha F^{\alpha\mu}\right)\left(u_\beta F^\beta{}_\mu\right). \tag{3.143}$$

Note that $\mathcal{E}$ is positive definite, since n^μ is a time-like vector; for example, in the rest frame of the heat bath, we find $\mathcal{E} = \boldsymbol{E}^2$. It is obvious that any gauge-invariant Lorentz scalar constructed from the field strength tensor is expressible in terms of this set of invariants (3.143). In Appendix E, we construct a convenient coordinate system in which even the components of any Lorentz vector or tensor of the problem can be expressed in terms of these invariants. There, we define the *vierbein* $e^{A\mu}$, which mediates between the given system, labeled by $\mu, \nu, \ldots = 0, 1, 2, 3$, and the desired system, labeled by the (Lorentz) indices $A, B, \ldots = 0, 1, 2, 3$:

$$e_0{}^\mu = u^\mu,$$

$$e_1{}^\mu = \frac{u_\alpha F^{\alpha\mu}}{\sqrt{\mathcal{E}}},$$

$$e_2{}^\mu = \frac{1}{\sqrt{d}}\left(u^\alpha F_{\alpha\beta}F^{\beta\mu} - \mathcal{E}\, e^{0\mu}\right),$$

$$e_3{}^\mu = \epsilon^{\alpha\beta\gamma\mu} e_{0\alpha}\, e_{1\beta}\, e_{2\gamma}, \tag{3.144}$$

where the quantity d is an abbreviation for the combination of invariants

---

[12] We employ the metric $g = \mathrm{diag}(-1, 1, 1, 1)$.

[13] For pure EM fields, the dimension of the little group would be 2, since boosts along and rotations around the field direction in a system where the $\boldsymbol{E}$ and $\boldsymbol{B}$ fields are parallel leave the fields invariant.

$$d = 2\mathcal{F}\mathcal{E} - \mathcal{G}^2 + \mathcal{E}^2. \tag{3.145}$$

The vierbein satisfies the identity

$$e_{A\mu}\,e_B{}^\mu = g_{AB} \equiv \mathrm{diag}(-1,1,1,1), \tag{3.146}$$

where $g_{AB} \sim g^{AB}$ denotes the metric which raises and lowers capital indices. By a direct computation, we can transform the field strength tensors and the heat-bath vector:

$$n^A = g^{AB}e_B{}^\mu\,n_\mu = (T,0,0,0),$$

$$F_{AB} = e_{A\mu}F^{\mu\nu}e_{B\nu} = \begin{pmatrix} 0 & \sqrt{\mathcal{E}} & 0 & 0 \\ -\sqrt{\mathcal{E}} & 0 & \sqrt{d/\mathcal{E}} & 0 \\ 0 & -\sqrt{d/\mathcal{E}} & 0 & -\mathcal{G}/\sqrt{\mathcal{E}} \\ 0 & 0 & \mathcal{G}/\sqrt{\mathcal{E}} & 0 \end{pmatrix},$$

$$^*F_{AB} = e_{A\mu}\,^*F^{\mu\nu}e_{B\nu} = \begin{pmatrix} 0 & -\mathcal{G}/\sqrt{\mathcal{E}} & 0 & \sqrt{d/\mathcal{E}} \\ \mathcal{G}/\sqrt{\mathcal{E}} & 0 & 0 & 0 \\ 0 & 0 & 0 & -\sqrt{\mathcal{E}} \\ -\sqrt{d/\mathcal{E}} & 0 & \sqrt{\mathcal{E}} & 0 \end{pmatrix}. \tag{3.147}$$

So, indeed, the components of these tensors can be completely expressed in terms of invariants. Hence, any tensor-algebraic manipulation involving the objects from (3.147) can immediately be performed on the level of gauge and Lorentz invariants.

It is worthwhile to point out at this stage that a duality transformation of the type $\boldsymbol{E} \to \boldsymbol{B}$ and $\boldsymbol{B} \to -\boldsymbol{E}$ not only implies an interchange of a and b (and a sign flip for $\mathcal{G}$) but also requires $\mathcal{E} \to \mathcal{E} + 2\mathcal{F} = \mathcal{E} + a^2 - b^2$. Hence, it is not sufficient in the finite-temperature case to perform a calculation for magnetic fields and then draw an analogy for electric fields by replacing $\boldsymbol{B}$ by $-\mathrm{i}\boldsymbol{E}$, in contrast to a zero-temperature calculation.

3.5.3 Effective Action

From (3.119), we can read off the definition of the effective Lagrangian at finite temperature:

$$\mathcal{L}^{1+1T} = \frac{\mathrm{i}}{2}\mathrm{tr}_\gamma \int_0^\infty \frac{\mathrm{d}s}{s}\,\mathrm{e}^{-\mathrm{i}sm^2}K^T(s|A), \tag{3.148}$$

where the superscript implies that $\mathcal{L}^{1+1T}$ consists of the zero-temperature as well as the finite-temperature one-loop part: $\mathcal{L}^{1+1\mathrm{T}} = \mathcal{L}^1 + \mathcal{L}^{1\mathrm{T}}$. The thermal transition amplitude in the integrand is obtained simply, by applying the modified Matsubara prescription (3.142) to the zero-temperature transition amplitude, (3.124)–(3.125), in the coincidence limit:

$$K^T(s; A) = \mathrm{i}T \sum_{n=-\infty}^\infty \int_V \frac{\mathrm{d}^3 p}{(2\pi)^3}\,\mathrm{e}^{\mathrm{i}\frac{e}{2}\sigma Fs}\mathrm{e}^{-Y(\mathrm{i}s)}\mathrm{e}^{-pX(\mathrm{i}s)p}\bigg|_{p_\mu = e\bar{A}_\mu - \mathrm{i}\pi T(2n+1)}, \tag{3.149}$$

where V denotes the three-space volume orthogonal to the u^μ direction. We now perform the computation of (3.148) and (3.149) within the system that we established in the previous subsection (and, in more detail, in Appendix E). With respect to the capital labels, this volume is related to the components $A, B, \ldots = 1, 2, 3$, whereas the components along the u^μ direction correspond to the label $A, B, \ldots = 0$.

In particular, we need an expression for the matrix X, as defined in (3.125), with respect to the new coordinate base. Of course, this is simply achieved by "replacing" the Greek indices by the capital Latin ones:

$$
X_{AB}(s) = e_{A\mu}\, X^{\mu\nu}\, e_{B\nu} = e_{A\mu} \left[(eF)^{-1} \tan(eFs) \right]^{\mu\nu} e_{B\nu}
$$
$$
= \sum_j \frac{\tan(ef_j s)}{ef_j} \left(e_{A\mu} A_j^{\mu\nu} e_{B\nu} \right) \equiv \sum_j \frac{\tan(ef_j s)}{ef_j} A_{j\,AB}. \quad (3.150)
$$

Here, we have made use of the spectral representation of the field strength tensor as described in Appendix B; the f_j's denote the eigenvalues as given in (B.13) and (B.14), while the projectors A_j read

$$
A_{j\,AB} = \frac{1}{2(f_j^2 - \bar{f}_j^2)} \left(-\bar{f}_j^2\, g_{AB} + f_j\, F_{AB} + F_{AB}^2 - i\bar{f}_j^{\,*}F_{AB} \right), \quad (3.151)
$$

where $F_{AB}^2 = F_A{}^C F_{CB}$. Note that X is an even function of the field strength tensor. Thus, it must be a symmetric tensor (antisymmetric parts would vanish anyway in the exponent pXp). Performing the sum over j in (3.150) and inserting (B.13) and (B.14), we arrive at

$$
X_{AB}(is) = i \left[\frac{\tan eas}{ea} \frac{(b^2 g_{AB} - F_{AB}^2)}{a^2 + b^2} + \frac{\tanh ebs}{eb} \frac{(a^2 g_{AB} + F_{AB}^2)}{a^2 + b^2} \right]. \quad (3.152)
$$

Incidentally, an identical equation also holds, of course, with the labels A, B replaced by μ, ν, but then the components are not related to gauge and Lorentz invariants.

The metric tensor is diagonal, and the structure of F_{AB}^2 can be evaluated from (3.147):

$$
F_{AB}^2 = \begin{pmatrix} -\mathcal{E} & 0 & \sqrt{d} & 0 \\ 0 & \mathcal{E} - d/\mathcal{E} & 0 & -\sqrt{d}\mathcal{G}/\mathcal{E} \\ \sqrt{d} & 0 & -\mathcal{G}^2/\mathcal{E} - d/\mathcal{E} & 0 \\ 0 & -\sqrt{d}\mathcal{G}/\mathcal{E} & 0 & -\mathcal{G}^2/\mathcal{E} \end{pmatrix}. \quad (3.153)
$$

Obviously, the only nonvanishing components of the symmetric tensor X_{AB} are the diagonal elements and X_{02} and X_{13}. We postpone the remaining calculation involving the explicit representation of X_{AB} to Appendix E and simply take advantage here of (E.49) and (E.55), which state

$$
X_{11} X_{33} - X_{13}^2 \equiv -X_{00} X_{22} - X_{02}^2, \quad (3.154)
$$
$$
X_{11} X_{33} - X_{13}^2 = -\frac{\tan eas}{ea} \frac{\tanh ebs}{eb}. \quad (3.155)
$$

The Gaussian momentum integration in (3.149) therefore results in

$$
\int_V \frac{d^3 p}{(2\pi)^3}\, e^{-p^A X_{AB} p^B} = \frac{1}{(4\pi)^{3/2}}\, \exp\left(\frac{X_{11}X_{33} - X_{13}^2}{X_{22}} p_0^2\right)
$$

$$
\times \left[(X_{11}X_{33} - X_{13}^2)X_{22}\right]^{-1/2}. \tag{3.156}
$$

Substituting the modified Matsubara frequencies $p_0 \equiv p_u = e\bar{A}_u - i\pi T(2n + 1)$ into the exponent of (3.156), the summation over n in (3.149) can be reorganized according to the Poisson formula (also "Jacobi's transformation formula"):

$$
\sum_{n=-\infty}^{\infty} e^{-\sigma(n-z)^2} = \sum_{n=-\infty}^{\infty} \left(\frac{\pi}{\sigma}\right)^{1/2} e^{-(\pi^2/\sigma)n^2 - 2\pi i\, zn}. \tag{3.157}
$$

In this case, we set $z = -1/2 - ie\bar{A}_u/(2\pi T)$ and $\sigma = (4\pi^2 T^2/X_{22})(X_{11}X_{33} - X_{13}^2)$. At this point, we have to mention that the formula (3.157) is not valid for $\mathrm{Re}\,\sigma < 0$, which would lead to a divergent behavior of the sum. We shall comment on the question of convergence in more detail later on.

The Poisson resummation serves the purpose of separating the zero-temperature part from the finite-temperature part, since the complete loop-momentum integration/summation in (3.149) now yields

$$
iT \sum_{n=-\infty}^{\infty} \int_V \frac{d^3 p}{(2\pi)^3}\, e^{-pXp}
$$

$$
= \frac{i}{16\pi^2}\left[(X_{11}X_{33} - X_{13}^2)X_{22}\right]^{-1} \tag{3.158}
$$

$$
\times \left[1 + 2\sum_{n=1}^{\infty}(-1)^n \exp\left(-\frac{X_{22}}{X_{11}X_{33} - X_{13}^2}\frac{n^2}{4T^2}\right)\cosh\frac{e\bar{A}_u n}{T}\right].
$$

Keeping only the "1" in the second factor in square brackets leads to the standard proper-time expression for the zero-temperature effective action, while the sum represents the thermal correction. For the remaining terms in (3.148) and (3.149), we make use of the results of (C.10) and (B.32) in Appendices C and B:

$$
e^{-Y(is)} = (\cos eas\, \cosh ebs)^{-1},
$$

$$
\mathrm{tr}_\gamma\, e^{i(e/2)\sigma Fs} = 4\cos eas\, \cosh ebs.
$$

As a nontrivial check, we first reconstruct the zero-temperature effective action by considering the terms which accompany the "1" in the second factor in square brackets in (3.158); inserting our findings into (3.148), we obtain for the zero-temperature contribution

$$
\mathcal{L}^1 = \frac{i}{2}\int_0^\infty \frac{ds}{s}\, e^{-ism^2}\, 4\,\frac{i}{16\pi^2}\left(X_{11}X_{33} - X_{13}^2\right)^{-1}
$$

$$\overset{(3.155)}{=} \frac{1}{8\pi^2} \int_0^\infty \frac{ds}{s^3} e^{-ism^2} \, eas \cot eas \, ebs \coth ebs. \tag{3.159}$$

Up to a charge and field strength renormalization, this corresponds exactly to Schwinger's well-known formula, which we cited in Sect. 2.1 (cf. (2.49)).

Taking the remaining terms of (3.158) into account, we arrive at the desired expression for the finite-temperature contribution to the one-loop QED effective action for constant fields:

$$
\begin{aligned}
\mathcal{L}^{1T} &= \frac{i}{2} \int_0^\infty \frac{ds}{s} e^{-ism^2} \, 4 \, \frac{i}{16\pi^2} \left(X_{11} X_{33} - X_{13}^2 \right)^{-1} \\
&\quad \times 2 \sum_{n=1}^\infty (-1)^n \exp\left(-\frac{X_{22}}{X_{11}X_{33} - X_{13}^2} \frac{n^2}{4T^2} \right) \cosh \frac{e\bar{A}_u n}{T} \\
&= \frac{1}{4\pi^2} \int_0^\infty \frac{ds}{s^3} e^{-ism^2} \, eas \cot eas \, ebs \coth ebs \\
&\quad \times \sum_{n=1}^\infty (-1)^n \exp\left[ih(s) \frac{n^2}{4T^2} \right] \cosh \frac{e\bar{A}_u n}{T}.
\end{aligned}
\tag{3.160}
$$

Here we have defined the function $h(s)$ in the exponent of $\mathcal{L}^{1T}$ by

$$
\begin{aligned}
h(s) &= \frac{iX_{22}}{X_{11}X_{33} - X_{13}^2} \\
&= \frac{b^2 - \mathcal{E}}{a^2 + b^2} \, ea \cot eas + \frac{a^2 + \mathcal{E}}{a^2 + b^2} \, eb \coth ebs,
\end{aligned}
\tag{3.161}
$$

where we have taken advantage of (E.58) in Appendix E. Equation (3.160) represents the central result of this section. In the rest frame of the heat bath, where $\mathcal{E} = \boldsymbol{E}^2$, we recover the findings of [72] for $h(s)$.

Note that $h(s)$ is strictly real, so there are no apparent convergence problems in employing the Poisson resummation (3.157). However, the situation is, unfortunately, a bit more subtle. Since $h(s)$ is real, the function σ in (3.157) is purely imaginary. In order to define the Poisson resummation for imaginary values of σ, one employs the usual machinery of analytical continuation. For example, one can give σ a small positive real value ϵ and finally consider the limit $\epsilon \to 0$. Such an ϵ is usually attached to one of the intrinsic parameters, e.g. $m^2 \to m^2 - i\epsilon$.

Taking a look at (3.160), one is tempted to shift the temperature by a small imaginary value. But the problem is that the function $h(s)$ is not strictly positive for certain background fields. Hence, the sign of the ϵ prescription for shifting the temperature depends on the sign of $h(s)$ and therefore on the values of the field invariants and the proper-time parameter. Obviously, this complicates the evaluation of $\mathcal{L}^{1T}$, and, in particular, prohibits interchanging summation and integration in some cases. Additionally, one has to be very careful in rotating the contour of the proper-time integration; nor is obtaining numerical estimates for (3.160) a straightforward exercise, owing to the wildly oscillatory behavior of the whole integrand, especially in the sum.

An unproblematic case is given by the limit of vanishing electric field (the limit is most conveniently taken for $\boldsymbol{E}$ and $\boldsymbol{B}$ (anti)parallel). Then, $h(s)$ reduces to $1/s$; and, further assuming $\bar{A}_u = 0$, we recover the findings of [54] for a purely magnetic field at finite temperature:

$$\mathcal{L}^{1T}(B,T) = -\frac{1}{4\pi^2}\int\limits_0^\infty \frac{\mathrm{d}s}{s}\, \mathrm{e}^{-\mathrm{i}sm^2}\, eBs \coth eBs \sum_{n=1}^\infty (-1)^n \mathrm{e}^{\mathrm{i}n^2/(4T^2 s)}. \quad (3.162)$$

Furthermore, the general form of (3.160) coincides with the representation found in the worldline approach [154] (in the heat-bath rest frame), except for the dependence on the gauge potential.[14]

The case of the effective Lagrangian of a constant magnetic field at nonzero chemical potential (which is related to the $\bar{A}_u$ field) has been discussed in [43]. A comprehensive study of this situation based on the real-time formalism, including finite temperature, can be found in [71], where astrophysical implications are also discussed. The same physical situation was investigated employing the imaginary-time formalism in [40], where high- and low-temperature expansions were approached in a more direct way. As is demonstrated in these references, the zero-temperature limit of (3.160) with a chemical potential obeying $\mu > m$ requires careful study.

In spite of all these technical difficulties, it is possible to construct an asymptotic weak-field expansion of the Lagrangian (3.160). Contrary to the zero-temperature case, where such an approximation can be achieved by a small-coupling e expansion of the integrand (cf. (2.52)), we have to rely on a standard Taylor expansion of the complete effective action. This is because the coupling also appears in the exponent, which must not be expanded, because of its convergence-ensuring properties. The advantage of a Taylor expansion of $\mathcal{L}^{1T}$ lies in the fact that the fields are set to zero after differentiation; therefore, the function $h(s)$ in the exponent always reduces to $1/s$ in each term. The resulting proper-time integrals can always be represented as n-sums over modified Bessel functions $K_\nu(nm/T)$. For the low-temperature domain, $T \ll m$, we may use the asymptotic representation of the Bessel functions and obtain, up to second order in the invariants $\mathcal{F}, \mathcal{G}, \mathcal{E}$ at zero chemical potential,

$$\mathcal{L}^{1T} \simeq \frac{\alpha}{\pi}\left(\frac{1}{6}\sqrt{\frac{2\pi m}{T}}\mathcal{E} + \frac{2}{3}\sqrt{\frac{2\pi T}{m}}\mathcal{F}\right)\mathrm{e}^{-m/T} - \alpha^2\left[\frac{1}{36}\frac{m^4}{T^4}\sqrt{\frac{\pi T}{2m}}\mathcal{E}^2\right.$$

$$\left. -\frac{\pi}{90}\frac{m^3}{T^3}(8\mathcal{F}\mathcal{E} + \mathcal{G}^2) + \frac{1}{45}\frac{m^2}{T^2}(8\mathcal{F}^2 - 8\mathcal{F}\mathcal{E} + 13\mathcal{G}^2)\right] \quad (3.163)$$

---

[14] The importance of the holonomy factor has been overlooked in [154].

$$+\frac{\pi}{30}\frac{m}{T}(4\mathcal{F}^2+7\mathcal{G}^2)+\frac{4}{45}(4\mathcal{F}^2+7\mathcal{G}^2)\Bigg]\frac{e^{-m/T}}{m^4},\quad T\ll m.$$

This agrees with the findings of [73]. In particular, we observe that the low-temperature contributions to the effective action are exponentially damped by the electron mass. This is plausible, since higher-lying fermion states beyond the mass gap m can hardly be excited at all for $T\ll m$.

The opposite limit, i.e. the behavior of $\mathcal{L}^{1T}$ in the high-temperature domain, can almost be guessed without calculation: for $T\gg m$, the thermally fluctuating electrons move ultrarelativistically so that the scale of the loop process is no longer set by the electron mass m, but by the temperature T. Therefore, in the limit $T\to\infty$, the loop particles become infinitely heavy and decouple from the theory. Consequently, vacuum polarization is strongly suppressed. This can also be inferred from dimensional analysis: any nonlinearity of the field strength induced by vacuum polarization must be of the form F^{2j+2}/T^{4j}, where F represents a field strength tensor contribution of mass dimension 2, and j is an arbitrary nonnegative integer. For any $j>0$, these terms are strongly suppressed for $T\to\infty$. Only $j=0$ terms can survive in this limit. As a consequence, $\mathcal{L}^{1T}$ must (and really does) contain a part identical to $-\mathcal{L}^1$ in order to cancel the effects of vacuum polarization [33,34]. Moreover, any remaining terms must be of second order in the field strength (first order in the invariants); these can be easily calculated and lead to

$$\mathcal{L}^{1T}(T\gg m)=-\frac{2\alpha}{3\pi}\mathcal{F}\ln\frac{T}{m}+\frac{\alpha}{6\pi}\mathcal{E}-\mathcal{L}^1. \tag{3.164}$$

This result can be viewed as a thermal renormalization of the Maxwell Lagrangian. In fact, the prefactor of $\mathcal{F}$ can be identified as the one-loop coefficient of the β function of QED; this implies a logarithmic increase of the coupling with increasing T/m. More calculational details of these formulas can be found in the next section, when we discuss light propagation in a thermal background.

3.5.4 The $\bar{A}_u$ Field

The physical interpretation of $\bar{A}_u$ can most easily be illuminated in the limiting case of vanishing field invariants, $a,b,\mathcal{E}=0$; this limit corresponds to switching off all the nonlinear self-interactions of the electromagnetic field, which are not at the center of interest at this stage. Under these circumstances we are able to rotate the contour $s\to-is$, and an interchange of integration and summation is permitted in (3.160):

$$\mathcal{L}^{1T}(\bar{A}_u)=-\frac{1}{4\pi^2}\int\limits_0^\infty\frac{ds}{s^3}e^{-m^2s}\sum_{n=1}^\infty(-1)^n\exp\left(-\frac{n^2}{4T^2s}\right)\cosh\frac{e\bar{A}_u n}{T}$$

$$= -\frac{1}{4\pi^2} \sum_{n=1}^{\infty} (-1)^n \cosh \frac{e\bar{A}_u n}{T} \int_0^\infty \frac{ds}{s^3} \exp\left[-m^2\left(s + \frac{n^2}{(2mT)^2 s}\right)\right]$$

$$= -\frac{2}{\pi^2} m^2 T^2 \sum_{n=1}^{\infty} \frac{(-1)^n}{n^2} \cosh \frac{e\bar{A}_u n}{T} K_2\left(\frac{m}{T}n\right), \tag{3.165}$$

where we have taken advantage of the representation

$$2\left(\frac{\mu}{2}\right)^\nu K_\nu(\mu) = \int_0^\infty du\, u^{\nu-1} \exp\left(-u - \frac{\mu^2}{4u}\right) \tag{3.166}$$

of the modified Bessel function. Decomposing the cosh term into exponential functions and employing a different representation of K_2 [91],

$$\frac{1}{n}\frac{T}{m} K_2\left(\frac{m}{T}n\right) = \int_1^\infty dq\, q\sqrt{q^2-1}\, e^{-(m/T)nq} = \int_0^\infty dk\, k^2 e^{-(n/T)\sqrt{k^2+m^2}}$$

$$= \frac{1}{4\pi} \int d^3k\, e^{-(1/T)\sqrt{k^2+m^2}n}, \tag{3.167}$$

where we have substituted $k = m\sqrt{q^2-1}$, we arrive at

$$\mathcal{L}^{1T}(\bar{A}_u) = T\,2 \int \frac{d^3k}{(2\pi)^3} \left[\ln\left(1 + e^{-\beta(E+e\bar{A}_u)}\right) + \ln\left(1 + e^{-\beta(E-e\bar{A}_u)}\right)\right]. \tag{3.168}$$

Here we have introduced the particle energy for the e^+e^- gas, $E = \sqrt{k^2 + m^2} \equiv |k_u|$. According to the relation

$$\mathcal{L}^{1T}(\bar{A}_u) = T\,\frac{\ln Z(T, \bar{A}_u)}{V},$$

we indeed find the general expression for the partition function Z of an ideal e^+e^- gas in which the $\bar{A}_u$ field plays the role of a chemical potential. If we had started the computation including a chemical potential, we would always have encountered the combination $-e\bar{A}_u - \mu \hat{=} e\bar{A}^0 - \mu$, which is therefore the only physical quantity. In other words, we can identify $e\bar{A}_u$ with a chemical potential during the complete calculation; hence, the additional information (compared with the zero-temperature case) which is required to define the correct choice of the background gauge potential A^μ is obtained from the value of the chemical potential of the system under consideration. If one wants to perform a gauge transformation beyond the class of periodic gauge functions Λ_p, one has to redefine the chemical potential to obtain the same physical system.

Going beyond the constant-field approximation, the effective Lagrangian in (3.159) and (3.160) can be viewed as the zeroth order of a gradient expansion of the one-loop effective action which governs the dynamics of the background gauge field $A^\mu(x)$. Beyond the meaning of a chemical potential, the $\bar{A}_u$ field will additionally carry dynamical information by being space-dependent. An immediate physical consequence of the fact that the $\bar{A}_u$ field appears explicitly in the Lagrangian is the well-known Debye screening of electric fields. A weak-field expansion will take the form $\mathcal{L}^{1T} = -(1/2)\partial_\mu \bar{A}_u \partial^\mu \bar{A}_u + (m_{\text{eff}}^2/2)(\bar{A}_u)^2 + \mathcal{O}(\bar{A}_u^4)$, where the effective photon mass (inverse Debye screening length) is given by

$$m_{\text{eff}}^2(T) = \left.\frac{\partial^2 \mathcal{L}^{1T}}{\partial \bar{A}_u^2}\right|_{\bar{A}_u=0}. \tag{3.169}$$

Considering the zero-field limit for simplicity, we find, according to (3.165),

$$\left.\frac{\partial^2 \mathcal{L}^{1T}}{\partial \bar{A}_u^2}\right|_{\bar{A}_u=0} = -\frac{2e^2}{\pi^2} m^2 \sum_{n=1}^{\infty} (-1)^n K_2\left(\frac{m}{T}n\right). \tag{3.170}$$

In the high-temperature limit, $T \gg m$, the sum can be expanded, for example by employing the techniques described in Appendix D, leading to $\sum_{n=1}^{\infty}(-1)^n K_2\left[(m/T)n\right] \simeq -\pi^2 T^2/(6m^2) + \mathcal{O}(1)$ (cf. (D.63)). We finally obtain

$$m_{\text{eff}}^2(T) = \frac{(eT)^2}{3}, \tag{3.171}$$

which is the well-known result for the Debye screening mass found in the literature. The leading corrections to the Debye mass in the high-density and high-temperature limit can be looked up in the erratum of [72].

Now, an attractive interpretation of Debye screening arises in this context: at zero temperature and vanishing chemical potential, the gauge symmetry is completely present in quantum electrodynamics. This symmetry prohibits the photon from acquiring a mass. An immediate consequence is the Coulomb law for the static potential. At finite temperature, the gauge symmetry is restricted to the class of periodic gauge transformations. Therefore, the masslessness of the photon is no longer ensured in the former rigorous way. It is the (integrated) time-like gauge field component $\bar{A}_u$ that acquires a mass; consequently, the Coulomb law for the static potential is modified by a screening mechanism, whereby the mass value characterizes the inverse screening length.

We would like to stress that there is no such interpretation in the real-time formalism, since there is no a priori reason for restricting the class of gauge transformations. At present, it is not known how the real-time formalism could resolve the problem of gauge invariance without additional postulates.

From a microscopic, more phenomenological viewpoint, the screening is, of course, due to the temperature-induced virtual pairs, which allow a polarization of the vacuum.

3.5.5 Temperature-Induced Pair Production at One Loop?

Next, we turn to the question of whether Schwinger's well-known formula for the pair-production probability provides finite-temperature corrections at one-loop order. Taking a closer look at the literature, this is perhaps the most puzzling question concerning the properties of the finite-temperature effective action for nonvanishing electric fields. In fact, most of the papers of the early 1990s have been motivated by the quest for a temperature-dependent imaginary part of the effective Lagrangian that would indicate pair production. The aim was to find a mechanism to explain the strongly correlated e^+e^- peaks in heavy-ion collisions[15] [101]. Incidentally, knowledge of a temperature-dependent part of the Heisenberg–Euler Lagrangian would also be useful for the study of $q\bar{q}$ production in strong chromo-electric fields, since fermion coupling to nonabelian fields does not dramatically differ from the abelian version.

While thermal contributions to the imaginary part of $\mathcal{L}^{1T}$ have been found neither within the real-time formalism [48,72] nor by employing the functional Schrödinger representation [96], an imaginary part seems to appear in the imaginary-time formalism [82]. The latter result was also computed in the real-time formalism in [121].

Indeed, comparing the finite-temperature version (3.160) with the one-loop Schwinger Lagrangian, there are subtle traps in the computation. In the following, we shall, as an example, argue against the calculation in [121]. Although our findings for the effective thermal Lagrangian (3.160) in the heat-bath rest frame formally coincide with those found in [121] (up to numerical prefactors and an interchange of proper-time integration and summation), we do not agree with their computation of the imaginary part, which follows the line of the zero-temperature calculation. Various obstacles are encountered when proceeding in this way in the finite-temperature case: since the function $h(s)$ in the exponential of (3.160) reduces to

$$h(s) = eE \coth eEs \tag{3.172}$$

for a purely electric field, $E = |\boldsymbol{E}|$, (i) a rotation of the contour, $s \to -is$, becomes useless owing to the coth term in the exponent of (3.160); (ii) each term in the sum of (3.160) exhibits an essential singularity at the poles of the coth term on the imaginary axis; the use of the rule $\cot z \to \mathrm{P} \cot z + i\pi \sum_{z_0} \delta(z - z_0)$ is therefore nonsensical (cf. [121]); (iii) proper-time integration and summation must not be interchanged, as we pointed out above. If they are, the imaginary parts of the successive terms in the sum diverge exponentially, as can be shown by evaluating the residues of the singularities on the imaginary s axis. Incidentally, we do not agree with the imaginary part computed in [82], simply because the expressions for the effective Lagrangians do not coincide.

---

[15] Meanwhile, the peaks have mysteriously vanished and joined the ranks of popular aberrants of modern physics.

However, we can give an indirect argument for the vanishing of the imaginary part following [72]: owing to the formal resemblance between our result (3.160) and the findings of Loewe and Rojas [121] for the effective Lagrangian (but not for its imaginary part!), we can follow their steps backwards and end up with the starting point of the real-time formalism,

$$\frac{\partial \Gamma^{1T}}{\partial m} = -i \operatorname{Tr} \left[f_F(k_u, \bar{A}_u) \left(\frac{1}{\not{\Pi} - m + i\epsilon} - \frac{1}{\not{\Pi} - m - i\epsilon} \right) \right], \qquad (3.173)$$

where $f_F(k_u, \bar{A}_u)$ denotes a (real) thermal distribution function for the fermions and $\Gamma = \int d^4 x \, \mathcal{L}^{1T}$. Obviously, since the right-hand side is purely real, there is no imaginary part in the thermal contribution to the effective one-loop action and hence no thermal correction to the Schwinger pair-production formula to this order of calculation. So the disagreement between the real- and imaginary-time formalisms is resolved in this respect.

In the next section, we shall see that the vanishing of the thermal imaginary part of the effective action is in fact a particularity of the one-loop contribution. There is no mechanism or algebraic constraint that prevents the existence of an imaginary part at higher-loop order. We shall prove this by calculating the two-loop contribution for low temperature explicitly.

3.5.6 Discussion

The derivation of the effective QED action to one-loop order in the presence of an arbitrary constant electromagnetic field at finite temperature reveals some delicate features in the imaginary-time formalism.

The gauge invariance of the classical action turns out to be restricted to periodic gauge transformations Λ_p on the quantum level in order to leave the boundary conditions of the functional integral unaltered. This implies the existence of further gauge-invariant quantities besides the field strength, which are constructed from the background field A^μ. Additional information about the system under consideration has to be employed to fix the form of the gauge potential A^μ. In the present case, the requirement of homogeneity (constant fields and constant chemical potential) gives rise to the additional gauge-invariant quantity $\bar{A}_u$.

The way in which $\bar{A}_u$ enters the effective action can be viewed as a topological effect that arises from the compactification of the finite-temperature configuration space in imaginary time; namely, the configuration space loses its property of being simply connected and allows infinitely many topologically inequivalent paths to connect two different points in space–time. Each path can be classified by its winding number around the space–time cylinder. The holonomy factor that carries the gauge dependence of the Green's function is sensitive to these inequivalent paths, since it represents a mapping of the paths in configuration space into the gauge group. A Poisson resummation of the sum over the winding number leads to a sum over Matsubara frequencies shifted by the $\bar{A}_u$ field. The quantity $e\bar{A}_u$ acts like a chemical potential in the partition function and therefore can be identified with μ.

With the aid of a simple Taylor expansion in terms of the invariants, we obtained the weak-field expansion in Heisenberg–Euler form with temperature-dependent coefficients ((3.163) and (3.164)). The cancellation of all nonlinear terms arising from vacuum polarization was first observed in [33, 34]. A detailed and rigorous weak-field expansion of the effective Lagrangian at finite temperature and finite chemical potential was suggested in [73], relying on the "real-time" representation of the effective action as given in [72].

3.6 QED Effective Action at Finite Temperature: Two-Loop Dominance

We calculate the two-loop effective action of QED for an arbitrary constant electromagnetic field at finite temperature T in the limit of T much smaller than the electron mass. It is shown that in this regime the influence of the two-loop contribution always exceeds that of the one-loop part owing to the thermal excitation of the internal photon. As an application, we study light propagation and photon splitting in the presence of a magnetic background field at low temperature. Furthermore, we discover a thermally induced contribution to pair production in an electric field.

The preceding investigations of the low-energy sector of QED have remained more or less on the one-loop level; radiative corrections have not been studied in a systematic way. For purely electromagnetically modified vacua, this program has successfully been carried out to two-loop order [55, 138, 140]. Such radiative corrections to the weak-field expansion of the Heisenberg–Euler Lagrangian are included in, for example, (3.39) and (3.40).

After the one-loop study of the preceding section, the present section is devoted to an investigation of the thermal QED effective action at the two-loop level. But, contrary to the zero-temperature case, where the two-loop contribution represents only a 1% correction to the one-loop effective action, we demonstrate that the thermal two-loop contribution is of a qualitatively different kind from the thermal one-loop part and exceeds the latter by far in the low-temperature domain [89].

The simple physical reason for this is the following: at one loop, one takes only the massive electrons and positrons as virtual loop particles into account (Fig. 3.5a). Owing to the mass gap in the fermion spectrum, a heat bath at temperatures much below the electron mass m can hardly excite higher fermion states at all. Hence, one expects thermal one-loop effects to be suppressed by the electron mass. In fact, in a low-temperature expansion of the thermal one-loop effective action [73], one finds that each term is accompanied by a factor of $\exp(-m/T)$, exhibiting an exponential damping for $T \to 0$ (see (3.163)).

On the other hand, the two-loop contribution to the thermal effective action involves a virtual photon within the fermion loop (Fig. 3.5b). Since the photon is massless, a heat bath of arbitrarily low temperature can easily excite higher photon states, implying a comparably strong influence of thermal effects on the effective action. In Sect. 2.2, we were able to show that the dominant contribution to the thermal two-loop effective action in the low-temperature limit is proportional to T^4/m^4. This power-law behavior always wins out over the exponential damping of the one-loop case, leading to a *two-loop dominance* in the low-temperature domain. One might ask whether this inversion of the loop hierarchy signals the failure of perturbation theory for finite-temperature QED. But, of course, this is not the case, since the inclusion of a virtual photon does not "amplify" the two-loop graph and higher ones. Rather, calculating the one-loop graph should be rated as only an inconsistent truncation of the theory, since the one-loop approximation does not include all species of particles as virtual particles. Besides, effective-field theory techniques indicate that the three-loop contribution is of the order of T^8/m^8 [118] for $T/m \ll 1$, thereby obeying the usual loop hierarchy.

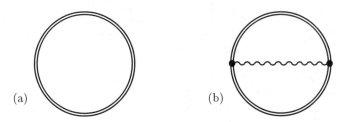

Fig. 3.5. Diagrammatic representation of (a) the one-loop and (b) the two-loop contribution to the effective QED action. The fermionic *double line* represents the coupling to all orders to the external electromagnetic field

The present section is organized as follows. First, we present the calculation of the two-loop effective QED action at finite temperature, employing the imaginary-time formalism and concentrating on the low-temperature limit at zero chemical potential. The outcome will be valid for slowly varying external fields of arbitrary strength.

Section 3.6.2 is devoted to an investigation of light propagation at finite temperature. The light cone condition for a thermal vacuum is derived, this time taking into account the dependence of the effective Lagrangian on the complete set of invariants. While, on the one hand, the well-known result for the velocity shift $\delta v \sim T^4/m^4$ is rediscovered [20, 60, 120, 157], we are also able to determine further contributions to the velocity shift arising from a nontrivial interplay between a temperature and an additional magnetic background field. By taking the one-loop contributions as derived in the preceding section into account, we can also comment on the intermediate-temperature domain $T \sim m$, where the velocity shift approaches a constant value.

In Sect. 3.6.3, we study aspects of thermally induced photon splitting. There, we point out that the thermal two-loop contribution to the splitting process exceeds the zero-temperature and one-loop contributions in the low-temperature and weak-field limit, but is negligible in comparison with other thermally induced scattering processes.

Sections 3.6.2 and 3.6.3 are mainly concerned with the limit of a weak magnetic background field and low-frequency photons ($\omega \ll m$), and therefore represent only a first glance at these extensive subjects. In fact, the quantitative results for this energy regime describe only tiny effects; any relevance to astrophysical topics such as pulsar physics has not been identified up to now. However, the intention of the present work is a more fundamental one, namely, to elucidate the mechanism of the violation of the usual loop hierarchy of perturbative thermal field theories involving virtual massless particles.

In Sect. 3.6.4, we calculate the thermal contribution to Schwinger's well-known pair-production formula [148] for a constant electric background field in the low-temperature limit. As mentioned before, a thermal one-loop contribution, surprisingly, does not exist [72, 87], since the thermal one-loop effective action is purely real by construction. Hence, the findings of Sect. 3.6.4 prove the existence of thermally induced pair production – an effect which has been searched for for 15 years [48, 82, 83, 96, 121]. In the low-temperature limit, we find that the situation for a strong electric field is dominated by the zero-temperature part (Schwinger formula), while the thermal contribution can become dominant for a weak electric field. Unfortunately, the experimentally more interesting high-temperature limit cannot be covered by our approach.

One last word of caution: the inclusion of electric background fields in finite-temperature QED is always plagued with the question of how violently this collides with assumptions on thermal equilibrium. In fact, electric fields and thermal equilibrium exclude each other, thus calling into question the physical meaning of the results of Sect. 3.6.4 at least quantitatively. However, it is reasonable to assume the existence of at least a small window of parameters in the low-temperature and weak-field domain for which the thermal-equilibrium calculation represents a good approximation. Moreover, knowledge of the effective Lagrangian, including a full dependence on all possible field configurations, is mandatory for deriving equations of motion for the fields, even in the limit of vanishing electric field.

3.6.1 Two-Loop Effective Action of QED at Low Temperature

In the following, we shall outline the calculation of the two-loop effective action, concentrating on the low-temperature limit where a *two-loop dom-*

inance is expected. The calculation is necessarily very technical; therefore, some details are left for the appendices.[16]

We employ the same notation as in Sect. 3.5.2; in addition to the standard gauge and Lorentz invariants, $\mathcal{F} = (1/4)F^{\mu\nu}F_{\mu\nu}$ and $\mathcal{G} = (1/4)F^{\mu\nu}{}^\star F_{\mu\nu}$ or, alternatively, to the *secular* invariants a and b (Appendix B), we can introduce one further invariant (besides the temperature itself):

$$\mathcal{E} = (u_\mu F^{\mu\alpha})(u_\nu F^\nu{}_\alpha), \tag{3.174}$$

where u^μ denotes the four-velocity of the heat bath. Let us recall that we have assumed without loss of generality that a Lorentz system exists in which the electric and magnetic fields are antiparallel. In this particular frame, the secular invariants can be identified with the field strengths: $a = B \equiv |\boldsymbol{B}|$, $b = E \equiv |\boldsymbol{E}|$. In the heat-bath rest frame, $\mathcal{E}$ simply reduces to $\mathcal{E} = \boldsymbol{E}^2$.

Contrary to the one-loop calculation, there will be a coupling between the loop momentum k^μ and the external field, so that we shall encounter another invariant in the intermediate stages of the calculation:

$$
\begin{aligned}
z_k &= (k_\mu F^{\mu\alpha})(k_\nu F^\nu{}_\alpha) \tag{3.175}\\
&= |\boldsymbol{k}|^2 B^2 \sin^2 \theta_B + |\boldsymbol{k}|^2 E^2 \sin^2 \theta_E - k^2 E^2 + 2k^0 \, \boldsymbol{E} \cdot (\boldsymbol{k} \times \boldsymbol{B}),
\end{aligned}
$$

where θ_B and θ_E denote the angle between the magnetic and electric field, respectively, and the three-space vector $\boldsymbol{k}$ (cf. (2.63)). Furthermore, the combination $k^\mu u_\mu$ filters out the momentum component that will be discretized by Matsubara frequencies at finite temperature: $(k^\mu u_\mu)^2 \to -\omega_n^2$.

Since the effective Lagrangian is a Lorentz covariant and gauge-invariant quantity, it can only be a function of the complete set of invariants of the system under consideration; hence, it will be of the form

$$\mathcal{L} = \mathcal{L}(\mathcal{E}, \mathcal{F}, \mathcal{G}, T), \tag{3.176}$$

as in the one-loop case. For simplicity, we work at zero chemical potential. Equipped with these conventions, we now turn to the calculation.

The two-loop contribution to the effective action/Lagrangian $\mathcal{L}^2$ is generally given by the diagram in Fig. 3.5b. This translates into the following formula in coordinate space [55]:

$$\mathcal{L}^2 = \frac{e^2}{2} \int \mathrm{d}^4x' \, \mathrm{tr} \left[\gamma^\mu \, G(x, x'|A) \, \gamma^\nu \, G(x', x|A) \right] D_{\mu\nu}(x - x'), \tag{3.177}$$

where $G(x, x'|A)$ represents the fermionic Green's function for the Dirac operator in the presence of an external electromagnetic field A. $D_{\mu\nu}$ denotes the photon propagator. Throughout this section, we assume the background field to be constant, or at most slowly varying compared with the scale of the Compton wavelength; therefore, the fermionic Green's function can be written as

---

[16] The primarily phenomenologically interested reader may simply take notice of the conventions used here, consult (3.202)–(3.209) directly, and skip the remainder of this subsection.

$$G(x, x'|A) = \Phi(x, x') \int \frac{\mathrm{d}^4 p}{(2\pi)^4} \, \mathrm{e}^{\mathrm{i}p(x-x')} \, g(p), \qquad (3.178)$$

where $g(p)$ denotes the Fourier transform of $G(x, x'|A)$, depending only on the field strength, and $\Phi(x, x')$ is the holonomy, carrying the complete gauge dependence of the Green's function. Inserting (3.178) into (3.177) leads us to the object $\Phi(x, x')\Phi(x', x) \equiv \Phi(\bigcirc)$, where the right-hand side represents the holonomy evaluated for a closed path. For a simply connected manifold such as the Minkowski space, $\Phi(\bigcirc) = 1$; hence, it does not contribute to the zero-temperature Lagrangian. For a nonsimply connected manifold such as the finite-temperature coordinate space $(\mathbb{R} \times S^1)$, $\Phi(\bigcirc)$ can pick up a winding number [87]. However, in the present case, we restrict our considerations to a situation with zero density, which implies the existence of a gauge in which $A_0 = 0$. Then, $\Phi(\bigcirc) = 1$ and the influence of the holonomy can be discarded.

This leads us to the representation

$$\mathcal{L}^2 = \frac{\mathrm{i}}{2} \int \frac{\mathrm{d}^4 k}{(2\pi)^4} \, D_{\mu\nu}(k) \, \Pi^{\mu\nu}(k) \qquad (3.179)$$

for the two-loop Lagrangian, where $D_{\mu\nu}(k)$ denotes the photon propagator in momentum space, and we have introduced the one-loop polarization tensor in an arbitrary constant external background field:

$$\Pi^{\mu\nu}(k) = -\mathrm{i}e^2 \int \frac{\mathrm{d}^4 p}{(2\pi)^4} \, \mathrm{tr} \left[\gamma^\mu \, g(p) \, \gamma^\nu \, g(p - k) \right]. \qquad (3.180)$$

So we have finally arrived at the well-known fact that the two-loop effective action can be obtained from the polarization tensor in an external field by gluing the external lines together.

The transition to finite-temperature field theory can now be made within the imaginary-time formalism by replacing the momentum integration over the zeroth component in (3.179) and (3.180) by a summation over bosonic and fermionic Matsubara frequencies, respectively. For example, performing this procedure in (3.180) corresponds to thermalizing the fermions in the loop. Now we come to an important point: confining ourselves to the low-temperature domain where $T \ll m$, we know from the one-loop calculations [73, 88] that thermal fermionic effects are suppressed by factors of $\mathrm{e}^{-m/T}$, indicating that the mass of the fermions suppresses thermal excitations. Hence, thermalizing the polarization tensor contributes at most terms of order $\mathrm{e}^{-m/T}$ to the two-loop Lagrangian for $T \ll m$; these are, furthermore, accompanied by an additional factor of the coupling constant α and can therefore be neglected compared with the one-loop terms. At low temperature, it is therefore sufficient to thermalize the internal photon only in order to obtain the leading T dependence of $\mathcal{L}^2$.

Since, in the Feynman gauge, the photon propagator reads

$$D_{\mu\nu}(k) = g_{\mu\nu} \frac{1}{k^2 - \mathrm{i}\epsilon}, \quad k^2 = -(k^0)^2 + \mathbf{k}^2, \quad g = (-, +, +, +), \quad (3.181)$$

the introduction of bosonic Matsubara frequencies $(k^0)^2 \to -\omega_n^2 = -(2\pi Tn)^2$, $n \in \mathbb{Z}$, leads us to[17]

$$\mathcal{L}^{2+2T} = \frac{\mathrm{i}}{2}\,\mathrm{i}T \sum_{\omega_n} \int \frac{\mathrm{d}^3 k}{(2\pi)^3}\, \frac{1}{k^2 - \mathrm{i}\epsilon}\, \Pi^\mu{}_\mu(k). \tag{3.182}$$

From now on, we write $\mathcal{L}^2$ for the zero-temperature two-loop Lagrangian, $\mathcal{L}^{2T}$ for the purely thermal part and $\mathcal{L}^{2+2T}$ for their sum. In (3.182), we need the trace of the polarization tensor in a constant but otherwise arbitrary electromagnetic field. Inserting the trace of the representation (2.75) for $\Pi^{\mu\nu}$ into (3.182), we obtain for the Lagrangian ($Q^\mu{}_\mu = 0$)

$$\mathcal{L}^{2+2T} = -\frac{T}{2}\frac{\alpha}{2\pi} \sum_{\omega_n} \int \frac{\mathrm{d}^3 k}{(2\pi)^3} \int_0^\infty \frac{\mathrm{d}s}{s} \int_{-1}^1 \frac{\mathrm{d}\nu}{2}\, \frac{e^{-\mathrm{i}s\phi_0}}{a^2 + b^2}\, \frac{eas\,ebs}{\sin eas \sinh ebs} \tag{3.183}$$

$$\times \left[\frac{z_k}{k^2 - \mathrm{i}\epsilon}(\tilde{N}_2 - \tilde{N}_1) + \left[2N_0(a^2 + b^2) + b^2\tilde{N}_2 + a^2\tilde{N}_1\right] \right]\Bigg|_{(k^0)^2 = -\omega_n^2},$$

where ϕ_0, N_0 and $\tilde{N}_i$ are functions of the integration variables s and ν and of the invariants a and b; only ϕ_0 depends additionally on z_k, as defined in (3.175). Their explicit form can be looked up in (2.77). In order to ensure convergence of the proper-time integrals, the causal prescription $m^2 \to m^2 - \mathrm{i}\epsilon$ for the mass term in ϕ_0 is understood; this agrees with deforming the s contour slightly below the real axis.

Now, the aim is to perform the k momentum integration/summation; note that the k dependence is contained in ϕ_0, z_k (and k^2, of course). Concentrating on this step, we encounter the integrals

$$I_1 = T \sum_{\omega_n} \int \frac{\mathrm{d}^3 k}{(2\pi)^3}\, e^{-\mathrm{i}s\phi_0}\Bigg|_{(k^0)^2 = -\omega_n^2},$$

$$I_2 = T \sum_{\omega_n} \int \frac{\mathrm{d}^3 k}{(2\pi)^3}\, \frac{z_k}{k^2 - \mathrm{i}\epsilon}\, e^{-\mathrm{i}s\phi_0}\Bigg|_{(k^0)^2 = -\omega_n^2}, \tag{3.184}$$

which allow us to write the Lagrangian (3.183) in terms of

$$\mathcal{L}^{2+2T} = -\frac{\alpha}{4\pi} \int_0^\infty \frac{\mathrm{d}s}{s} \int_{-1}^1 \frac{\mathrm{d}\nu}{2}\, \frac{eas\,ebs}{(a^2 + b^2)\sin eas \sinh ebs} \tag{3.185}$$

$$\times \left\{ (\tilde{N}_2 - \tilde{N}_1)\,I_2 + \left[2N_0(a^2 + b^2) + b^2\tilde{N}_2 + a^2\tilde{N}_1\right] I_1 \right\}.$$

[17] Of course, the present calculation does not necessarily have to be performed in the imaginary-time formalism. For example, instead of (3.181), we could just as well work with the real-time representation of the thermal photon propagator. We could even use only the one-component formalism, since we merely consider the photon to be thermalized. However, from our viewpoint, the calculations in the imaginary-time formalism appear a bit simpler since the momentum integrals remain Gaussian. Of course, this might be just a matter of taste.

The explicit k dependence of the function ϕ_0 in terms of the variables k^2 and z_k can be isolated by a reorganization of the first line of (2.77):

$$e^{-is\phi_0} = e^{-im^2 s}\, e^{-A_z\, z_k}\, e^{-A_k\, k^2}, \tag{3.186}$$

where we have implicitly defined

$$A_z = \frac{is}{2}\,\frac{1}{a^2+b^2}\left(\frac{\cos\nu eas - \cos eas}{eas\sin eas} + \frac{\cosh\nu ebs - \cosh ebs}{ebs\sinh ebs}\right), \tag{3.187}$$

$$A_k = \frac{is}{2}\,\frac{1}{a^2+b^2}\left(b^2\frac{\cos\nu eas - \cos eas}{eas\sin eas} - a^2\frac{\cosh\nu ebs - \cosh ebs}{ebs\sinh ebs}\right).$$

Employing (3.186) for ϕ_0, we can reduce the evaluation of I_2 to that of I_1:

$$I_2 = T\sum_{\omega_n}\int\frac{d^3k}{(2\pi)^3}\frac{z_k}{k^2 - i\epsilon}\, e^{-im^2 s}e^{-A_z z_k}\, e^{-A_k k^2}\Bigg|_{(k^0)^2=-\omega_n^2}$$

$$= -\frac{\partial}{\partial A_z}\int_{A_k}^{\infty} dA'_k\, I_1, \tag{3.188}$$

where A_z and A_k are again functions of the integration variables s and ν and of the invariants a and b, as defined in (3.187). In view of (3.188), it is sufficient to consider the momentum integration/summation for I_1 only:

$$I_1 \stackrel{(3.186)}{=} T e^{-im^2 s}\sum_{\omega_n}\int\frac{d^3k}{(2\pi)^3}\, e^{-A_z z_k}\, e^{-A_k k^2}\Bigg|_{(k^0)^2=-\omega_n^2}. \tag{3.189}$$

At this stage, the *finite-temperature coordinate frame* as introduced in Appendix E becomes extremely useful, since it enables us to perform the calculation in terms of the invariants. This coordinate system is adapted to the situation of an electromagnetic field at finite temperature in such a way that the components of any tensor-valued function of the field strength can be expressed in terms of the invariants $\mathcal{E}$, $\mathcal{F}$ and $\mathcal{G}$. Details are presented in Appendix E, from where we take the final formula for the exponent of (3.189) (see (E.60)):

$$A_z z_k + A_k k^2$$

$$= \left[A_k + (a^2 - b^2 + \mathcal{E})A_z\right]\left(k^2 - \frac{A_z\sqrt{d}}{A_z(2\mathcal{F}+\mathcal{E})+A_k}\,k^0\right)^2$$

$$- \frac{(A_k + a^2 A_z)(A_k - b^2 A_z)}{A_k + (a^2-b^2+\mathcal{E})A_z}(k^0)^2 + \frac{(A_k + a^2 A_z)(A_k - b^2 A_z)}{A_k\frac{a^2 b^2}{\mathcal{E}} + A_k}(k^1)^2$$

$$+ \left(A_z\frac{a^2 b^2}{\mathcal{E}} + A_k\right)\left(k^3 + \frac{A_z\frac{\sqrt{d}\mathcal{G}}{\mathcal{E}}}{A_z\frac{\mathcal{G}^2}{\mathcal{E}} + A_k}\,k^1\right), \tag{3.190}$$

where k^0, k^1, k^2, k^3 represent the components of the rotated momentum vector $k^A = e^A{}_\mu k^\mu$, and $e^A{}_\mu$ denotes the vierbein which mediates between the given coordinate system and the finite-temperature coordinate frame (see Appendix E). Since the transformation into the new reference frame is only a rigid rotation in Minkowski space, no Jacobian arises for the measure of the momentum integral. Hence, only integrals of Gaussian type are present in (3.189), which can easily be evaluated to give

$$I_1 = T \frac{e^{-im^2 s}}{(4\pi)^{3/2}} \frac{1}{\sqrt{p\, q_a\, q_b}} \sum_{\omega_n} \exp\left(-\frac{q_a\, a_b}{p} \omega_n^2\right), \qquad (3.191)$$

where it was convenient to introduce the short forms

$$q_a = A_k + a^2 A_z, \quad q_b = A_k - b^2 A_z, \quad p = A_k + (a^2 - b^2 + \mathcal{E}) A_z. \quad (3.192)$$

The sum in (3.191) can be rewritten with the aid of Jacobi's transformation formula:

$$\sum_{n=-\infty}^{\infty} \exp\left[-\sigma(n - z)^2\right] = \sum_{n=-\infty}^{\infty} \sqrt{\frac{\pi}{\sigma}} \exp\left(-\frac{\pi^2}{\sigma} n^2 - 2\pi i z n\right). \quad (3.193)$$

With $z = 0$ and $\sigma = (2\pi T)^2 q_a q_b / p$, we obtain for (3.191)

$$
\begin{aligned}
I_1 &\equiv I_1^{T=0} + I_1^T \\
&= \frac{e^{-im^2 s}}{16\pi^2} \frac{1}{q_a q_b} + \frac{e^{-im^2 s}}{8\pi^2} \frac{1}{q_a q_b} \sum_{n=1}^{\infty} \exp\left(-\frac{p}{q_a q_b} \frac{n^2}{4T^2}\right),
\end{aligned}
\qquad (3.194)
$$

where we have separated the ($n = 0$) term from the remaining sum in order to find the ($T = 0$) contribution. The first term in (3.194) (($n = 0$) term) is independent of T and $\mathcal{E}$, while the second term vanishes in the limit $T \to 0$ exponentially. In Appendix F, we check explicitly that the first term of (3.194) indeed leads to the (unrenormalized) two-loop Lagrangian for an arbitrary constant electromagnetic field at zero temperature. For example, for a purely magnetic field, the representation of Dittrich and Reuter [55] is rediscovered.

For our finite-temperature considerations, we shall keep only the second term of (3.194), which we denote by I_1^T in the following. Concerning the formula for $\mathcal{L}^{2T}$ in (3.185), I_1^T is already in its final form (it will turn out later that this term is subdominant in the low-T limit and only I_2^T contains the important contributions). Hence, let us turn to the evaluation of I_2^T, i.e. the thermal part of (3.188); for this, we have to interpret I_1^T as a function of A_z and A_k (remember that q_a, q_b and p are functions of A_z and A_k):

$$
\begin{aligned}
I_2^T &= -\frac{\partial}{\partial A_z} \int_{A_k}^{\infty} dA_k'\, I_1^T(A_k', A_z) = -\frac{\partial}{\partial A_z} \int_0^{\infty} ds'\, I_1^T(s' + A_k, A_z) \\
&= -\frac{e^{-im^2 s}}{8\pi^2} \sum_{n=1}^{\infty} \frac{\partial}{\partial A_z} J(A_z),
\end{aligned}
\qquad (3.195)
$$

where we have defined the auxiliary integral

$$J(A_z) = \int\limits_0^\infty \frac{ds'}{(s' + q_a)(s' + q_b)} \exp\left(-\frac{s' + p}{(s' + q_a)(s' + q_b)} \frac{n^2}{4T^2}\right). \quad (3.196)$$

Upon substitution of the integration variable[18]

$$u = \frac{q_a q_b}{p} \frac{s' + p}{(s' + q_a)(s' + q_b)}, \quad (3.197)$$

$$\frac{ds'}{(s'+q_a)(s'+q_b)} = -du \left[\frac{q_a^2 q_b^2}{p^2} + \frac{2q_a q_b}{p}(2p - q_a - q_b)u + (q_a - q_b)^2 u^2\right]^{-1/2},$$

the auxiliary integral becomes

$$J(A_z) = \int\limits_0^1 du \left[\frac{q_a^2 a_b^2}{p^2} + \frac{2q_a q_b}{p}(2p - q_a - q_b)u + (q_a - q_b)^2 u^2\right]^{-1/2}$$

$$\times \exp\left(-\frac{n^2}{4T^2}\frac{p}{q_a q_b}u\right). \quad (3.198)$$

Now we come to an important point: since we have only thermalized the photons, our effective Lagrangian $\mathcal{L}^{2T}$ is only valid for $T \ll m$ anyway. Nevertheless, our formulas also contain information about the high-temperature domain which we should discard, since it is incomplete. Regarding (3.198), the exponential function causes the integrand to be extremely small for small values of T, except where u is also small. Hence, the auxiliary integral is mainly determined by the lower end of the integration interval.

Taking these considerations into account, we expand the square root for small values of u and then extend the integration interval to infinity (in fact, remaining 1 as the upper bound creates only terms of the order $\exp[-(2nm)/T]$, which are subdominant in the low-temperature limit). The remaining u integration can then easily be performed for each order in the u expansion; up to u^2, we obtain

$$J(A_z) = 4\frac{T^2}{n^2} - 16\frac{T^4}{n^4}(2p - q_a - q_b) \quad (3.199)$$

$$-64\frac{T^6}{n^6}\left[(q_a - q_b)^2 - 3(2p - q_a - q_b)^2\right] + \mathcal{O}(T^8/n^8).$$

Upon differentiation, the T^2 dependence drops out, and we obtain (cf. (3.192))

$$\frac{\partial}{\partial A_z} J(A_z) = -2^5 \frac{T^4}{n^4}(\mathcal{F} + \mathcal{E}) - 2^9 \frac{T^6}{n^6}\left[\mathcal{F}^2 + \mathcal{G}^2 - 3(\mathcal{F} + \mathcal{E})^2\right]A_z$$

$$+\mathcal{O}(T^8/n^8). \quad (3.200)$$

In this equation, we indeed discover a power-law dependence on the temperature, which will directly translate into a power-law dependence of the

---

[18] Solving for $s' = s'(u)$ leads to a quadratic equation, from which the positive root has to be taken in order to take care of the integral boundaries.

two-loop effective action after insertion into (3.195) and (3.185). Technically speaking, this arises from the fact that the omnipresent exponential factor $\exp[-(n^2/4T^2)(p/q_a q_b)u]$, which finally causes exponential damping for $T/m \to 0$, becomes equal to 1 after the u integration at the lower bound at $u = 0$.

At this stage, it is important to observe that the u integration appears only in I_2^T (via the A_k' integration in (3.184)) and not in I_1^T. Therefore, I_1^T will always contain exponential damping factors in the limit $T \to 0$. Even the remaining proper-time integrations do not provide a mechanism similar to the u integration, since for large s, the mass factor $\exp(-im^2 s)$ with the causal prescription $m^2 \to m^2 - i\epsilon$ causes the integrand to vanish, and, for small s, the combination $p/q_a q_b$ in the exponent becomes

$$\frac{p}{q_a q_b} = -\frac{4i}{1 - \nu^2} \frac{1}{s} + \mathcal{O}(s). \tag{3.201}$$

Obviously, inserting (3.201) into the exponent leads to an exponential fall-off (bearing in mind that the s contour will run slightly below the real axis). Similar conclusions can be drawn for the ν integration. To summarize these technical considerations, we conclude that only the term containing I_2^T (the thermal part of I_2) in (3.185) contributes dominantly to $\mathcal{L}^{2T}$ in the low-temperature limit.

Inserting the first and second term of $(\partial/\partial A_z)J(A_z)$ in (3.200) successively into (3.195) and then into (3.185), we obtain the dominant terms of order T^4 and T^6 of the two-loop effective QED Lagrangian at low temperature; in particular, for the T^4 term, various useful representations can be given:

$$\mathcal{L}^{2T}\Big|_{T^4} = -\frac{\alpha\pi}{90} T^4 (\mathcal{F} + \mathcal{E}) \int_0^\infty \frac{ds}{s} \int_{-1}^1 \frac{d\nu}{2} e^{-im^2 s} \frac{eas\, ebs}{\sin eas \sinh ebs} \frac{\tilde{N}_2 - \tilde{N}_1}{a^2 + b^2}$$

$$= -\frac{\alpha\pi}{45} T^4 (\mathcal{F} + \mathcal{E}) \int_0^\infty \frac{ds}{s} \frac{1}{a^2 + b^2} e^{-im^2 s} \tag{3.202}$$

$$\times \left(ebs \coth ebs \frac{1 - eas \cot eas}{\sin^2 eas} + eas \cot eas \frac{1 - ebs \coth ebs}{\sinh^2 ebs} \right)$$

$$= \frac{\pi^2}{45} T^4 (\mathcal{F} + \mathcal{E}) \left(\frac{1}{a^2 + b^2} (\partial_a^2 + \partial_b^2) \right)$$

$$\times \left[\frac{1}{8\pi^2} \int_0^\infty \frac{ds}{s^3} e^{-im^2 s} eas \cot eas\, ebs \coth ebs \right]. \tag{3.203}$$

The term proportional to T^6 reads

$$\mathcal{L}^{2T}\Big|_{T^6} = -\frac{16\alpha\pi^3}{945} T^6 \left[\mathcal{F}^2 + \mathcal{G}^2 - 3(\mathcal{F} + \mathcal{E})^2 \right]$$

$$\times \int\limits_0^\infty \frac{ds}{s} \int\limits_{-1}^1 \frac{d\nu}{2} \frac{e^{-im^2 s}}{a^2 + b^2} \frac{eas\,ebs}{\sin eas\,\sinh ebs} \left(\tilde{N}_2 - \tilde{N}_1\right) A_z, \quad (3.204)$$

where $\tilde{N}_i$ and A_z are functions of the integration variables and the invariants a and b (but not of $\mathcal{E}$), and are defined in (2.77) and (3.187). The ν integration can be performed analytically, but the extensive result does not provide for new insights; hence we do not bother to write it down.

These equations represent the central result of the present work; therefore, a few of their properties should be stressed:

(1) While we have worked explicitly in the low-temperature approximation $T \ll m$, we have put no restrictions on the strength of the electromagnetic field.

(2) The low-temperature Lagrangians contain arbitrary powers of the invariants a and b (or, equivalently, $\mathcal{F}$ and $\mathcal{G}$), but the additional invariant at finite temperature $\mathcal{E}$ appears only linearly in the T^4 term and quadratically in the T^6 term. The small-T expansion thus corresponds to a small-$\mathcal{E}$ expansion.

(3) The fact that only the integral I_2^T, with the prefactor $(\tilde{N}_2 - \tilde{N}_1)$, contributes to the low-temperature Lagrangian in (3.185) implies that only the spatially transverse modes $\Pi_\parallel$ and $\Pi_\perp$ of the polarization tensor (2.76) play a role in this thermalized virtual two-loop process. The timelike or longitudinal mode Π_0 (depending on the character of k^μ) might become important at higher values of temperature.

(4) The fact that the invariant $\mathcal{E}$ always appears in the combination $\mathcal{F} + \mathcal{E}$ ensures a kind of dual invariance of the Lagrangian. Under the replacement $\boldsymbol{E} \to \boldsymbol{B}$ and $\boldsymbol{B} \to -\boldsymbol{E}$, the invariants change such that $\mathcal{F} \to -\mathcal{F}$, $\mathcal{G} \to -\mathcal{G}$ and $\mathcal{E} \to \mathcal{E} + 2\mathcal{F}$, so that $\mathcal{F} + \mathcal{E}$ remains invariant.

(5) The T^4 term of $\mathcal{L}^{2T}$ as exhibited in (3.203) possesses the peculiarity of being derivable from the one-loop zero-temperature Lagrangian, which we have marked by square brackets in (3.203) after the derivative terms. This will be elucidated further in the following subsection.

(6) The entire thermal contribution to the effective action is finite. This reflects the well-known fact that the counterterms which are necessary and sufficient in order to renormalize the zero-temperature effective action are also necessary and sufficient for the finite-temperature action. Even more conveniently, we were able to separate the zero-temperature parts from the thermal parts, implicitly assuming that the renormalization of the zero-temperature parts is performed without any reference to the finite-temperature system. As a consequence, we are dealing with the same renormalization point as at zero temperature, which is naturally given by the zero-temperature electron mass. At finite temperature, this does not have to be, and indeed is not, identical to the physical electron mass, which undergoes further renormalization by finite-temperature effects.

For example, from a one-loop calculation of the mass operator one finds $m_{\mathrm{phys}}^2 = m^2 + (2/3)\alpha\pi T^2$ for $T \ll m$ [64]. Therefore, the thermal effective action given above must be viewed as "off-shell" renormalized. Nevertheless, since the physics is independent of the renormalization point, we can work with the zero-temperature as well as the physical electron mass.[19] The "off-shell", i.e. zero-temperature, renormalization is, of course, more transparent, since all temperature dependence is explicitly displayed; it would otherwise be partly hidden in the physical electron mass.

For the remainder of this section, we shall discuss certain limiting cases of the two-loop low-temperature Lagrangian. First, let us concentrate on a weak-field expansion, which corresponds to a small-s expansion of the proper-time integral owing to the exponential mass factor. Expanding the integrands for small values of s (except the mass factor) and integrating over ν and s leads us to the dominant terms in the weak-field limit:

$$\left.\mathcal{L}^{2T}\right|_{T^4} = \frac{44\alpha^2\pi^2}{2025}\frac{T^4}{m^4}(\mathcal{F}+\mathcal{E}) - \frac{2^6 \times 37\alpha^3\pi^3}{3^4 \times 5^2 \times 7}\frac{T^4}{m^4}\frac{\mathcal{F}(\mathcal{F}+\mathcal{E})}{m^4} + \mathcal{O}(3), \quad (3.205)$$

$$\left.\mathcal{L}^{2T}\right|_{T^6} = \frac{2^{13}\alpha^3\pi^5}{3^6 \times 5 \times 7^2}\frac{T^6}{m^6}\left(2\mathcal{F}^2 + 6\mathcal{E}\mathcal{F} + 3\mathcal{E}^2 - \mathcal{G}^2\right)\frac{1}{m^4} + \mathcal{O}(3), \quad (3.206)$$

where $\mathcal{O}(3)$ signals that we have omitted terms of third order in the field invariants (sixth order in the field strength). Note that no linear term in the field invariants exists to order T^6. For the terms of quadratic order, the T^6 term is subdominant for $T/m \leq 0.05$, and amounts to a 10% correction to the T^4 term for $T/m \sim 0.1$. For larger values of temperature, we expect a failure of the low-temperature approximation.

Finally, we consider $\left.\mathcal{L}^{2T}\right|_{T^4}$ in the limit of a purely magnetic background field: $b \to 0$, $a \to B$, $\mathcal{F} + \mathcal{E} \to (1/2)B^2$. The T^4 term in (3.202) then reduces to

$$\left.\mathcal{L}^{2T}(B)\right|_{T^4} = \frac{\alpha\pi}{90}T^4 \int_0^\infty \frac{dz}{z}e^{-m^2 z/eB}\left(\frac{1 - z\coth z}{\sinh^2 z} + \frac{1}{3}z\coth z\right), \quad (3.207)$$

where we have performed the substitution $eas \to -iz$ in concordance with the causal prescription $m^2 \to m^2 - i\epsilon$. Incidentally, the limit of a purely electric field can simply be obtained by replacing B by $-iE$ and multiplying (3.207) by -1.

Introducing the critical field strength $B_{\mathrm{cr}} = m^2/e$, we can evaluate the integral in (3.207) analytically [60], and obtain

$$\left.\mathcal{L}^{2T}(B)\right|_{T^4} = \frac{\alpha\pi}{90}T^4\left[\left(\frac{B_{\mathrm{cr}}^2}{2B^2} - \frac{1}{3}\right)\psi\left(1 + \frac{B_{\mathrm{cr}}}{2B}\right) - \frac{2B_{\mathrm{cr}}}{B}\ln\Gamma\left(\frac{B_{\mathrm{cr}}}{2B}\right) - \frac{3B_{\mathrm{cr}}^2}{4B^2}\right.$$

---

[19] In the case of an "on-shell" renormalization, first, m has to be replaced by m_{phys}, and, secondly, we obtain an additional term $-(2/3)\alpha\pi T^2(\partial\mathcal{L}^1/\partial m^2)$ from the mass renormalization at one-loop order.

$$-\frac{B_{\mathrm{cr}}}{2B} + \frac{B_{\mathrm{cr}}}{B}\ln 2\pi + \frac{1}{6} + 4\zeta'\left(-1, \frac{B_{\mathrm{cr}}}{2B}\right) + \frac{B}{3B_{\mathrm{cr}}}\bigg], \quad (3.208)$$

where $\psi(x)$ denotes the logarithmic derivative of the Γ function, and $\zeta'(s, q)$ is the first derivative of the Hurwitz ζ function with respect to its first argument.

For strong magnetic fields, $B \gg B_{\mathrm{cr}}$, the last term in the square brackets in (3.208) dominates the whole expression, and we find a linear increase of the effective Lagrangian:

$$\mathcal{L}^{2T}(B \gg B_{\mathrm{cr}})\bigg|_{T^4} = \frac{\alpha\pi}{270}\,T^4\,\frac{eB}{m^2}. \quad (3.209)$$

This contribution remains subdominant compared with the one arising from pure vacuum polarization, $\sim B^2 \ln(eB/m^2)$, which is not astonishing, since the magnetization of (real) thermalized plasma particles is bounded: the spins can at maximum be completely aligned. In contrast, the nonlinearities of vacuum polarization set no such upper bound. Quantitatively, the same result was found for the thermal one-loop contribution [71].

3.6.2 Light Propagation

As a first application of the one- and two-loop QED effective action at finite temperature, we reinvestigate the modification of the velocity of light in the presence of a heat bath [88,89].

As we reported in Sect. 3.3, the most important contribution to the velocity shift in the low-temperature region ($T \ll m$), $\delta v \sim -T^4/m^4$ (cf. (3.77)), can be obtained from the two-loop polarization tensor, in which the radiative photon within the fermion loop is considered to be thermalized [120,157]. We rediscovered the same result by means of an effective one-loop calculation employing the Heisenberg–Euler Lagrangian [20,60].

In the present subsection, we enlarge the effective-Lagrangian approach to the light cone condition as applied in Sect. 3.3 by taking into account the complete set of gauge and Lorentz invariants of the given situation. Our intention is to avoid two problems of the former approach: first, by working with the complete set of invariants, we do not have to rely on approximate methods for the high-temperature case [60], i.e. within the framework of the effective-action approach, the procedure is well defined in a field-theoretical sense. Secondly, the one-loop contribution is taken into account for the first time; in particular, in the intermediate-temperature region $T \sim m$, the one-loop contribution is expected to be the dominant one.

Our knowledge about the two-loop thermal effective action will not only reproduce the well-known velocity shift in the low-temperature region, $\delta v \sim -T^4/m^4$, serving as an independent check of our computation, but will moreover allow a study of combined temperature- and field-modified vacua.

Light Cone Condition

Following the lines of Sect. 3.2, the system considered here is a propagating photon (plane wave) field in the presence of a vacuum that is modified by an external constant electromagnetic field, at finite temperature and zero density. We assume the propagating photons to be soft, and neglect any vacuum modifications caused by the propagating light itself. Thus, we can linearize the field equations with respect to the photon field.

This time, the effective Lagrangian (3.160) depends not only on the field invariants $\mathcal{F}$ and $\mathcal{G}$, but also on the invariant $\mathcal{E}$ and the temperature T (see (3.176)). Within the framework of these assumptions, the field equations are obtained from the Euler–Lagrange equations of motion of the effective thermal QED Lagrangian:

$$0 = 2\partial_\mu \frac{\partial \mathcal{L}}{\partial F_{\mu\nu}} = 2\partial_\mu \left(\partial_{\mathcal{F}}\mathcal{L} \frac{\partial \mathcal{F}}{\partial F_{\mu\nu}} + \partial_{\mathcal{G}}\mathcal{L} \frac{\partial \mathcal{G}}{\partial F_{\mu\nu}} + \partial_{\mathcal{E}}\mathcal{L} \frac{\partial \mathcal{E}}{\partial F_{\mu\nu}} \right). \qquad (3.210)$$

The Lagrangian $\mathcal{L}$ that we are going to insert into (3.210) contains the classical Maxwell part as well as the zero-temperature and thermal one-loop contributions: $\mathcal{L} = \mathcal{L}_{\mathrm{M}} + \mathcal{L}^1 + \mathcal{L}^{1T}$; however, the following derivation of the light cone condition does not rely on any perturbative approximation to $\mathcal{L}$. It is only necessary that the field dependence of the Lagrangian be completely contained in a set of three linearly independent invariants, for which we take the standard invariants $\mathcal{F}$, $\mathcal{G}$ and the invariant $\mathcal{E}$ as defined in (3.174).

Note that the differentiation with respect to $F_{\mu\nu}$ has to be performed with regard to its antisymmetry properties. It is convenient to introduce the following notation for antisymmetrizing a tensor $A^{\mu\nu}$:

$$A^{[\mu\nu]} = \frac{1}{2}\left(A^{\mu\nu} - A^{\nu\mu} \right). \qquad (3.211)$$

Obviously, we find for the field strength tensor that $F^{[\mu\nu]} \equiv F^{\mu\nu}$. This implies the following for the partial derivative of $\mathcal{E}$ with respect to the field strength tensor:

$$\frac{\partial \mathcal{E}}{\partial F_{\mu\nu}} \overset{(3.143)}{=} \frac{\partial}{\partial F_{\mu\nu}} \left(u_\alpha F^{\alpha\gamma} u_\beta F^\beta{}_\gamma \right) = 2u_\alpha F^{\alpha\gamma} u_\beta g^{\beta[\mu} \delta^{\nu]}_\gamma,$$

$$= 2u_\alpha F^{\alpha[\nu} u^{\mu]}. \qquad (3.212)$$

Taking the identities (3.3) into account and moving the space–time derivative in (3.210) to the right, the field equation reads

$$0 = \partial_\mu \left[\partial_{\mathcal{F}}\mathcal{L}\, F^{\mu\nu} + \partial_{\mathcal{G}}\mathcal{L}\, {}^*F^{\mu\nu} + 4\partial_{\mathcal{E}}\mathcal{L}\, u^{[\mu}(uF)^{\nu]} \right] \qquad (3.213)$$

$$= \partial_{\mathcal{F}}\mathcal{L}\, \partial_\mu F^{\mu\nu} + \partial_{\mathcal{G}}\mathcal{L} \underbrace{\partial_\mu {}^*F^{\mu\nu}}_{=0} + 4\partial_{\mathcal{E}}\mathcal{L}\, \partial_\mu \left[u^{[\mu}(uF)^{\nu]} \right]$$

$$+ F^{\mu\nu}\left(\partial_{\mathcal{F}}^2\mathcal{L}\, \partial_\mu \mathcal{F} + \partial_{\mathcal{F}\mathcal{G}}\mathcal{L}\, \partial_\mu \mathcal{G} + \partial_{\mathcal{F}\mathcal{E}}\mathcal{L}\, \partial_\mu \mathcal{E} \right)$$

$$+ {}^\star F^{\mu\nu}\left(\partial_{\mathcal{F}\mathcal{G}}\mathcal{L}\,\partial_\mu\mathcal{F} + \partial_{\mathcal{G}}^2\mathcal{L}\,\partial_\mu\mathcal{G} + \partial_{\mathcal{G}\mathcal{E}}\mathcal{L}\,\partial_\mu\mathcal{E}\right)$$

$$+ 4u^{[\mu}(u\mathrm{F})^{\nu]}\left(\partial_{\mathcal{F}\mathcal{E}}\mathcal{L}\,\partial_\mu\mathcal{F} + \partial_{\mathcal{G}\mathcal{E}}\mathcal{L}\,\partial_\mu\mathcal{G} + \partial_{\mathcal{E}}^2\mathcal{L}\,\partial_\mu\mathcal{E}\right).$$

Here, we have employed the Bianchi identity and matrix notation, e.g. $(u\mathrm{F})^\nu = u_\alpha F^{\alpha\nu}$. For further evaluation, we need the following identities:

$$\partial_\mu\mathcal{F} = \frac{1}{2}F_{\alpha\beta}\,\partial_\mu F^{\alpha\beta}, \qquad\qquad \partial_\mu\mathcal{G} = \frac{1}{2}{}^\star F_{\alpha\beta}\partial_\mu F^{\alpha\beta},$$

$$\partial_\mu\mathcal{E} = \frac{\partial\mathcal{E}}{\partial F^{[\alpha\beta]}}\,\partial_\mu F^{\alpha\beta} \overset{(3.212)}{=} 2u_{[\alpha}(u\mathrm{F})_{\beta]}\,\partial_\mu F^{\alpha\beta}$$

$$= 2(u\mathrm{F})_\beta\,\partial_\mu(u\mathrm{F})^\beta,$$

$$\partial_\mu\big[u^\mu(u\mathrm{F})^\nu\big] = u^\mu\,\partial_\mu(u\mathrm{F})^\nu. \tag{3.214}$$

Inserting these identities into the field equation (3.213), we finally obtain

$$0 = \partial_{\mathcal{F}}\mathcal{L}\,\partial_\mu F^{\mu\nu} + \partial_\mu F^{\alpha\beta}\bigg\{4\partial_{\mathcal{E}}\mathcal{L}\,u^{[\mu}u_{[\alpha}\delta_{\beta]}^{\nu]}$$

$$+ F^{\mu\nu}\left[\frac{1}{2}\partial_{\mathcal{F}}^2\mathcal{L}\,F_{\alpha\beta} + \frac{1}{2}\partial_{\mathcal{F}\mathcal{G}}\mathcal{L}\,{}^\star F_{\alpha\beta} + 2\partial_{\mathcal{F}\mathcal{E}}\mathcal{L}\,u_{[\alpha}(u\mathrm{F})_{\beta]}\right]$$

$$+ {}^\star F^{\mu\nu}\left[\frac{1}{2}\partial_{\mathcal{F}\mathcal{G}}\mathcal{L}\,F_{\alpha\beta} + \frac{1}{2}\partial_{\mathcal{G}}^2\mathcal{L}\,{}^\star F_{\alpha\beta} + 2\partial_{\mathcal{G}\mathcal{E}}\mathcal{L}\,u_{[\alpha}(u\mathrm{F})_{\beta]}\right]$$

$$+ 4u^{[\mu}(u\mathrm{F})^{\nu]}\left[\frac{1}{2}\partial_{\mathcal{F}\mathcal{E}}\mathcal{L}\,F_{\alpha\beta} + \frac{1}{2}\partial_{\mathcal{G}\mathcal{E}}\mathcal{L}\,{}^\star F_{\alpha\beta} + 2\partial_{\mathcal{E}}^2\mathcal{L}\,u_{[\alpha}(u\mathrm{F})_{\beta]}\right]\bigg\}$$

$$= \partial_{\mathcal{F}}\mathcal{L}\,\partial_\mu F^{\mu\nu} + \partial_\mu F^{\alpha\beta}\left(\frac{1}{2}M_{\alpha\beta}^{\mu\nu}\right). \tag{3.215}$$

In the last line, we have introduced the tensor $M_{\alpha\beta}^{\mu\nu}$, which is equal to twice the terms in the curly brackets. Note that (3.215) is formally identical to the field equation (3.17) for the purely electromagnetic case.

Hence, we can easily obtain the light cone condition for the present case by repeating the steps following (3.17). Then, we can read off the (polarization-state-averaged) light cone condition from (3.20),

$$0 = 2(\partial_{\mathcal{F}}\mathcal{L})\,k^2 + M_{\alpha\nu}^{\mu\nu}\,k_\mu k^\alpha, \tag{3.216}$$

which leads us to the partially contracted tensor $M_{\alpha\nu}^{\mu\nu}$. With the aid of the fundamental algebraic relations (B.3) and (B.4), the contractions can be simplified; inserting the contracted tensor into (3.216), we may write the light cone condition as

$$0 = \left(\partial_{\mathcal{F}}\mathcal{L} + \mathcal{G}\partial_{\mathcal{F}\mathcal{G}}\mathcal{L} - \mathcal{F}\partial_{\mathcal{G}}^2\mathcal{L}\right)k^2 + \frac{1}{2}\left(\partial_{\mathcal{F}}^2 + \partial_{\mathcal{G}}^2\right)\mathcal{L}\,(\mathrm{F}k)^\nu(\mathrm{F}k)_\nu$$

$$- \partial_{\mathcal{E}}\mathcal{L}\,k^2 + 2\big[\partial_{\mathcal{E}}\mathcal{L} + \mathcal{E}\partial_{\mathcal{E}}^2\mathcal{L}\big]\,(ku)^2 \tag{3.217}$$

$$+ 2\partial_{\mathcal{F}\mathcal{E}}\mathcal{L}\,(ku)(\mathrm{F}k)^\nu(\mathrm{F}u)_\nu + 2\partial_{\mathcal{G}\mathcal{E}}\mathcal{L}\,(ku)({}^\star\mathrm{F}k)^\nu(\mathrm{F}u)_\nu$$

$$+ 2(\partial_{\mathcal{F}\mathcal{E}}\mathcal{L} - \partial_{\mathcal{E}}^2\mathcal{L})\,(u\mathrm{F}k)^2 + 2\partial_{\mathcal{G}\mathcal{E}}\mathcal{L}\,(u\mathrm{F}k)(u{}^\star\mathrm{F}k),$$

where we have employed matrix notation again: $(\mathsf{F}k)^\nu \equiv F^\nu{}_\alpha k^\alpha$, or $(ku) = k^\alpha u_\alpha$. The first line in (3.217) corresponds to the purely field-modified light cone condition as derived in Sect. 3.2, which can conveniently be expressed in terms of the energy–momentum tensor (or its vacuum expectation value) of the electromagnetic field. In the present case, the introduction of the vacuum expectation value of the energy–momentum tensor would appear to be artificial at this stage, but we shall come back to this point later on. In fact, (3.217) is the most transparent representation of the light cone condition for electromagnetically modified and temperature-modified vacua.

In the following, we confine ourselves to the case of vanishing external field: $F^{\mu\nu}, \mathcal{F}, \mathcal{G}, \mathcal{E} \to 0$. Since it is more than reasonable to assume that the effective Lagrangian behaves smoothly at this origin in field space, the derivatives of the Lagrangian are bounded in this limit. So the light cone condition becomes greatly simplified, and, except for the linear term in $\mathcal{F}$, which is filtered out by $\partial_{\mathcal{F}}\mathcal{L}|_{\mathcal{F},\mathcal{G},\mathcal{E}=0}$, only the first term in the square brackets in (3.217) survives:

$$0 = (\partial_{\mathcal{F}}\mathcal{L} - \partial_{\mathcal{E}}\mathcal{L})k^2 + 2\partial_{\mathcal{E}}\mathcal{L}\,(ku)^2, \quad \text{for} \quad \mathcal{F},\mathcal{G},\mathcal{E} \to 0. \tag{3.218}$$

Introducing the phase velocity $v = k^0/|\mathbf{k}|$, which is identical to the group velocity in the soft-photon limit, we may rewrite the light cone condition in the heat-bath rest frame $((ku)^2 = (k^0)^2)$ in terms of the squared velocity:

$$v^2 = \left(1 + \frac{2\partial_{\mathcal{E}}\mathcal{L}}{-\partial_{\mathcal{F}}\mathcal{L} + \partial_{\mathcal{E}}\mathcal{L}}\right)^{-1}. \tag{3.219}$$

In order to maintain $v \leq 1$, the fraction in the denominator of (3.219) should always be positive. Note that this statement is independent of any loop approximation.

Finally, we transcribe the light cone condition into the form that we derived in the preceding sections. For the present purpose, it is sufficient to consider the zero-field case only. Then, the effective Lagrangian simply depends on the temperature, $\mathcal{L} \equiv \mathcal{L}(T)$, which is related to the Lorentz-invariant norm of a vector n^μ (cf. (3.143)).

We are interested in the vacuum expectation value of the energy–momentum tensor for the present system. With the aid of its definition in (3.23), the vacuum expectation value of $\langle T^{\mu\nu} \rangle$ yields

$$\langle T^{\mu\nu} \rangle = \mathcal{L}\,g^{\mu\nu} - 2\frac{\partial\mathcal{L}}{\partial g_{\mu\nu}}. \tag{3.220}$$

Here, we need

$$\frac{\partial\mathcal{L}}{\partial g_{\mu\nu}} = \frac{\partial\mathcal{L}}{\partial(T^2)}\frac{\partial(T^2)}{\partial g_{\mu\nu}} = \frac{1}{2T}\frac{\partial\mathcal{L}}{\partial T}\frac{\partial}{\partial g_{\mu\nu}}\left(-n^\alpha n^\beta g_{\alpha\beta}\right)$$

$$= -\frac{1}{2}T\frac{\partial\mathcal{L}}{\partial T}u^\mu u^\nu. \tag{3.221}$$

Upon insertion of this into (3.220), we arrive at

$$\langle T^{\mu\nu}\rangle = \mathcal{L}\, g^{\mu\nu} + T\, \partial_T \mathcal{L}\, u^\mu u^\nu. \tag{3.222}$$

It is convenient to rewrite the last equation as

$$u^\mu u^\nu = \frac{1}{T\partial_T \mathcal{L}}\left(\langle T^{\mu\nu}\rangle - \mathcal{L}\, g^{\mu\nu}\right). \tag{3.223}$$

This form can now be substituted into the light cone condition (3.218), which leads us to

$$0 = (\partial_{\mathcal{F}}\mathcal{L} - \partial_\varepsilon \mathcal{L})k^2 + \frac{2\partial_\varepsilon \mathcal{L}}{T\partial_T \mathcal{L}}\left(\langle T^{\mu\nu}\rangle k_\mu k_\nu - \mathcal{L}\, k^2\right). \tag{3.224}$$

We obtain from (3.224)

$$k^2 = Q_T \langle T^{\mu\nu}\rangle k_\mu k_\nu, \tag{3.225}$$

where the effective-action charge Q_T for the present system is given by

$$Q_T = \frac{(2/T)(\partial_\varepsilon \mathcal{L}/\partial_T \mathcal{L})}{-\partial_{\mathcal{F}}\mathcal{L} + \partial_\varepsilon \mathcal{L} + (2/T)(\partial_\varepsilon \mathcal{L}/\partial_T \mathcal{L})\,\mathcal{L}}. \tag{3.226}$$

The form of (3.225) is identical to our findings in (3.29), where we considered the case of a purely magnetically modified vacuum. Of course, the explicit form of the effective-action charge in (3.226) differs from the magnetic case, since we are dealing with a different set of invariants. In fact, there is not even a formal resemblance to a Poisson equation in the $(\mathcal{E}, T)$ parameter space; nevertheless, we should expect a localized distribution in parameter space in order to ensure a bounded velocity shift.

However, since the structure of (3.218) (or (3.219)) furnishes a more convenient computation, this form of the light cone condition will be employed in the following.

One-Loop Results

As an application, we insert the one-loop effective action into the light cone condition. This will finally give us physical results for the intermediate-temperature domain, $T \sim m$. Let us first consider a thermalized vacuum with vanishing background fields. Here, the velocity shift stems from the thermal contribution $\mathcal{L}^{1T}$ only, and we need the derivative $\partial_\varepsilon \mathcal{L} = \partial_\varepsilon \mathcal{L}^{1T}$ of $\mathcal{L}^{1T}$ (3.160) in the zero-field limit. For this, we find at zero chemical potential

$$\partial_\varepsilon \mathcal{L}^{1T} = -\frac{\alpha}{3\pi}\sum_{n=1}^{\infty}(-1)^n\left(\frac{m}{T}n\right)K_1\left(\frac{m}{T}n\right). \tag{3.227}$$

Here, we have performed a Wick rotation $s \to -is$ in accordance with the causal prescription $m^2 \to m^2 - i\epsilon$ and have employed the special representation of the modified Bessel function given in (3.166). Also, note that summation and integration can be interchanged, since the essential singularities in the complex s plane vanish in the zero-field limit.

Regarding (3.219), we also need $\partial_{\mathcal{F}}\mathcal{L}$; for this, note that the zero-temperature one-loop part is renormalized in such a way that the term linear in $\mathcal{F}$ vanishes, in order to recover the pure Maxwell theory in the weak-field limit. Hence, it is only the thermal contribution $\mathcal{L}^{1T}$ which provides an additional term linear in $\mathcal{F}$: $\partial_{\mathcal{F}}\mathcal{L} = -1+\partial_{\mathcal{F}}\mathcal{L}^{1T}$, where the (-1) stems from the Maxwell part $\mathcal{L}_M$. In the zero-field limit, we get

$$\partial_{\mathcal{F}}\mathcal{L}^{1T} = -\frac{4\alpha}{3\pi}\sum_{n=1}^{\infty}(-1)^n K_0\left(\frac{m}{T}n\right). \tag{3.228}$$

In the limiting cases of low, intermediate and high temperature, (3.227) and (3.228) can be expanded in terms of the parameter (T/m). Let us first consider (3.228): at low temperature, only the first term of the sum needs to be taken into account, which leads to

$$\partial_{\mathcal{F}}\mathcal{L}^{1T}(T \ll m) \simeq \frac{2\alpha}{3\pi}\sqrt{\frac{2\pi T}{m}}\,\mathrm{e}^{-m/T}. \tag{3.229}$$

For an intermediate- or high-temperature expansion, the infinite sum must be completely taken into account; this can be achieved by employing the techniques proposed at the end of Appendix D. The result for (3.228) when $T \sim m$ is

$$\partial_{\mathcal{F}}\mathcal{L}^{1T} = \frac{2\alpha}{3\pi}\left[(0.666\ldots) + (0.814\ldots)\ln\frac{T}{m}\right], \tag{3.230}$$

where the numbers stem from pure integrals over analytic functions. Incidentally, the expansion for $T \gg m$ is formally identical to (3.230) with the factor of $0.814\ldots$ replaced by 1, as found in (3.164).

For our purpose of investigating the low- and intermediate-temperature domains, we observe that $\partial_{\mathcal{F}}\mathcal{L}^{1T}(T) = \mathcal{O}(\alpha/\pi) \ll 1$. For the calculation of the velocity shifts to order α/π, it is thus sufficient to employ simply

$$v^2 \simeq 1 - 2\partial_{\mathcal{E}}\mathcal{L}, \tag{3.231}$$

i.e. we neglect $\partial_{\mathcal{F}}\mathcal{L}^{1T}$ compared with $\partial_{\mathcal{F}}\mathcal{L}_M = -1$ in (3.219).

Turning to the low-temperature expansion of (3.227), we have to take into account only the first term of the sum:

$$\partial_{\mathcal{E}}\mathcal{L}^{1T}(T \ll m) \simeq \frac{\alpha}{3\pi}\sqrt{\frac{\pi}{2}}\sqrt{\frac{m}{T}}\,\mathrm{e}^{-(m/T)}. \tag{3.232}$$

Inserting this into (3.231), we find for the squared velocity at low temperature an exponentially decreasing modification:

$$v^2(T \ll m) \simeq 1 - \frac{\alpha}{3}\sqrt{\frac{2}{\pi}}\sqrt{\frac{m}{T}}\,\mathrm{e}^{-(m/T)}. \tag{3.233}$$

The high-temperature expansion of (3.227) is worked out in Appendix D, leading to

$$\partial_{\mathcal{E}} \mathcal{L}^{1T}(T \gg m) \simeq \frac{\alpha}{6\pi}\left[1 - k_2 \frac{m^2}{T^2} + \mathcal{O}\left(\frac{m^4}{T^4}\right)\right], \tag{3.234}$$

where k_2 denotes the number $0.213139\ldots$ and arises from a parameter-independent integral over analytic functions during the expansion. The light cone condition then yields the following for the squared velocities at high temperature:

$$v^2(T \gg m) \simeq 1 - \frac{\alpha}{3\pi}\left(1 + k_2 \frac{m^2}{T^2}\right) + \ldots, \tag{3.235}$$

which implies a maximum velocity shift of $\delta v_{\mathrm{max}}^2 = -\alpha/(3\pi) \simeq -1/1291$. It should, however, be noted that an expansion for $T/m \gg 1$ is a formal trick to extract analytical results. For identifying the values of temperature to which the light cone condition (3.217) is applicable, we notice that the frequency of the plane wave should, on the one hand, be smaller than the electron mass in order to justify the assumption of slowly varying fields and, on the other hand, should be larger than the plasma frequency in order to ensure the existence of such a propagating mode:

$$\omega_{\mathrm{p}} \ll \omega \ll m. \tag{3.236}$$

For a plasma frequency corresponding to the Debye screening mass, we employ the representation found in (3.169) and (3.170):

$$\omega_{\mathrm{p}}^2 = -8\frac{\alpha}{\pi}m^2 \sum_{n=1}^{\infty}(-1)^n K_2\left(\frac{m}{T}n\right). \tag{3.237}$$

The maximum value of temperature up to which the low-frequency assumption can formally be justified is determined by $\omega_{\mathrm{p}}(T_{\mathrm{max}}) = m$. Numerically, one finds $T_{\mathrm{max}}/m \simeq 5.74\ldots$. Of course, this is just a formal value; in order to obtain reasonable results, i.e. in order to satisfy (3.236), the actual temperature should be kept smaller than this maximum value.

Nevertheless, the numerical results given below confirm that the formal expansion for $T/m \gg 1$ is already appropriate for $T \sim m$, which justifies extracting physical conclusions from this analytical procedure.

For the remainder of the section, we additionally allow a further perturbation of the vacuum in the form of an external weak magnetic field, $B \ll B_{\mathrm{cr}} = m^2/e$. In terms of the invariants, a purely magnetic field implies that $\mathcal{G}, \mathcal{E} \to 0$, $\mathcal{F} \to (1/2)B^2$, where the latter expression is understood to be valid in the heat-bath rest frame. Owing to the simple form of the four-velocity vector $u^\mu = (1, 0, 0, 0)$ in this special frame, terms containing $F^{\mu\nu}u_\nu$ vanish, since this product is proportional to the electric field, which is assumed to be zero.

Taking these considerations into account, the light cone condition (3.217) is reduced to

$$0 = \left(\partial_{\mathcal{F}}\mathcal{L} - \partial_{\mathcal{E}}\mathcal{L} - \mathcal{F}\partial_{\mathcal{G}}^2\mathcal{L}\right)k^2 + \frac{1}{2}\left(\partial_{\mathcal{F}}^2 + \partial_{\mathcal{G}}^2\right)\mathcal{L}(Fk)^\nu(Fk)_\nu + 2\partial_{\mathcal{E}}\mathcal{L}(ku)^2. \tag{3.238}$$

We shall restrict the following investigations to experimentally more accessible parameter values, i.e. weak fields and low temperature (compared with the electron mass). Here we would like to answer the question of whether there is an additional induced velocity shift for the case of combined magnetic *and* thermal vacuum modifications. For each single case, the velocity shifts are known; for example, the polarization-summed velocity shift of light propagating in a magnetic background field is given by (3.12):

$$\delta v_B^2 = -\frac{22}{45}\frac{\alpha^2}{m^4} B^2 \sin^2\Theta, \tag{3.239}$$

where Θ denotes the angle between the propagation direction and the magnetic field. The velocity shift for the purely thermal part is found in (3.233). In order to find a contribution to the velocity shift for the combined case apart from the trivial sum of the single cases, we make use of the expansion of the thermal one-loop Lagrangian $\mathcal{L}^{1T}$ for weak fields and low temperature as given in (3.163). In the desired limit, the required terms, which are quadratic in the field invariants and dominant at low temperatures, read

$$\Delta\mathcal{L}_{BT}(\mathcal{F},\mathcal{G},\mathcal{E},T) = \frac{\alpha^2}{45}\left(\frac{8\mathcal{F}\mathcal{E}-8\mathcal{F}^2-13\mathcal{G}^2}{m^2T^2} + \frac{\pi}{2}\frac{8\mathcal{F}\mathcal{E}+\mathcal{G}^2}{mT^3}\right)e^{-m/T}.$$
$$\tag{3.240}$$

These terms can be inserted into (3.238) together with the classical, one-loop and purely thermal parts of the effective Lagrangian, $\mathcal{L}_M, \mathcal{L}^1$ and $\mathcal{L}^{1T}$. Introducing the phase velocity $v = k^0/|\boldsymbol{k}|$ again and omitting the terms of higher order in T/m and eB/m^2, we arrive at the following polarization sum rule describing the modification of light propagation in a QED vacuum at low temperature and in a weak magnetic background field:

$$v^2 = 1 - \frac{\alpha}{3}\sqrt{\frac{2}{\pi}}\sqrt{m/T}e^{-\frac{m}{T}} - \frac{22}{45}\frac{\alpha^2}{m^4}B^2\sin^2\Theta \tag{3.241}$$
$$-\frac{22}{45}\frac{\alpha^2}{m^4}\left[\left(\frac{\pi}{44}\frac{m^3}{T^3} - \frac{21}{22}\frac{m^2}{T^2}\right)e^{-m/T}B^2\sin^2\Theta\right.$$
$$\left.+\left(\frac{2\pi}{11}\frac{m^3}{T^3} + \frac{4}{11}\frac{m^2}{T^2}\right)e^{-m/T}B^2\right].$$

In the first line of this equation, we encounter the familiar classical, purely thermal and purely magnetic contributions. The second and third line contain the correction terms due to the simultaneous presence of a heat bath and a magnetic background field, indicating that the combination of both perturbations leads to more than a simple sum of the single effects.

Two-Loop Results

As is obvious from the preceding results, the one-loop contributions to the velocity shift in the low-temperature domain are exponentially damped by the electron mass. Hence, the most important contributions for $T \ll m$ can be expected from inserting the two-loop thermal Lagrangian (3.203) into the light cone condition.

Let us first consider the situation of a thermalized QED vacuum without an additional background field. In the low-temperature domain, this vacuum is characterized by the Lagrangian $\mathcal{L} = -\mathcal{F} + \mathcal{L}^{2T}$, where $-\mathcal{F}$ represents the classical Maxwell term. Inserting (3.205) and (3.206) into (3.219) leads us to

$$v^2 = \frac{1}{1 + 2(44/2025)\alpha^2\pi^2(T^4/m^4)}$$
$$\simeq 1 - 2\frac{44}{2025}\alpha^2\pi^2\frac{T^4}{m^4} + \mathcal{O}(T^8/m^8). \tag{3.242}$$

Note that there is no T^6 term, since $\mathcal{L}^{2T}|_{T^6}$ is at least quadratic in the field invariants. In (3.242), we have rediscovered the well-known velocity shifts for light propagation in a thermal background, as found in [120,157] via the two-loop polarization operator and in [20,60] via considering vacuum expectation values of field bilinears in a thermal background. The rederivation within the effective-action approach from first principles presented here can thus be viewed as an independent check of our calculations of $\mathcal{L}^{2T}$ and of the light cone condition as derived above.

But we can go one step further and take a weak external magnetic field additionally into account. The Lagrangian describing a thermal QED vacuum with a weak magnetic background field at finite temperature is given by $\mathcal{L} = -\mathcal{F} + \mathcal{L}^1 + \mathcal{L}^{2T}$, where $\mathcal{L}^1$ denotes the (zero-T) Heisenberg–Euler Lagrangian $\mathcal{L}^1$; inserting this into (3.238) gives, to lowest order in the parameters T and B,

$$v^2 = 1 - \frac{22}{45}\frac{\alpha^2}{m^4}B^2\sin^2\theta_B - 2\frac{44}{2025}\alpha^2\pi^2\frac{T^4}{m^4}$$
$$+ \frac{22}{45}\frac{\alpha^2}{m^4}\left(\frac{2^5 \times 37}{3^2 \times 5 \times 7 \times 11}\alpha\pi^3\frac{T^4}{m^4}\right)B^2(1 + \sin^2\theta_B). \tag{3.243}$$

The second and third terms are the well-known velocity shifts for purely magnetic [4,28] and purely thermal (cf. (3.242)) vacua, respectively. The last term describes a nontrivial interplay between these two vacuum modifications. The latter can best be elucidated in the various limits of the angle θ_B; for propagation orthogonal to the magnetic field, $\theta_B = \pi/2$, we obtain

$$v^2 = 1 - 2\frac{44}{2025}\alpha^2\pi^2\frac{T^4}{m^4} - \frac{22}{45}\frac{\alpha^2}{m^4}B^2\left(1 - (0.15...)\frac{T^4}{m^4}\right). \tag{3.244}$$

For propagation parallel to the magnetic field, $\theta_B = 0$, we find

$$v^2 = 1 - 2\frac{44}{2025}\alpha^2\pi^2\frac{T^4}{m^4}\left[1 - (0.96...)\left(\frac{eB}{m^2}\right)^2\right]. \tag{3.245}$$

Since T/m and eB/m^2 are considered to be small in each case, the corrections to the pure effects in the mixed situation are comparably small. Note that the mixed thermal and magnetic corrections always diminish the values of the velocity shift of the pure magnetic or thermal situation. Finally, let us remind the reader that the velocities given here hold for low-frequency light ($\omega \ll m$) only, and represent averages over the two possible polarization modes. While for the purely thermal case the polarization modes cannot be distinguished, a situation involving an electromagnetic field generally leads to birefringence owing to the existence of a preferred direction of the field lines.

Discussion

Low Temperature. At temperatures well below the fermion mass, the one-loop modification of the velocity of light as described by (3.241) vanishes exponentially, demonstrating the two-loop dominance. In the present "first-principles" calculation on the level of the two-loop effective action, we have reproduced the low-temperature velocity shift $\delta v = -(44\pi^2/2025)\alpha^2(T^4/m^4)$ (cf. (3.77)). From this fundamental point of view, we can now comment on the derivation of this shift in Sect. 3.3. The philosophy there was to calculate the velocity shifts in a (weak) purely electromagnetic background first, and then take thermal vacuum expectation values of the field bilinears, leading to the formula

$$v^2 = 1 - \frac{2}{3}(\partial_{\mathcal{F}}^2 + \partial_{\mathcal{G}}^2)\mathcal{L} \langle T^{00}\rangle, \tag{3.246}$$

where $\langle T^{00}\rangle^T = (\pi^2/15)T^4$ denotes the vacuum expectation value of the energy–momentum tensor of the modified vacuum. From the present viewpoint, the correctness of this approach arises from the special form of the low-temperature two-loop Lagrangian $\mathcal{L}^{2T}|_{T^4}$ as given in (3.203). Since (B.12)

$$\frac{1}{a^2 + b^2}(\partial_a^2 + \partial_b^2) = \partial_{\mathcal{F}}^2 + \partial_{\mathcal{G}}^2, \tag{3.247}$$

(3.203) can also be written as

$$\partial_{\mathcal{E}}\mathcal{L}^{2T}\Big|_{T^4} = \frac{2}{3}\langle T^{00}\rangle^T \frac{1}{2}(\partial_{\mathcal{F}}^2 + \partial_{\mathcal{G}}^2)\mathcal{L}^1. \tag{3.248}$$

Incidentally, (3.248) holds for arbitrary field strength, but for this line of argument, it is required for weak fields only. Inserting (3.248) into the correct light cone condition at finite temperature, i.e. (3.219), we obtain to lowest order

$$v^2 \simeq 1 - 2\partial_{\mathcal{E}}\mathcal{L} = 1 - \frac{2}{3}(\partial_{\mathcal{F}}^2 + \partial_{\mathcal{G}}^2)\mathcal{L} \langle T^{00}\rangle^T, \tag{3.249}$$

which is equal to the light cone condition deduced above for a thermal QED vacuum.

When the temperature increases, one finds that the two-loop dominance does not hold over the complete range of $T < m$. As can be discovered numerically, we find significant one-loop modifications of the phase velocity for comparatively small values of the temperature. The increasing influence of the one-loop term is due to the factor of $T^{-1/2}$ and, of course, the proportionality to α (and not α^2).

In Fig. 3.6, we compare the one-loop velocity shift with the two-loop T^4 behavior. Obviously, the one-loop contribution becomes dominant for comparatively small values of temperature, $T/m \geq 0.058$. Of course, this statement should be handled with care because here we are comparing a one-loop result including thermalized fermions with a two-loop result without thermalized fermions. Nevertheless, the increasing two-loop contribution from thermalized fermions will always be suppressed by a factor of α, which could only be compensated for by an unexpected conspiracy of numerical prefactors.

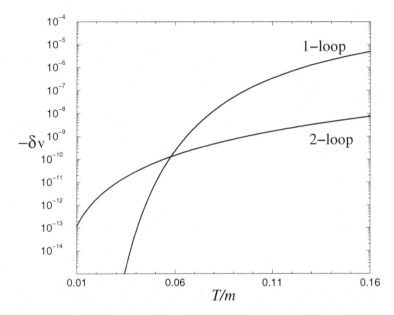

Fig. 3.6. Low-temperature velocity shift δv in units of the vacuum velocity $c = 1$ versus the dimensionless temperature scale T/m; the one-loop contribution exceeds the well-known two-loop result $\sim T^4/m^4$ for comparatively small values of temperature

Intermediate Temperature. In the present context, the results of the high-temperature expansion can be applied to values of temperature of the order of the electron mass, $T \sim m$, where plasma effects do not yet dominate the physical properties of the modified vacuum. At even higher temperatures, $T \gg m$, the e^+e^- gas constitutes a plasma in which soft photons simply no longer exist, owing to the appearance of a plasma frequency [113, 114]; then our formalism becomes meaningless, because the low-frequency modes simply do not propagate.

It is remarkable that a maximally possible velocity shift, to this order of calculation, which is simply given by $\delta v_{\text{max}}^2 = -\alpha/(3\pi)$ (Fig. 3.7) exists. Of course, this maximal velocity shift is only reached asymptotically and, therefore, strictly speaking, lies beyond the scope of the present formalism; nevertheless, the actual velocity shift already comes close to the maximum value in the intermediate-temperature domain where $T \sim m$ (Fig. 3.7). To complete the discussion, it should be mentioned that the inclusion of the contributions from (3.228) increases the negative velocity shift proportionally to $(\alpha^2/\pi^2) \ln(T/m)$ for $T \geq m$. For reasonable values of temperature, this contribution is in fact completely negligible.

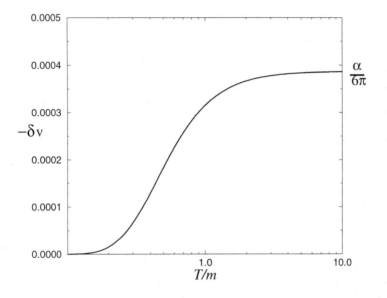

Fig. 3.7. High-temperature velocity shift $-\delta v$ in units of the vacuum velocity $c = 1$ versus the dimensionless temperature scale T/m; the maximum velocity shift $-\delta v = \alpha/(6\pi)$ is approached at comparatively low temperatures

Effective-Action Charge. In (3.225), we were able to formulate the thermally induced deformation of the light cone in terms of the vacuum expectation value of the energy–momentum tensor times a factor Q_T called the effective-action charge. The charge concept was introduced in Sect. 3.2.1 as a useful picture which provides for an intuitive understanding of this proportionality factor. Since any perturbed QED vacuum can be expected to control the magnitude of the velocity shift, even for vacuum modifications of high energy density, this factor Q_T has to decrease sufficiently fast for increasing energy scale parameters (in this case, the temperature). Therefore, the factor Q_T should exhibit a localized distribution in this parameter space.

For calculating the effective-action charge Q_T according to (3.226) for a thermalized QED vacuum, we not only consider the one-loop contribution from the thermalized fermions $\mathcal{L}^{1T}$, but also take the free photonic part $\mathcal{L}_\gamma = (\pi^2/45)T^4$ into account. Although the latter does not exert an influence on the velocity shift and drops out of (3.225), we have to include it in the present considerations in order to work with the complete vacuum expectation value of the energy–momentum tensor for the thermalized QED vacuum.

To lowest order in α/π, the formula for the effective-action charge (3.226) reduces to

$$Q_T \simeq \frac{2}{T}\frac{\partial_\varepsilon\mathcal{L}}{\partial_T\mathcal{L}} = \frac{2\partial_\varepsilon\mathcal{L}}{T\partial_T\mathcal{L}_\gamma + T\partial_T\mathcal{L}^{1T}}, \tag{3.250}$$

where $\partial_\varepsilon\mathcal{L}^{1T}$ is given in (3.227), and the terms in the denominator can be written as

$$T\partial_T\mathcal{L}_\gamma(T) = \frac{4\pi^2}{45}T^4,$$

$$T\partial_T\mathcal{L}^{1T}(T) = -\frac{2}{\pi^2}m^4\frac{T}{m}\sum_{n=1}^{\infty}\frac{(-1)^n}{n}K_3\left(\frac{m}{T}n\right). \tag{3.251}$$

Note that the appearance of $\partial_T\mathcal{L}_\gamma$ and $\partial_T\mathcal{L}^{1T}$ in the denominator of (3.250) corresponds to the appearance of a photonic and a fermionic part in the energy–momentum tensor. One usually expects the fermionic part to become important only for high temperature, $T \gg m$, where the fermions become ultrarelativistic.

On the one hand, we indeed find the expected localized behavior, as can be seen in Fig. 3.8: the effective-action charge vanishes for high as well as low temperatures; in between, it develops a maximum at $T/m \simeq 0.22$. On the other hand, it is interesting to note that the inclusion of the fermionic contributions $\sim T\partial_T\mathcal{L}^{1T}$ (solid line) in the denominator of (3.250) becomes important for values of the temperature close to the maximum of Q_T. This again indicates that the thermalization of the fermions becomes important even for comparatively low values of temperature.

Contrary to the cases discussed in Sect. (3.2), the effective-action charge is not centered at the origin. This is because the electron mass damps the fermionic thermal fluctuations exponentially for small values of temperature at the one-loop level. To complete the picture at the origin, we have to include the two-loop contribution $(22/45)(\alpha^2/m^4)$, corresponding to $2.6 \times 10^{-5}/m^4$ for T/m close to zero.

Concluding Remarks. First, we would like to stress that the one-loop contributions to the velocity shift as calculated in the present work do not fit into the scheme of the "unified formula" proposed in [120]; the latter connects the velocity shift with a shift of the vacuum energy density caused by external influences, with a universal constant coefficient as the proportionality

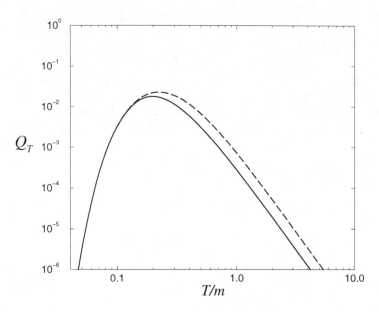

Fig. 3.8. One-loop contribution to the effective-action charge Q_T in units of $1/m^4$; for high temperature, Q_T decreases proportionally to $1/T^4$. The *solid line* corresponds to Q_T as given in (3.250); for the *dashed line*, the fermionic contributions $\sim T\partial_T\mathcal{L}^{1T}$ have been omitted. In the low-temperature limit, the effective-action charge at the one-loop level vanishes owing to the influence of a finite electron mass

factor. In the language of (3.225), this coefficient has to be identified with the effective-action charge Q_T, which in the present case is not constant at all but carries almost the complete physical information of the problem. This misfit indicates that the "unified formula" is an artifact of an approximation scheme rather than a fundamental principle [60].

Furthermore, it should be pointed out that the maximum velocity shift, $-\delta v_{max}^2 = \alpha/(3\pi)$, cannot be viewed as an experimentally significant limiting value, since light propagation will be increasingly dominated by plasma effects for increasing temperature for $T \sim m$. Nevertheless, the existence and amount of such a maximum shift are at least interesting from a theoretical viewpoint, since they characterize the classically forbidden interaction between the modified vacuum and a photon which is exposed to the effects of vacuum polarization. If there were no collective excitations constituting a plasmon at high temperature, then QED would not allow an arbitrarily strong influence of a heat bath on the propagation of light for reasonable values of the temperature. Of course, for an extremely high temperature, the logarithmic corrections from $\partial_{\mathcal{F}}\mathcal{L}$ in (3.219) would also significantly increase the negative velocity shift, slowing down the speed of light; for example, another shift of $\alpha/3\pi$ would be reached at $T/m \simeq 6.4 \times 10^7$.

3.6.3 Photon Splitting

In Sect. 3.4, we discussed the basics of photon splitting for low-frequency photons in a weak magnetic field, for which the effective-action approach is most appropriate. There,[20] we showed that the hexagon graph (three propagating photons and three couplings to the external field) represents the first nonvanishing contribution; in particular, the box graph (one coupling to the external field) vanishes owing to the invariant structure of the splitting matrix element.

At finite temperature, the situation changes qualitatively, since there is the additional invariant $\mathcal{E}$ of the external field. In calculating the splitting matrix element $\mathcal{M}$ according to (3.102), we encounter the following additional derivatives of $\mathcal{E}$:

$$\frac{\partial \mathcal{E}}{\partial F_r^{\pm}} = \mp \frac{1}{2}\left(F_r^{+} - F_r^{-}\right), \quad \frac{\partial^2 \mathcal{E}}{\partial F_r^{\mp} \partial F_s^{\pm}} = \frac{1}{2}\delta_{rs}, \quad \frac{\partial^2 \mathcal{E}}{\partial F_r^{\pm} \partial F_s^{\pm}} = -\frac{1}{2}\delta_{rs}, \tag{3.252}$$

which should be read side by side with the derivatives of $\mathcal{F}$ in (3.105). The essential difference is the nonvanishing right-hand side of the second equality in (3.252), which is responsible for the nonvanishing of the box graph for photon splitting at finite temperature.

The question of thermally induced photon splitting was first investigated by Elmfors and Skagerstam [73], with the aid of the thermal one-loop effective QED Lagrangian; their studies were motivated by the fact that a vacuum may be a bad approximation to the surroundings of some astrophysical compact objects, but a thermalized environment at zero or finite density might be more appropriate. In the following, we shall complete this one-loop study of thermally induced photon splitting with the dominant low-temperature contributions stemming from the two-loop process. Here, we again concentrate on the splitting process $\perp \rightarrow \|_1 + \|_2$, where a photon with its electric field vector orthogonal ($\perp$) to the plane spanned by the external magnetic field and the propagation direction splits into two photons with their electric field vectors within ($\|$) that plane.[21] This is the only allowed process when dispersion effects are taken into account.

With reference to Sect. 3.4, we recall that the splitting amplitude is obtained by attaching the external photon legs to the fermion loop, i.e. by differentiating the effective action (which is represented by the loop) three times with respect to the fields and then contracting the result with the field strengths of the photons involved. Following the lines of (3.102)–(3.110), the thermal amplitude arising from the box graph finally yields

$$\mathcal{M}(\perp \rightarrow \|_1 + \|_2) = 2\omega\omega_1\omega_2\, B \sin\theta_B\, \partial_{\mathcal{E}\mathcal{F}}\mathcal{L}, \tag{3.253}$$

---

[20] For notation and conventions concerning photon splitting, see Sect. 3.4.

[21] Note that Adler's definitions of the $\|$, $\perp$ modes are based on the direction of the magnetic field vector of the photon and thus are opposite to ours.

where $\omega, \omega_1, \omega_2$ denote the frequencies of the incoming and two outgoing photons, respectively, and θ_B again represents the angle between the propagation direction and the magnetic field. From the splitting amplitude, we obtain the absorption coefficient κ via (3.111):

$$\kappa = \frac{1}{32\pi\omega^2} \int_0^\omega d\omega_1 \int_0^\omega d\omega_2\, \delta(\omega - \omega_1 - \omega_2)\, \mathcal{M}^2. \tag{3.254}$$

Inserting (3.253) for the thermal splitting amplitude into (3.254) leads us to

$$\frac{\kappa}{m} = \frac{1}{2^6 \times 3 \times 5\pi^2} \left(\frac{eB}{m^2}\right)^2 \sin^2\theta_B \left(\frac{\omega}{m}\right)^5 (\partial_{\mathcal{EF}}\mathcal{L})^2\, m^8. \tag{3.255}$$

Here, we encounter the typical $(\omega/m)^5$ dependence of the photon-splitting absorption coefficient for low-frequency photons. The appearance of the magnetic field to the second power is directly related to the fact that the box graph exhibits only one coupling to the external field. In contrast, Adler's result for the absorption coefficient at zero temperature arising from the hexagon graph reads (see (3.114))

$$\frac{\kappa^{T=0}}{m} = \frac{13^2}{3^5 \times 5^3 \times 7^2} \frac{\alpha^3}{\pi^2} \left(\frac{eB}{m^2}\right)^6 \sin^6\theta_B \left(\frac{\omega}{m}\right)^5. \tag{3.256}$$

Here, the three couplings to the external magnetic field produce a B^6 dependence of the absorption coefficient. Therefore, any finite-temperature contribution will exceed the zero-temperature one for small enough magnetic fields; but, of course, the absorption coefficient may then become very tiny.

In order to obtain the one-loop and two-loop absorption coefficients for thermally induced photon splitting at low temperature, the derivatives of the corresponding Lagrangian are required in (3.255):

$$\partial_{\mathcal{EF}}\mathcal{L}^{1T} = \left[\frac{8\alpha^2}{45}\left(\frac{m}{T}\right)^2 + \frac{4\pi\alpha^2}{45}\left(\frac{m}{T}\right)^3\right] \frac{e^{-m/T}}{m^4}, \tag{3.257}$$

$$\partial_{\mathcal{EF}}\mathcal{L}^{2T} = \left[-\frac{2^6 \times 37\alpha^3\pi^3}{3^4 \times 5^2 \times 7}\left(\frac{T}{m}\right)^4 + \frac{2^{14}\alpha^3\pi^5}{3^5 \times 5 \times 7^2}\left(\frac{T}{m}\right)^6\right] \frac{1}{m^4}, \tag{3.258}$$

where we have made use of the results of (3.163) for the low-temperature/ weak-field approximation of the one-loop Lagrangian $\mathcal{L}^{1T}$, and employed (3.205) and (3.206) for the two-loop Lagrangian. Obviously, inserting the two-loop terms from (3.258) into (3.255) leads to a power-law dependence of the absorption coefficient $\sim T^8/m^8$, while the one-loop terms from (3.257) imply an exponential mass damping $\exp(-2m/T)$ for $T \to 0$.

As mentioned above, photons of frequency below the pair-production threshold not only are exposed to splitting at finite temperature, but can also scatter directly with the plasma of electrons and positrons. Following [73], the absorption coefficient for the Compton process is given by

$$\frac{\kappa_C}{m} = \frac{\sigma_C}{m} \frac{2}{\pi^2} \int\limits_0^\infty dp \frac{p^2}{e^{\omega_e/T} + 1},$$ (3.259)

where ω_e denotes the fermion energy $\omega_e = \sqrt{p^2 + m^2}$, and the cross section σ_C for unpolarized photons at $\omega/m \simeq 1$ is approximately given by

$$\sigma_C \simeq \frac{4\pi\alpha^2}{3m^2}.$$ (3.260)

Although $\omega/m \simeq 1$ formally represents the maximal limit of validity of our constant-field approximation for the effective action, we shall continue to consider photons of such a frequency in the following, since, on the one hand, this circumvents a suppression of the absorption coefficients by the common factor $(\omega/m)^5$ and, on the other hand, it has been shown for the hexagon graph in [4] that the difference between calculations for $\omega/m = 1$ and $\omega/m \sim 0$ is negligible for weak magnetic fields.

Finally, we have to consider another scattering process which arises from the presence of a heat bath: photon–photon scattering between the propagating photon and the black-body radiation of the thermal background. We estimate the absorption coefficient for this process by

$$\frac{\kappa_{\gamma\gamma}}{m} = \frac{\sigma_{\gamma\gamma} n_\gamma}{m},$$ (3.261)

where n_γ denotes the density of photons and is given by

$$n_\gamma = 2 \int \frac{d^3 p}{(2\pi)^3} \frac{1}{e^{\sqrt{p^2}/T} - 1} = \frac{2\zeta(3)}{\pi^2} T^3.$$ (3.262)

Here we encounter the Riemann ζ function, for which $\zeta(3) \simeq 1.202$. The total polarization-averaged cross section for photon–photon scattering at low frequencies, as can be obtained, for example, from the Heisenberg–Euler Lagrangian [76], reads

$$\sigma_{\gamma\gamma} = \frac{973}{10125} \frac{\alpha^2}{\pi} \frac{\alpha^2}{m^2} \left(\frac{\omega_{\mathrm{CM}}}{m}\right)^6,$$ (3.263)

where ω_{CM} denotes the frequency of the two photons in the center-of-mass (CM) frame. In order to determine ω_{CM}, we first have to find the mean frequency at temperature T. Averaging over the thermal probability distribution for the photons, we find the mean value $\omega_T = [\pi^4/30\zeta(3)]T \simeq 2.701T$. According to relativistic kinematics, the average value of the CM frequency ω_{CM} is given by $\omega_{\mathrm{CM}} = \sqrt{\omega\omega_T/2} \simeq 1.16\sqrt{T\omega}$, where we have averaged over the propagation direction of the thermal photons. Putting everything together, we obtain the following for the absorption coefficient for photon–photon scattering with the thermal background:

$$\frac{\kappa_{\gamma\gamma}}{m} = \frac{7 \times 139}{2^5 \times 3^7 \times 5^6} \frac{\pi^9}{\zeta(3)^2} \alpha^4 \left(\frac{T}{m}\right)^6 \left(\frac{\omega}{m}\right)^3$$

$$\simeq 5.21 \times 10^{-11} \left(\frac{T}{m}\right)^6 \left(\frac{\omega}{m}\right)^3. \tag{3.264}$$

Since the average frequency of the heat-bath photons is proportional to the temperature, this formula becomes invalid for $T \sim m$ and above, because we have employed the low-frequency cross section in (3.261).

It is already clear from a qualitative viewpoint that there must be a domain where the two-loop splitting process exceeds the one-loop and Compton contributions, owing to the power-law dependence on the temperature. But since $\kappa^{2T} \sim (T/m)^8$ and $\kappa_{\gamma\gamma} \sim (T/m)^6$, the two-loop contribution will eventually be surpassed by the photon–photon scattering for $T \to 0$.

However, quantitative results can only be revealed by numerical studies. In fact, as shown in Fig. 3.9a, the two-loop contribution is completely irrelevant for those parameter values which may be appropriate for a neutron star system, and which are close to the upper bound of validity of our approximation: $eB/m^2 = 0.2$, $\omega/m = 1$, $\sin\theta_B = 1$ and $T/m = 0.05$–0.1. Even the one-loop contribution is small compared with the zero-temperature result; but both contributions are negligible compared with the Compton process.

Concentrating on the relative strengths of the thermal splitting processes, the one-loop contribution loses its major role for $T/m \leq 0.041$, where its exponential decrease causes it to be surpassed by the two-loop power law.

In order to find a domain in which the two-loop splitting wins out over the zero-temperature process, we have to look at smaller values of the magnetic field strength; for example, at a value of temperature $T/m = 0.025$, the two-loop process exceeds the zero-temperature one for $eB/m^2 \leq 2.1 \times 10^{-4}$. Since this means a more moderate field strength, the absorption coefficient naturally becomes very small: $\kappa/m \sim 10^{-34}$–10^{-33}. Hence, in order for us to be able to measure the splitting rate, the extension of the magnetic field in which the photon propagates must be comparable with galactic scales.

Finally, we have plotted the Compton and photon–photon absorption coefficients, κ_C and $\kappa_{\gamma\gamma}$, and the two-loop coefficient κ^{2T} for a weak magnetic field $eB/m^2 = 10^{-4}$ and $T/m = 0.001$–0.1 in Fig. 3.9b. Obviously, the Compton process loses its dominant role for $T/m \leq 0.03$; in this range, the absorption coefficient is governed by the photon–photon scattering as long as the temperature does not become so small that only the zero-temperature amplitude remains. As is also visible in Fig. 3.9b, the two-loop contribution does not exceed that of the photon–photon process, owing to the weaker temperature dependence of the latter. Hence, we may say in summary that the photon absorption coefficient in the low-temperature domain is dominated either by the zero-temperature contribution for strong magnetic fields or by the photon–photon scattering with the thermal background for weak fields. So the two-loop contribution always belongs to the top flight but is never ranked first.

In the intermediate-temperature domain, $T \sim m$, the one-loop contribution can exceed the zero-temperature part (without any physical consequence, because Compton scattering greatly dominates). But for larger values of temperature, the thermal constribution cancels the zero-temperature part owing to coherent summation [73]. This reflects the suppression of vacuum polarization effects for $T \gg m$ as described in (3.164).

In order to account for realistic astrophysical systems, it is necessary to include a finite chemical potential. Initial estimates of the effect of a chemical potential can be found in [73] to one-loop order, where signs have been found that a finite chemical potential of $\mu \simeq m$ may induce an increase of the thermal splitting amplitude at low temperatures. The present work shows that an investigation of these systems should take the two-loop contributions into account in order to settle this question properly. Some initial progress in this direction has been achieved in [168], in which a two-loop calculation with an external magnetic field at finite density has been performed.

Let us conclude this section with the remark that in order to obtain the sum of the zero-temperature and thermal contributions to the photon splitting absorption coefficient, the amplitudes must be summed coherently, since the final states of the processes coincide, and a thermal vacuum with a constant background field does not provide a mechanism of decoherence. While the zero-temperature amplitude and the thermal one-loop amplitude are strictly positive for $T \ll m$, the T^4 term in (3.258) contributes with a negative sign. Hence, an exceptional curve in the parameter space of eB/m^2 and T/m exists where the thermal two-loop amplitude interferes with the thermal one-loop and zero-temperature amplitudes destructively so that the photon splitting vanishes.

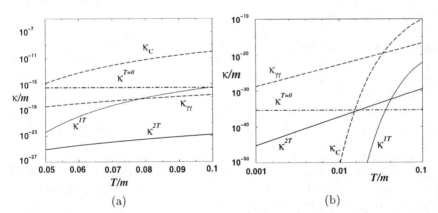

Fig. 3.9. Absorption coefficient κ in units of the electron mass versus temperature T in units of the electron mass. In (**a**), the various contributions are plotted for parameter values of a realistic astrophysical system ($eB/m^2 = 0.2$, $\omega/m = 1 = \sin\theta_B$). In (**b**), the parameters are chosen in such a way that the dominance of the two-loop contribution over the one-loop contribution and the Compton process is revealed ($eB/m^2 = 10^{-4}$, $\omega/m = 1 = \sin\theta_B$); the photon–photon scattering contribution cannot be overtaken in the low-temperature limit

3.6.4 Pair Production

In Sect. 3.5.5, we reported on the futile search for thermally induced pair production in electric fields at the one-loop level, which is absent by construction. With the aid of the thermal two-loop Lagrangian, we can now show that an imaginary part nevertheless exists, and thermally induced pair production can become an important process for certain parameter values. As already mentioned in the introduction of Sect. 3.6, drawing conclusions from an imaginary part of the thermal effective action about pair production is not as immediate and straightforward as at zero temperature, since the presence of an electric pair-producing field and the thermal-equilibrium assumption which is inherent to our approach contradict each other.

In the following, we simply assume that on the one hand, the timescale of pair production is much shorter than the timescale of the depletion of the electric field so that dynamic back-reactions can be neglected (this assumption is familiar from the zero-temperature Schwinger formula). On the other hand, we also assume that the state of the plasma can be appropriately approximated by a thermal equilibrium although it is exposed to an electric field. Whether the assumption of thermal equilibrium is justified in concrete experimental situations, such as heavy-ion collisions, is still under discussion.

Recently, pair production has been studied with the aid of a quantum kinetic equation (including nonhomogeneous electric-field configurations, back-reactions and collisions), revealing the non-Markovian character of the creation process [145]. In these works, the Schwinger formula is rediscovered in the low-density limit for constant fields. We expect that our results hold in the same limit at finite temperature.

Let us now turn to the computation of the imaginary part of the two-loop thermal effective action for external electric fields. For this, we concentrate on the T^4 contribution as given in (3.202). For purely electric fields, $a \to 0$, $b \to E$, $\mathcal{E} + \mathcal{F} \to (1/2)E^2$, this reads

$$\mathcal{L}^{2T}(E)\Big|_{T^4} = -\frac{\alpha\pi}{90} T^4 \int_0^\infty \frac{dz}{z} \, e^{-i(m^2/eE)z} \left(\frac{1}{3} z \coth z + \frac{1 - z \coth z}{\sinh^2 z}\right), \quad (3.265)$$

where we have substituted $z = eEs$. For reasons of convenience, it is useful to define the symbol $\eta = eE/m^2$, which denotes the dimensionless ratio between the electric field and the critical field strength $E_{cr} = m^2/e$. Integrating the $1/\sinh^2 z$ term by parts leads us to

$$\mathcal{L}^{2T}(E)\Big|_{T^4} = -\frac{\alpha\pi}{90} T^4 \lim_{\epsilon \to 0}\left\{\frac{1}{2\epsilon^2} + \frac{1}{2} + \frac{1}{4\eta^2}\right. \qquad (3.266)$$

$$\left. + \int_\epsilon^\infty dz \, e^{-iz/\eta}\left(\frac{1}{3} - \frac{i}{\eta z} - \frac{1}{z^2} + \frac{1}{2\eta^2}\right)\coth z\right\}.$$

Here, it should be pointed out that the isolated pole in the first term in the curly brackets does not signal a divergence, but simply cancels the pole at the lower bound of the integral; the whole expression is still finite. Our aim is to evaluate the imaginary part of (3.266); for this, the behavior of the integral at the lower bound is of no interest. An imaginary part $\mathrm{Im}\, \mathcal{L}^{2T}(E)|_{T^4}$ arises from the poles of the $\coth z$ term on the imaginary axis at $z = \pm i n\pi$, $n = 1, 2, \ldots$.

Decomposing the exponential function into $\cos + \mathrm{i}\sin$, it becomes obvious that the imaginary parts of the integrand are even functions of z, while the real parts are odd. Thus, extending the integration interval from $-\infty$ to ∞ exactly cancels the real parts and simply doubles the imaginary parts. We finally obtain

$$\mathrm{Im}\, \mathcal{L}^{2T}(E)\bigg|_{T^4} = -\frac{\alpha\pi}{90}\frac{T^4}{2\mathrm{i}}\int_{-\infty}^{\infty} dz\, e^{-\mathrm{i}z/\eta}\left(\frac{1}{3} - \frac{\mathrm{i}}{\eta z} - \frac{1}{z^2} + \frac{1}{2\eta^2}\right)\coth z. \quad (3.267)$$

Now we can close the contour in the lower complex half plane, which is in agreement with the causal prescription $m^2 \to m^2 - \mathrm{i}\epsilon$. The value of the integral is then simply given by the sum of the residues of the $\coth z$ poles at $z = -\mathrm{i}\pi n$, $n = 1, 2, \ldots$. Hence, we arrive at

$$\mathrm{Im}\, \mathcal{L}^{2T}(E)\bigg|_{T^4} = \frac{\alpha\pi^2}{90}T^4\sum_{n=1}^{\infty} e^{-n\pi/\eta}\left(\frac{1}{3} + \frac{1}{n\pi\eta} + \frac{1}{n^2\pi^2} + \frac{1}{2\eta^2}\right), \quad (3.268)$$

where $\eta = eE/m^2$. Equation (3.268) represents our final result for the imaginary part of the thermal effective QED action at low temperature, and should be read side by side with Schwinger's one-loop result:

$$\mathrm{Im}\, \mathcal{L}^1(E) = \frac{m^4}{8\pi^3}\eta^2\sum_{n=1}^{\infty}\frac{e^{-n\pi/\eta}}{n^2}. \quad (3.269)$$

The sum in (3.268) and (3.269) can be evaluated analytically; but here, it should be sufficient to consider the limiting cases of weak and strong electric fields.

In the weak-field limit, i.e. for small values of η, the sum over n in (3.268) is dominated by the first term, $n = 1$. Furthermore, of the terms in parentheses, it is the last term which is the most important one. These considerations then lead us to

$$\mathrm{Im}\, \mathcal{L}^{2T}(eE \ll m^2) \simeq \frac{\alpha\pi^2}{180}T^4\frac{e^{-\pi/\eta}}{\eta^2}. \quad (3.270)$$

Combining this with the weak-field approximation of (3.269), we obtain roughly, for the total imaginary part of the effective Lagrangian

$$\mathrm{Im}\, \mathcal{L}(eE \ll m^2) = m^4 e^{-\pi/\eta}\left(\frac{\eta^2}{8\pi^3} + \frac{\alpha\pi^2}{180}\frac{1}{\eta^2}\frac{T^4}{m^4}\right)$$

$$\simeq m^4 e^{-\pi/\eta}\left(4\times 10^{-3}\eta^2 + 4\times 10^{-4}\frac{T^4/m^4}{\eta^2}\right). \quad (3.271)$$

For example, for $T/m \simeq 0.1$, where the present low-temperature approximation should still be appropriate, the thermal contribution can be neglected for $\eta \geq 0.1$; both contributions become roughly equal for $\eta \simeq 0.056$ (and $T/m = 0.1$). For weaker fields and $T/m \simeq 0.1$, the thermal contribution becomes the dominant one.

In the opposite limit, where $\eta \gg 1$, i.e. for strong electric fields beyond the critical field strength, the $1/3$ in the parentheses dominates the expression in (3.268), which then gives

$$\mathrm{Im}\,\mathcal{L}^{2T}(eE \gg m^2) = \frac{\alpha\pi^2}{270}\,T^4\sum_{n=1}^{\infty}\left(e^{-\pi/\eta}\right)^n = \frac{\alpha\pi^2}{270}\,T^4\,\frac{e^{-\pi/\eta}}{1-e^{-\pi/\eta}}$$

$$= \frac{\alpha\pi}{270}\,T^4\,\eta + \mathcal{O}(\eta^0). \tag{3.272}$$

Together with the strong-field approximation of the Schwinger formula, this gives

$$\mathrm{Im}\,\mathcal{L}(eE \gg m^2) = m^4\,\eta\left(\frac{\eta}{48\pi} + \frac{\alpha\pi}{270}\,\frac{T^4}{m^4}\right)$$

$$\simeq m^4\,\eta\left(6.6\times10^{-3}\eta + 8.5\times10^{-5}\frac{T^4}{m^4}\right). \tag{3.273}$$

Since (3.273) is valid for $\eta \gg 1$ and $T/m \ll 1$, the low-temperature contribution to $\mathrm{Im}\,\mathcal{L}(E)$ can be neglected for strong electric fields. Similarly to the case of strong magnetic fields, we find that the nonlinearities of the pure (zero-T) vacuum polarization exceed the polarizability of the thermally induced real plasma by far in the strong-field limit.

Nevertheless, in the limit of weak electric fields, thermal effects can increase the pair-production probability $P = 1 - \exp[-2\,\mathrm{Im}\,\mathcal{L}(E)]$ significantly, as was shown in (3.271). Of course, for such values of η, the total imaginary part is very small owing to the inverse power of η in the exponential.

Since we did not consider thermalized fermions, our approach is not capable of describing high-temperature pair production, which would be desirable for the forthcoming heavy-ion collision experiments. However, as can be read off from our results for light propagation and photon splitting, extrapolating the power-law behavior to higher temperature scales of $T \sim m$ or even $T/m \gg 1$ overestimates any possible two-loop contribution by far, since, for these values of temperature, the one-loop contribution can be expected to be the dominant one. The latter increases at most logarithmically with T.

Therefore, it is reasonable to assume that the pair-production probability also increases at most logarithmically with T. In view of these considerations, a power-law growth as suggested in [82, 83, 121] does not appear plausible. Of course, in order to decide this question, the two-loop calculation has to be carried out for arbitrary values of temperature.

3.6.5 Discussion

In the preceding subsections, we calculated the thermal two-loop contribution to the effective QED action for arbitrary constant electromagnetic fields in the low-temperature limit, $T/m \ll 1$. Contrary to the usual loop hierarchy in a perturbation theory with small coupling, the thermal two-loop part is found to be dominant over the thermal one-loop part in the low-temperature limit, since the former exhibits a power-law behavior in T/m, while the latter is exponentially suppressed by factors of $\exp(-m/T)$. The physical reason behind this is that the one-loop approximation does not involve virtual photons, which, because they are massless, can be more easily excited at low temperatures than massive fermions; thus, the one-loop approximation should be rated as an inconsistent truncation of finite-temperature QED for T much below the electron mass.

The power-law dependence of the thermal effective action to two loops starting with T^4/m^4, implies a *two-loop dominance* in the low-energy domain of thermal QED, which holds up to roughly $T/m \simeq 0.05$.

In the case of light propagation at finite temperature, this two-loop dominance has been known for some time from studies of the polarization tensor [120, 157]. Moreover, in the case of QED in a Casimir vacuum such as the parallel-plate configuration, the two-loop dominance is very natural and well known, since the fermions are not considered to be subject to the periodic boundary conditions anyway. This gives rise to a nontrivial check of our results, since Casimir and finite-temperature calculations highly resemble each other. Replacing, as usual, T by $(2a)^{-1}$ in (3.270) for the weak-field limit of the imaginary part of the effective Lagrangian, where a denotes the separation of the Casimir plates, we obtain

$$\mathrm{Im}\, \mathcal{L}^{2a}(E)\Big|_{a^{-4}} = \frac{\pi e^2}{2^8 \times 45} \frac{1}{a^4} \left(\frac{m^2}{eE}\right)^2 e^{-\pi m^2/eE}, \tag{3.274}$$

which agrees precisely with the findings of [142] for the Casimir corrections to the Schwinger formula.[22]

In order to illustrate the two-loop dominance, we studied light propagation and photon splitting in a weak magnetic background at low temperature. Since we are dealing with the two-loop level, the effects considered here are naturally very tiny, and a significant influence on, for example, photon physics near astrophysical compact objects appears not to be very probable. One should, rather, take a closer look at photon physics on large, galactic scales.

Furthermore, we calculated the imaginary part of the thermal two-loop effective action for electric background fields at low temperature. Under mild

---

[22] Actually, (3.274) agrees with the findings of [142] except for a global sign; however, as was pointed out by one of the authors of [142] in a footnote to [143], the expression in [142] is wrong by a minus sign, which saves the day.

assumptions, this result can be related to a thermally induced production probability of electron–positron pairs. In particular, in the weak-field limit, the thermal contribution has a significant influence on the production rate. Since no thermal one-loop imaginary part exists, any finite two-loop result automatically dominates at any temperature scale.

In the cases of light propagation and photon splitting, the loop hierarchy is restored above $T/m \simeq 0.05$. Even at this comparatively low value of temperature, the thermal excitation of the fermions begins to compete with that of the virtual photon. Hence, a calculation of the two-loop thermal Lagrangian for intermediate or high temperatures would appear as an imposition, were it not for the high-temperature pair-production probability, which is beyond the range of the one-loop approximation and of great interest for, for example, heavy-ion collisions.

4. QED in Two Spatial Dimensions

We briefly discuss general, though unfamiliar, aspects of 2+1-dimensional QED including a Chern–Simons term, distinguishing between a one-fermion (irreducible) and a two-fermion (reducible) formulation of the theory; each formulation exhibits different symmetry properties with respect to, for example, parity or chiral symmetry. The quantum aspects of these symmetries are investigated in detail to the one-loop level; we describe, as examples, the formation of a magnetically induced chiral condensate in the parity-preserving reducible formulation, and the perturbative generation of a parity-odd Chern–Simons term in the irreducible formulation. The relation between symmetry-breaking patterns and fermionic zero modes is outlined.

Furthermore, we perform a derivative expansion of the QED_{2+1} effective action for inhomogeneous fields, employing heat-kernel methods. As a by-product, we obtain the 2+1-dimensional analogue of the Uehling potential.

Models in relativistic field theory in low-dimensional space–time have been a subject of considerable interest over the whole history of quantum field theory. In the first place, they serve as "toy" models for gaining a better understanding of the rich collection of phenomena in field theory. In some cases, these models are even exactly soluble, such as the Schwinger model [149] and the Gross–Neveu model [93] in 1+1 dimensions. But beyond this pedagogical interest, 2+1-dimensional models in particular have been successfully applied to real physical systems. A prominent example is the application of 2+1-dimensional gauge theories to planar condensed-matter phenomena, such as the fractional quantum Hall effect [6, 108] and high-temperature superconductivity [65, 119, 151]. Finally, three-dimensional gauge theories can be matched to the high-temperature limit of four-dimensional ones, which is known as "temperature-induced dimensional reduction".

Our intention is to investigate 2+1-dimensional quantum electrodynamics with a classical background field in the effective-action approach. Similarities to the real 3+1-dimensional world are pointed out, but we shall mainly focus on the differences, i.e. the theoretical novelties of the lower-dimensional theory.

4.1 General Features of QED$_{2+1}$

4.1.1 Classical Properties

The Lagrangian for the abelian U(1) gauge theory for one (irreducible) fermion field reads

$$\mathcal{L} = \mathcal{L}_G + \mathcal{L}_{Gm} + \mathcal{L}_F + \mathcal{L}_{Fm} + \mathcal{L}_I, \tag{4.1}$$

where the various gauge field and fermion contributions are given by

$$\mathcal{L}_G = -(1/4)F^{\mu\nu}F_{\mu\nu}, \qquad \mathcal{L}_{Gm} = (\mu/4)\epsilon^{\mu\nu\alpha}F_{\mu\nu}A_\alpha,$$

$$\mathcal{L}_F = -\bar{\psi}\left(-i\gamma^\mu\partial_\mu\right)\psi, \qquad \mathcal{L}_{Fm} = -m\bar{\psi}\psi, \tag{4.2}$$

$$\mathcal{L}_I = e\bar{\psi}\gamma^\mu\psi A_\mu.$$

Note that the coupling constant e has dimension [mass]$^{1/2}$. The Euler–Lagrange equations of motion are given by

$$\partial_\mu F^{\mu\nu} + \frac{\mu}{2}\epsilon^{\nu\alpha\beta}F_{\alpha\beta} = J^\nu \equiv e\bar{\psi}\gamma^\nu\psi, \tag{4.3}$$

$$(-i\partial\!\!\!/ - eA\!\!\!/ + m)\psi = 0. \tag{4.4}$$

These are obviously invariant under gauge transformations:

$$A_\mu \to A_\mu + \partial_\mu\Lambda, \qquad \psi \to e^{ie\Lambda}\psi. \tag{4.5}$$

Note that the major difference from 3+1-dimensional QED is marked by the mass term for the gauge field $\mathcal{L}_{Gm}$. From (4.3), we can indeed read off that the parameter μ denotes a mass for the gauge field. Since the Lagrangian $\mathcal{L}_{Gm}$ changes by only a total derivative under gauge transformations,

$$\mathcal{L}_{Gm} \to \mathcal{L}_{Gm} + \partial_\alpha\left(\frac{\mu}{4}\epsilon^{\alpha\mu\nu}F_{\mu\nu}\Lambda\right), \tag{4.6}$$

even the quantum theory is gauge-invariant.[1] Owing to the topological origin of $\mathcal{L}_{Gm}$, which is the Chern–Simons secondary characteristic class, the theory is often called "topologically massive spinor electrodynamics" [52].

The Dirac algebra can be realized in a two-dimensional representation:

$$\gamma^0 = \sigma^3, \qquad \gamma^1 = i\sigma^1, \qquad \gamma^2 = i\sigma^2,$$

$$\gamma^\mu\gamma^\nu = -g^{\mu\nu} + i\epsilon^{\mu\nu\alpha}\gamma_\alpha, \qquad g^{\mu\nu} = \mathrm{diag}(-1,1,1). \tag{4.7}$$

The σ^i's are the Pauli matrices, and ψ represents a two-component spinor.

To elucidate the difference between the two-component spinor theories and the usual Dirac spinor theory, it is helpful to recall the discrete symmetries of the lower-dimensional theory [52, 127]. Charge conjugation,

---

[1] In the nonabelian version, the analogue to $\mathcal{L}_{Gm}$ additionally changes by an integer value under large gauge transformations; hence, gauge invariance enforces a topological quantization of the mass parameter μ in order to leave $\exp i \int dx\, \mathcal{L}$ invariant. This is not the case in the abelian theory.

$$\mathsf{C}A_\mu \mathsf{C}^{-1} = -A_\mu, \qquad \mathsf{C}\psi \mathsf{C}^{-1} = \sigma^1 \psi^\dagger, \tag{4.8}$$

leaves the equations of motion invariant. This is not the case for the parity transformation and time inversion. In 2+1 dimensions, parity corresponds to inverting one axis, say the x axis, since inversion of both axes could be undone by a rotation by π: $\mathsf{P}(x,y) = (x,y)_\mathsf{P} = (-x,y)$. The corresponding operation on the two-component spinor and on the gauge field is

$$\mathsf{P}\psi(t,\boldsymbol{r})\mathsf{P}^{-1} = \sigma^1 \psi(t,\boldsymbol{r}_\mathsf{P}), \tag{4.9}$$
$$\mathsf{P}A^0(t,\boldsymbol{r})\mathsf{P}^{-1} = A^0(t,\boldsymbol{r}_\mathsf{P}),$$
$$\mathsf{P}A^1(t,\boldsymbol{r})\mathsf{P}^{-1} = -A^1(t,\boldsymbol{r}_\mathsf{P}), \tag{4.10}$$
$$\mathsf{P}A^2(t,\boldsymbol{r})\mathsf{P}^{-1} = A^2(t,\boldsymbol{r}_\mathsf{P}).$$

Finally, time inversion leads to

$$\mathsf{T}\psi(t,\boldsymbol{r})\mathsf{T}^{-1} = \sigma^2 \psi(-t,\boldsymbol{r}), \tag{4.11}$$
$$\mathsf{T}A^0(t,\boldsymbol{r})\mathsf{T}^{-1} = A^0(-t,\boldsymbol{r}),$$
$$\mathsf{T}\boldsymbol{A}(t,\boldsymbol{r})\mathsf{T}^{-1} = -\boldsymbol{A}(-t,\boldsymbol{r}). \tag{4.12}$$

Operating with P as well as T on the equations of motion reverses the sign of both of the masses μ and m. The same conclusion holds for the mass terms in the Lagrangian, which are odd under the parity transformation. Therefore, the one-fermion (irreducible) formulation of QED$_{2+1}$ is not parity-invariant as long as there are finite mass terms. Of course, the theory is symmetric under CPT, which corresponds to a double sign flip of the mass terms.

Nevertheless, it is possible to construct a formulation of QED$_{2+1}$ which preserves parity (and consequently time inversion) without taking the zero-mass limit. For this, the inclusion of an additional fermion species (additional "flavor") is required [9, 10, 132]. But it is not sufficient to simply double the fermions by enlarging the Dirac space by another copy of the original one: on the one hand, the sign flip of the mass terms under parity would persist; on the other hand, we would still be confronted with the usual 2×2 Pauli algebra, which does not allow for another matrix that anticommutes with the γ^μ's. Therefore, there would be nothing to generate a chiral symmetry. In conclusion, such a theory would not be richer than the original one.

Therefore, the three 4×4 γ matrices have to be taken to be unitarily equivalent to

$$\gamma^\mu_{4\times4} = \begin{pmatrix} \gamma^\mu_{2\times2} & 0 \\ 0 & -\gamma^\mu_{2\times2} \end{pmatrix}. \tag{4.13}$$

Then, a four-component spinor is constructed from the two different fermions as

$$\psi = \begin{pmatrix} \psi_+ \\ \psi_- \end{pmatrix}, \tag{4.14}$$

and the four-component mass term becomes

$$m\bar{\psi}\psi = m\psi_+^\dagger \sigma^3 \psi_+ - m\psi_-^\dagger \sigma^3 \psi_-. \tag{4.15}$$

The mass term is obviously conserved under the parity transformation:

$$\psi_+ \to \sigma^1 \psi_-, \qquad\qquad \psi_- \to \sigma^1 \psi_+. \tag{4.16}$$

This theory is called reducible, since the fermionic part of the Lagrangian can be decomposed according to [129]

$$\mathcal{L} = -\bar{\psi}\Big[m + \gamma_{4\times4}(-i\partial - eA)\Big]\psi$$

$$= -\sum_{\lambda=\pm} \bar{\psi}_\lambda \Big[\lambda m + \gamma_{2\times2}(-i\partial - eA)\Big]\psi_\lambda. \tag{4.17}$$

(Note that there is a γ^0 in $\bar{\psi}$.) From (4.17), we can read off that one two-component fermion carries the mass m, while the other carries $-m$. This theory preserves parity (at least at the classical level) and additionally provides for a chiral U(2) symmetry in the limit of vanishing masses [9].

Of course, we can add more pairs of fermions to the theory without modifying the feature of parity conservation. In the limit of vanishing masses, the theory possesses a chiral U(2N) symmetry for N pairs of fermions.

4.1.2 Quantum Theory

Before we analyze the quantum theory in the effective-action approach in detail, the question arises as to whether the various classical symmetries of the different formulations of the theory persist at the quantum level.

In the two-component irreducible formulation, the gauge and fermion mass terms are intimately connected, since they are both parity-violating. It is to be expected that one can be generated from the other in perturbation theory. This will be demonstrated explicitly in Sect. 4.3, where we begin with a massive fermion field, omitting the Chern–Simons term in the bare Lagrangian. Whether a parity-violating term is induced in perturbation theory when starting from a bare Lagrangian with vanishing masses depends on the order of the limiting processes. If one uses a parity-preserving regularization method, parity cannot be spontaneously broken in the effective theory. On the other hand, if one employs a regulator mass (e.g. Pauli–Villars regularization), the regularization process introduces parity-violating terms which persist even in the limit when the regulator drops out of the theory. The same is true if one starts with a massive theory and takes the limit of vanishing mass at the end of the calculation.

The four-component reducible representation of QED_{2+1} requires a more careful study, since we have parity conservation as well as a chiral symmetry in the massless case. In the following, we assume that the massless theory is achieved as a limit of a massive theory. In this case, we can employ fermion masses as (parity-preserving) regulators and finally take the zero-mass limit.

We also restrict ourselves to static external-field configurations in the $A_0 = 0$ gauge.

The key to a simple understanding of the patterns of symmetry breaking [133] is the observation that the Dirac operator in two spatial dimensions has zero modes for magnetic background fields [7, 108]. Let $\Phi = (e/2\pi) \int d^2x\, B(x,y)$ be the total magnetic flux of the space; then, the number of fermion zero modes is given by $[|\Phi|]$, where $[\ ]$ denotes the largest integer[2] less than Φ. Since these zero modes can be either filled or empty in the vacuum state without any change of energy, the vacuum is degenerate, with degeneracy $2^{2N[|\Phi|]}$ for N pairs of fermions.

Whether one particular zero mode is filled or empty is arbitrary for an a priori massless theory. But, starting with a massive theory, the zero modes are shifted by the fermion masses m_i. If $m_i > 0$, the corresponding zero modes are empty in the vacuum state, and they will remain empty in the limit $m_i \to 0^+$. The reverse is true for $m_i < 0$ in the limit $m_i \to 0^-$; these zero modes will be filled. Defining the massless theory from a massive one in this way, it is clear that the $[|\Phi|]$ zero modes for each fermion species (flavor) are either all filled or all empty. Hence, we are dealing with a vacuum of degeneracy 2^{2N} only.

As a simple example, let us now discuss the case $N = 1$, i.e. the four-component theory with one pair of two-component fermions. Thus, we have to consider two zero modes – one for each two-component fermion. In the massive formulation given above, ψ_+ has positive and ψ_- negative mass. Hence, in the massless limit, one zero mode $(-)$ is filled and the other $(+)$ is empty. Since a chiral transformation is a rotation in flavor space, this asymmetric vacuum state (filled–empty) is not invariant under chiral transformations. We have to conclude that chiral symmetry is spontaneously broken at the quantum level and a fermion condensate arises for a sufficiently large magnetic flux. This will be demonstrated explicitly in the following section.

To see whether parity is conserved in this case, we have to know how the parity transformation acts on the zero modes. Inserting the definition (4.16) of the parity transformation into (4.15), it turns out that P reverses the sign of the mass term, $m \to -m$, and additionally interchanges the fermion flavors, $\psi_\pm \to \psi_\mp$. The mass sign reversal is related to changing the status of a zero mode, empty↔filled, when taking the zero-mass limit. The interchange of the fermion flavors also interchanges the zero modes. To conclude, the asymmetric vacuum state (filled–empty) is invariant under parity (similarly for the (empty–filled) state), and integrating out the fermions preserves parity at the quantum level.

---

[2] Instead of $\mathbb{R}^2$, we can equally consider a compact boundaryless two-dimensional manifold for convenience. Then, the flux Φ is a topological integer. This relation between Φ and the number of zero modes is a simple example of the Atiyah–Singer index theorem applied to the Euclidean two-dimensional Dirac operator [12].

If we want chiral symmetry to be preserved, a chirally invariant vacuum state is required, i.e. either both zero modes filled or both empty. Then, parity is not conserved, since, for example, the (empty–empty) state goes to (filled–filled) under parity. Clearly, in order to formulate a chirally symmetric theory, the signs of the masses have to be equal; for example, positive mass terms would lead to the (empty–empty) vacuum state in the zero-mass limit. It is indeed possible to construct such a mass term in the four-component formulation [9], which (in the two-component notation) reads $m\psi_+\sigma^3\psi_+ + m\psi_-\sigma^3\psi_-$. Obviously, this mass term violates parity even at the classical level, but also, as we have demonstrated, it violates parity at the quantum level in the zero-mass limit.

We have to conclude that either parity or chiral symmetry is broken at the quantum level. One can easily generalize these considerations to arbitrary numbers of flavors [133]. On the one hand, chiral symmetry can only be preserved if the zero modes are all filled or all empty; otherwise, the symmetry breaks down to direct products of symmetry groups for the either set of zero modes, the filled or the empty ones. On the other hand, parity can only be conserved if there is an even number of flavors, with one half of the zero modes filled and the other half empty. It is easy to show that, if one symmetry is exactly conserved, the other is maximally broken.

Finally, we have to stress that the present considerations depend strongly on the assumption that the massless theory arises from a massive one by taking the zero-mass limit at the end of the calculation. If one starts with a massless theory, the results of perturbation theory depend on how the infrared divergences are removed. If one is careful to preserve chiral symmetry and parity during the regularization, no symmetry-breaking terms can be produced perturbatively. Therefore, one has to employ nonperturbative methods to answer the question of whether a vacuum state with broken symmetry is favored. From a nonperturbative Dyson–Schwinger analysis [9, 134], it is believed that fermion masses are generated dynamically even in the absence of a magnetic field as long as N is smaller than a critical number of flavor pairs (see [10]). The allowed vacuum configurations preserve parity, i.e. the masses appear in pairs with opposite sign. This confirms energetic arguments [163] which state that these are the only types of mass terms that are allowed to be generated dynamically in vector-like theories.

In a well-known series of papers, Gusynin, Miransky and Shovkovy [94] have shown that chiral symmetry is dynamically broken in the presence of a magnetic field. To be precise, an (enormously strong[3]) magnetic field acts as a catalyst for the symmetry breaking by allowing those fermions with an energy much less than that of the Landau gap to move only in 1+1 dimensions. Owing to this dimensional reduction, the critical coupling for dynamical symmetry breaking vanishes; therefore, the latter occurs for any arbitrarily weak

---

[3] The necessary field strength has been estimated to amount to $B > 10^{38}$ T [102], which is much stronger than any known field in the universe.

attraction. Lattice results confirm these nonperturbative findings [77]. However, a detailed analysis of this effect, which appears to be model-independent and is present in 3+1 dimensions as well (see also [102]), is beyond the scope of this work.

4.2 Parity-Invariant QED$_{2+1}$

In this section, we analyze the four-component formulation of QED$_{2+1}$ in detail, employing the effective-action approach to one loop [59]. We confine ourselves mainly to constant external magnetic fields. In the sense of the preceding considerations, we look for spontaneous flavor/chiral symmetry breaking by studying the fermion condensate. Furthermore, we investigate the field dependence of the trace of the energy–momentum tensor.

Before going into details of the low-dimensional theory, we take the opportunity to briefly recall the quantities of interest in the real 3+1-dimensional world in order to point out the similarities to and differences from the 2+1-dimensional system.

Consider the fermionic part of the Lagrangian of QED$_{3+1}$,

$$\mathcal{L} = -\bar{\psi}\left[m + \gamma\left(\frac{1}{i}\partial - eA\right)\right]\psi, \tag{4.18}$$

where the γ matrices are the "usual" ones defined in Appendix A. When we take the derivative of the generating functional,

$$Z = e^{iW} = \int \mathcal{D}\psi \mathcal{D}\bar{\psi}\mathcal{D}A \, \exp\left\{-i\int d^4x\, \bar{\psi}\left[m + \gamma\left(\frac{1}{i}\partial - eA\right)\right]\psi\right\}, \tag{4.19}$$

i.e.

$$\begin{aligned}
\frac{\partial \ln Z}{\partial m} &= \frac{1}{Z}\frac{\partial Z}{\partial m} \\
&= \frac{-i}{Z}\int \mathcal{D}\psi \mathcal{D}\bar{\psi}\mathcal{D}A \left(\int d^4x\, \bar{\psi}(x)\psi(x)\right) e^{-i\int d^4x\, \bar{\psi}\{m+\gamma[-i\partial-eA]\}\psi} \\
&= -i\int d^4x\, \langle\bar{\psi}(x)\psi(x)\rangle,
\end{aligned}$$

or with $\ln Z = iW$,

$$-\frac{\partial W}{\partial m} = \int d^4x \langle\bar{\psi}(x)\psi(x)\rangle,$$

we obtain

$$\langle\bar{\psi}\psi\rangle = -\frac{\partial \mathcal{L}^{(1)}}{\partial m}. \tag{4.20}$$

Here we have introduced the one-loop effective Lagrangian $\mathcal{L}^{(1)}$ via $\Gamma^{(1)} = \int d^4x\, \mathcal{L}^{(1)}(x)$. When we employ the Green's function

$$G(x, x'|A) = i\langle T[\psi(x)\bar{\psi}(x')]\rangle,$$

we can produce the following useful equalities:

$$\frac{\partial \mathcal{L}^{(1)}(x)}{\partial m} = -i\,\mathrm{tr}\,G(x, x|A) = -\langle\bar{\psi}(x)\,\psi(x)\rangle. \tag{4.21}$$

This equation contains the same physical information as is described by (2.7)–(2.9).

A variant of the above procedure implies for the trace of the energy–momentum tensor

$$\langle T_\mu{}^\mu(x)\rangle = -i\,m\,\mathrm{tr}\,G(x, x|A) = m\frac{\partial\mathcal{L}^{(1)}(x)}{\partial m}$$

$$= -m\langle\bar{\psi}(x)\psi(x)\rangle\,. \tag{4.22}$$

With the aid of Schwinger's proper-time technique [148], we find, for a constant-magnetic-field configuration, $B = B_3 = \mathrm{const.}$,

$$\langle T_\mu{}^\mu\rangle(B) = \frac{eBm^2}{4\pi^2}\int_0^\infty \frac{dz}{z^2}\,e^{-(m^2/eB)z}\left(z\coth z - 1 - \frac{1}{3}z^2\right). \tag{4.23}$$

After performing the integration, we obtain [56]

$$\langle T_\mu{}^\mu\rangle(B) = -\frac{e^2B^2}{12\pi^2} - \frac{m^4}{4\pi^2}\ln\frac{m^2}{2eB} + \frac{eBm^2}{4\pi^2}\ln\frac{m^2}{2eB} + \frac{m^4}{4\pi^2}$$

$$+ \frac{eBm^2}{2\pi^2}\left[\ln\Gamma\left(\frac{m^2}{2eB}\right) - \frac{1}{2}\ln 2\pi\right]\,. \tag{4.24}$$

We can also identify the fermion condensate:

$$\langle\bar{\psi}\psi\rangle(B) = \frac{m^3}{4\pi^2}\ln\frac{m^2}{2eB} - \frac{eBm}{4\pi^2}\ln\frac{m^2}{2eB} - \frac{m^3}{4\pi^2}$$

$$- \frac{eBm}{2\pi^2}\left[\ln\Gamma\left(\frac{m^2}{2eB}\right) - \frac{1}{2}\ln 2\pi\right]. \tag{4.25}$$

First, note that the fermion condensate vanishes proportionally to $m\ln m$ for vanishing fermion mass m. Therefore, if we introduce additional fermions to obtain a flavor symmetry in the zero-mass limit, we find no spontaneous symmetry breakdown in QED$_{3+1}$.

As a consistency check against Schwinger's formula for pair production [148], we obtain the following by replacing B with $-iE$:

$$\mathrm{Im}\langle T_\mu{}^\mu\rangle(E) = m\frac{\partial}{\partial m}\mathrm{Im}\,\mathcal{L}^{(1)}(E) = -\frac{m^2eE}{4\pi^2}\sum_{n=1}^\infty \frac{e^{-(\pi m^2/eE)n}}{n}. \tag{4.26}$$

Now we turn to the parity-invariant four-component model of (2+1)-dimensional QED with two flavors ($\pm$) as defined in (4.17). The dynamics is contained in the Lagrangian

$$\mathcal{L} = \sum_{\lambda=\pm} \mathcal{L}_\lambda = -\sum_{\lambda=\pm} \bar{\psi}_\lambda \Big[\lambda m + \gamma_{2\times2}\big(-i\partial - eA\big)\Big]\psi_\lambda, \qquad (4.27)$$

where we have employed the reduced two-component representation, and the 2×2 γ matrices are related to the Pauli matrices by (4.7). The fermion condensate can be written as

$$S(x; m) \equiv \langle \bar{\psi}(x)\psi(x)\rangle \overset{(4.21)}{=} -\frac{\partial \mathcal{L}^{(1)}}{\partial m} = -\sum_{\lambda=\pm} \frac{\partial \mathcal{L}^{(1)}_\lambda}{\partial m}$$

$$= -\sum_{\lambda=\pm} \lambda \frac{\partial \mathcal{L}^{(1)}_\lambda}{\partial(\lambda m)} = \sum_{\lambda=\pm} \lambda S_\lambda(x; m)$$

$$= \sum_{\lambda=\pm} \lambda S_+(x; \lambda m). \qquad (4.28)$$

In the last step, we have made use of the fact that $S_-(x; m) = S_+(x; -m)$. Equation (4.28) clearly demonstrates that $S(x; -m) = -S(x; m)$; hence, performing the calculation for, for example, $m > 0$ requires us to multiply the even terms in m by a factor of $\text{sign}(m)$ at the end of the calculation.

Returning to the four-component formulation, we compute $\langle \bar{\psi}(x)\psi(x)\rangle = \langle \psi^\dagger(x)\gamma^0_{4\times4}\psi(x)\rangle$, for the moment assuming that $m > 0$. Employing the 2+1-dimensional analogue of the formula (2.47) of [55] together with (4.21), we arrive at

$$\langle \bar{\psi}(x)\psi(x)\rangle = i \operatorname{tr} G(x, x|A) = i \operatorname{tr} \int \frac{d^3k}{(2\pi)^3} \mathcal{G}(k) \qquad (4.29)$$

$$= -\frac{1}{(2\pi)^3} \int d^3k \int_0^\infty ds \, \exp\Big[-is\Big(m^2 - k^{0^2} + \frac{\tan z}{z} k_\perp^2\Big)\Big]$$

$$\times \operatorname{tr}\Big[\frac{e^{i\sigma_3 z}}{\cos z}\big(m + \gamma^0 k^0\big) - \frac{e^{-i\sigma_3 z}}{\cos z}(\gamma \cdot k)_\perp\Big], \quad z = eBs.$$

The momentum integral and the traces can readily be evaluated, with the intermediate result

$$\langle \bar{\psi}\psi\rangle = -\int_0^\infty \frac{ds}{8(\pi s)^{3/2}} \, 4m(eBs)\cot(eBs)e^{-i(m^2 s + \pi/4)}. \qquad (4.30)$$

To obtain a convergent expression on the right-hand side, we need to subtract the zero-field part, which changes (4.30) into ($s = -it$)

$$\langle \bar{\psi}\psi\rangle(B) = -\frac{m}{2\pi^{3/2}} \int_0^\infty dt \, e^{-m^2 t} t^{-1/2} \Big[eB\coth(eBt) - \frac{1}{t}\Big]. \qquad (4.31)$$

At this stage, we need the value of the following two integrals, which, after regularization with the ϵ-integration technique (see Appendix D), yield

$$I_1 = \int_0^\infty \mathrm{d}t\, e^{-m^2 t} t^{-1/2} (eB) \coth(eBt)$$

$$= \sqrt{eB}\sqrt{\pi}\left[\sqrt{2}\,\zeta\left(\frac{1}{2}, \frac{m^2}{2eB}\right) - \frac{\sqrt{eB}}{m}\right],$$

$$I_2 = \int_0^\infty \mathrm{d}t\, e^{-m^2 t} t^{-3/2} = 2\sqrt{\pi}m.$$

The use of these integrals produces

$$\langle\bar{\psi}\psi\rangle(B) = -\frac{1}{2\pi}\left[m\sqrt{2eB}\,\zeta\left(\frac{1}{2}, 1 + \frac{m^2}{2eB}\right) + eB - 2m^2\right]. \tag{4.32}$$

We observe that, for $m > 0$ and $eB > 0$, we obtain

$$\lim_{m \to 0_+} \langle\bar{\psi}\psi\rangle(B) = -\frac{eB}{2\pi}, \tag{4.33}$$

and, more generally,

$$\lim_{m \to 0} \langle\bar{\psi}\psi\rangle(B) = -\mathrm{sgn}(m)\frac{eB}{2\pi}. \tag{4.34}$$

Note that, in contrast to 3+1 dimensions (cf. (4.21)), (4.34) reveals spontaneous flavor symmetry breaking in QED_{2+1} in the four-component formulation with an even number of two-component fermions. This was expected from our general considerations in the preceding section.

Employing (4.22), we can make contact with the trace of the energy–momentum tensor in QED_{2+1}:

$$\langle T_\mu{}^\mu\rangle = -m\langle\bar{\psi}\psi\rangle$$

$$= \frac{1}{2\pi}\left[m^2\sqrt{2eB}\,\zeta\left(\frac{1}{2}, 1 + \frac{m^2}{2eB}\right) + meB - 2m^3\right]. \tag{4.35}$$

Unlike the result in four-dimensional QED [56],

$$\lim_{m \to 0} \langle T_\mu{}^\mu\rangle_{(4)} = -\frac{1}{12\pi^2}e^2 B^2 = -\frac{2\alpha}{3\pi}\frac{1}{4}F_{\mu\nu}F^{\mu\nu},$$

which reflects the breaking of scale invariance, there is no trace anomaly in three-dimensional QED:

$$\lim_{m \to 0} \langle T_\mu{}^\mu\rangle = 0, \tag{4.36}$$

as it should; namely, QED_{2+1} is super-renormalizable and the perturbative β function vanishes. Recall that the coefficient of $(1/4)F^{\mu\nu}F_{\mu\nu}$ in the right-hand side of the trace of $T^{\mu\nu}$ is related to the β function.

Incidentally, the formula for the pair production rate in QED_{2+1}, which should be read together with (4.26), is given by

$$2\,\mathrm{Im}\,\mathcal{L}^{(1)}(E) = \frac{1}{4\pi^2}\,(eE)^{3/2}\sum_{n=1}^{\infty}\frac{e^{-(\pi m^2/eE)n}}{n^{3/2}}. \tag{4.37}$$

We would like to conclude with the remark that the fermion condensate in QED$_{2+1}$ has also been investigated for spatially nonconstant magnetic background fields $B(x)$ [68,129]. At least for large magnetic flux, it turns out that (4.34) is also valid as a local statement:

$$\lim_{m\to 0}\langle\bar\psi(x)\psi(x)\rangle \sim -\mathrm{sign}(m)\frac{eB(x)}{2\pi}. \tag{4.38}$$

4.3 The Effective Action for Irreducible QED$_{2+1}$

In this section, we evaluate the QED$_{2+1}$ effective action for the irreducible two-component formulation of the theory containing only one fermion [137]. As argued in Sect. 4.1, parity is conserved neither in the massive classical theory nor in the quantum theory even in the zero-mass limit (if this limit is taken at the end of the computation).

By integrating out the fermions, we shall demonstrate in detail that the parity-violating term is of the form of $\mathcal{L}_{\mathrm{Gm}}$ as given in (4.2) and, therefore, is of topological origin.

First, we work in Euclidean space with $g = \mathrm{diag}(1,1,1)$, $\{\gamma^\mu,\gamma^\nu\} = -2\delta^{\mu\nu}$ and $(\gamma^\mu)^\dagger = -\gamma^\mu$ for all μ. A representation for the γ matrices which satisfies these conditions is given by $\gamma^\mu = i\sigma^\mu$, $\sigma^0 \equiv \sigma^3$. Then, the effective action to one loop is proportional to the logarithm of the fermion determinant:

$$\ln\det\big[m - i\slashed{D}\,[A]\big]_{\mathrm{E}} \equiv \mathrm{Tr}\,\ln\big[m - i\slashed{D}\,[A]\big]_{\mathrm{E}} = \mathrm{Tr}\,\ln\big[m + \slashed{\Pi}\,\big]_{\mathrm{E}}. \tag{4.39}$$

Here we have introduced $\Pi_\mu = p_\mu - eA_\mu$ and $p_\mu = -i\partial_\mu$, so that $D_\mu = i\Pi_\mu$. The operator $\slashed{D}$ is Hermitian, with real eigenvalues λ_n, while $\slashed{\Pi}$ is anti-Hermitian, so that its eigenvalues are purely imaginary:

$$\big[m + \slashed{\Pi}\,\big]_{\mathrm{E}}|m,i\lambda_n\rangle = (m - i\lambda_n)|m,i\lambda_n\rangle, \qquad \lambda_n \in \mathbb{R}. \tag{4.40}$$

The spectral representation of (4.39) can then be written as

$$\begin{aligned}
\ln\det\big[m - i\slashed{D}\,\big]_{\mathrm{E}} &= \mathrm{Tr}\,\ln\big[m + \slashed{\Pi}\,\big]_{\mathrm{E}} \\
&= \mathrm{Tr}\,\sum_n \ln(m - i\lambda_n)\,|m,i\lambda_n\rangle\langle m,i\lambda_n| \\
&= \mathrm{Tr}\,\sum_n \ln(m - i\lambda_n)\,\mathsf{P}_{m+\slashed{\Pi}}\,.
\end{aligned} \tag{4.41}$$

In the continuum, (4.41) reads

$$\ln\det\big[m + \slashed{\Pi}\,\big]_{\mathrm{E}} = \int_{-\infty}^{\infty} d\lambda\,\ln(m - i\lambda)\,\mathrm{Tr}\,\big[\mathsf{P}_{m+\slashed{\Pi}}\big]. \tag{4.42}$$

Writing $\ln(m - i\lambda) = \ln|m - i\lambda| - i\phi$, where the phase angle ϕ is introduced by $m - i\lambda = |m - i\lambda|e^{-i\phi}$, so that $\phi = \arctan(\lambda/m)$, we obtain

$$\ln(m - i\lambda) = \ln\sqrt{m^2 + \lambda^2} - i\arctan\frac{\lambda}{m}. \tag{4.43}$$

With the aid of the relation (4.43), (4.42) can be separated into two parts:

$$\ln\det\left[m + \slashed{\Pi}\right]_{\mathrm{E}} = \frac{1}{2}\int d\lambda\,\ln(m^2 + \lambda^2)\,\mathrm{Tr}\left[P_{m+\slashed{\Pi}}\right]$$
$$-i\int d\lambda\,\arctan\frac{\lambda}{m}\,\mathrm{Tr}\left[P_{m+\slashed{\Pi}}\right]. \tag{4.44}$$

Since

$$\lambda|m, i\lambda\rangle = i\slashed{\Pi}\,|m, i\lambda\rangle \qquad \text{or} \qquad \lambda^2|m, i\lambda\rangle = -\slashed{\Pi}^2|m, i\lambda\rangle, \tag{4.45}$$

we obtain the following representation for the one-loop effective action (4.39):

$$\ln\det\left[m + \slashed{\Pi}\right]_{\mathrm{E}} = \frac{1}{2}\ln\det\left[m^2 - \slashed{\Pi}^2\right]_{\mathrm{E}} - i\frac{\pi}{2}\eta_{m+\slashed{\Pi}}, \tag{4.46}$$

where

$$\eta_{m+\slashed{\Pi}} = \frac{2}{\pi}\int d\lambda\,\arctan\frac{\lambda}{m}\,\mathrm{Tr}\left[P_{m+\slashed{\Pi}}\right] \tag{4.47}$$

is a measure of the spectral asymmetry of the Dirac operator and is called the Atiyah–Patodi–Singer η invariant [13]. For massless fermions (4.47) reduces to

$$\eta_{\slashed{\Pi}} = \int d\lambda\,\frac{\lambda}{|\lambda|}\,\mathrm{Tr}\left[P_{\slashed{\Pi}}\right]. \tag{4.48}$$

Now we can write for the effective action

$$\Gamma^{(1)} \equiv -i\int d^3x\,\langle x|\mathrm{tr}\,\ln(-i\slashed{D} + m)|x\rangle,$$
$$= \int d^3x\,\mathcal{L}^{(1)}, \qquad d^3x = -i d^3x_{\mathrm{E}}, \quad \mathcal{L}^{(1)}(x) = -\mathcal{L}_{\mathrm{E}}^{(1)}(x_{\mathrm{E}})$$
$$= i\int d^3x_{\mathrm{E}}\,\mathcal{L}_{\mathrm{E}}^{(1)}(x_{\mathrm{E}})$$
$$= -i\int d^3x_{\mathrm{E}}\,{}_{\mathrm{E}}\langle x|\mathrm{tr}\,\ln[-i\slashed{D} + m]_{\mathrm{E}}|x\rangle_{\mathrm{E}} = i\Gamma_{\mathrm{E}}^{(1)} \tag{4.49}$$
$$= -i\int d^3x_{\mathrm{E}}\left[\frac{1}{2}\mathrm{tr}\,{}_{\mathrm{E}}\langle x|\mathrm{tr}\,\ln[m^2 - \slashed{\Pi}^2]_{\mathrm{E}}|x\rangle_{\mathrm{E}} - i\frac{\pi}{2}{}_{\mathrm{E}}\langle x|\eta_{m+\slashed{\Pi}}|x\rangle_{\mathrm{E}}\right].$$

The first term in (4.49) represents the parity-even part, and is closely related to the usual effective action of QED_{3+1}, with respect to its structure as well as its origin. A standard calculation (e.g. [30, 137]; a version including N fermion flavors and $\mathcal{N}$ gauge field colors can be found in [85]) yields

$$\Gamma_{\mathrm{even}}^{(1)} = \frac{\sqrt{2}}{2\pi}(eB)^{3/2}\,\zeta\left(-\frac{1}{2}, \frac{m^2}{2eB}\right) - \frac{meB}{4\pi}. \tag{4.50}$$

Using the identity $\zeta(z, q \to 0) = [2/(2\pi)^{1-z}]\Gamma(1-z)\sin(z\pi/2)\zeta(1-z)$ (cf. [91]), we can take the zero-mass limit of (4.50):

$$\Gamma^{(1)}_{\text{even}} = -\frac{1}{2\pi^2}\zeta(3/2)\int d^3x \left(\frac{e|{}^*F|}{2}\right)^{3/2},$$

(4.51)

where we have employed an identity for the Riemann ζ function and replaced the magnetic field B by the gauge- and Lorentz-invariant (${}^*F^\mu = (1/2)\epsilon^{\mu\alpha\beta}F_{\alpha\beta}$):

$$|{}^*F| = \sqrt{|{}^*F^\mu {}^*F_\mu|} = \sqrt{\frac{1}{4}|\epsilon^{\mu\alpha\beta}F_{\alpha\beta}\epsilon_{\mu\kappa\lambda}F^{\kappa\lambda}|} = \sqrt{\frac{1}{2}F^{\alpha\beta}F_{\alpha\beta}}$$

$$= \sqrt{B^2 - E^2}.$$

(4.52)

This result belongs to the class of "similarities" to standard QED$_{3+1}$ and does not reveal any surprising features, except for the polynomial dependence on the field strength instead of a logarithmic one, which is more familiar in field theory.

To obtain the parity-odd part of the effective Lagrangian for *massless* QED$_3$, we need to calculate

$$\Gamma^{(1)}_{\text{E,odd}} = i\frac{\pi}{2}\eta_{\slashed{\partial}},$$

(4.53)

where $\eta_{\slashed{\partial}}$ is given in (4.48). The question of how a nonvanishing fermion mass modifies the result will be answered afterwards, since it requires the full machinery of proper-time techniques.

Employing the following formula where $M = \lambda^2 > 0$,

$$M^{-s} = \frac{1}{\Gamma(s)}\int\limits_0^\infty d\tau\, \tau^{s-1}\, e^{-M\tau},$$

(4.54)

we obtain, by substituting $|\lambda| = \sqrt{\lambda^2} = \sqrt{M}$ and $s = 1/2$,

$$\eta_{\slashed{\partial}} = \frac{1}{\sqrt{\pi}}\int d\lambda \int\limits_0^\infty \frac{d\tau}{\sqrt{\tau}}\lambda e^{-\lambda^2\tau}\,\text{Tr}\left[P_{\slashed{\partial}}\right]$$

$$= \frac{1}{\sqrt{\pi}}\int\limits_0^\infty \frac{d\tau}{\sqrt{\tau}}\,\text{Tr}\left(i\slashed{\partial}\, e^{-(i\slashed{\partial})^2\tau}\right).$$

(4.55)

We shall now take the functional derivative of (4.53) with respect to A_μ to deduce the Chern–Simons term in the effective action. Equation (4.53), together with (4.54), produces

$$\frac{\delta\Gamma^{(1)}_{\text{E,odd}}}{\delta A_\mu} = i\frac{\pi}{2}\frac{1}{\sqrt{\pi}}\int\limits_0^\infty \frac{d\tau}{\sqrt{\tau}}\,\text{Tr}\,\frac{\delta}{\delta A_\mu}\left(i\slashed{\partial}\, e^{-(i\slashed{\partial})^2\tau}\right).$$

(4.56)

Here we need the operator identity

$$\frac{\delta}{\delta A_\mu}\left(i\slashed{\Pi}\, e^{-(i\slashed{\Pi})^2\tau}\right) = \frac{\delta(i\slashed{\Pi})}{\delta A_\mu}e^{-(i\slashed{\Pi})^2\tau} \tag{4.57}$$

$$+(i\slashed{\Pi})i(i\tau)\int_0^1 d\sigma\, e^{i(i\tau)(i\slashed{\Pi})^2(1-\sigma)}\frac{\delta(i\slashed{\Pi})^2}{\delta A_\mu}e^{i(i\tau)(i\slashed{\Pi})^2\sigma}.$$

The first term on the right-hand side of (4.57) can be rewritten in terms of a total derivative according to

$$\frac{\delta}{\delta A_\mu}\left(i\slashed{\Pi}\ e^{-(i\slashed{\Pi})^2\tau}\right)$$

$$= \sqrt{\tau}\frac{d}{d\tau}\left[2\sqrt{\tau}\frac{\delta(i\slashed{\Pi})}{\delta A_\mu}e^{-(i\slashed{\Pi})^2\tau}\right] - 2\tau\frac{\delta(i\slashed{\Pi})}{\delta A_\mu}\left[-(i\slashed{\Pi})^2\right]e^{-(i\slashed{\Pi})^2\tau}$$

$$-(i\slashed{\Pi})\tau\int_0^1 d\sigma\, e^{i(i\tau)(i\slashed{\Pi})^2(1-\sigma)}\frac{\delta(i\slashed{\Pi})^2}{\delta A_\mu}e^{i(i\tau)(i\slashed{\Pi})^2\sigma}$$

$$= 2\sqrt{\tau}\frac{d}{d\tau}\left[\sqrt{\tau}\frac{\delta(i\slashed{\Pi})}{\delta A_\mu}e^{-(i\slashed{\Pi})^2\tau}\right] + \left[2\tau\frac{\delta(i\slashed{\Pi})}{\delta A_\mu}(i\slashed{\Pi})^2e^{-(i\slashed{\Pi})^2\tau}\right. \tag{4.58}$$

$$\left.-2\tau(i\slashed{\Pi})\int_0^1 d\sigma\, e^{i(i\tau)(i\slashed{\Pi})^2(1-\sigma)}\frac{\delta(i\slashed{\Pi})}{\delta A_\mu}(i\slashed{\Pi})e^{i(i\tau)(i\slashed{\Pi})^2\sigma}\right].$$

After performing the Tr operation, the last two terms in (4.58) cancel, and we are left with

$$\frac{\delta\Gamma^{(1)}_{E,odd}}{\delta A_\mu} = i\frac{\sqrt{\pi}}{2}\int_0^\infty \frac{d\tau}{\sqrt{\tau}}\,\mathrm{Tr}\left\{\sqrt{\tau}\frac{d}{d\tau}\left[2\sqrt{\tau}\frac{\delta(i\slashed{\Pi})}{\delta A_\mu}e^{-(i\slashed{\Pi})^2\tau}\right]\right\}$$

$$= i\sqrt{\pi}\left[\sqrt{\tau}\,\mathrm{Tr}\left[(-ie\gamma^\mu)e^{-(i\slashed{\Pi})^2\tau}\right]\right]_{\tau=0}^{\tau=\infty}. \tag{4.59}$$

Here we recall that $(i\slashed{\Pi})$ has real eigenvalues so that, according to (4.45), $(i\slashed{\Pi})^2$ is a positive real number. Hence, the upper limit in (4.59) vanishes.

We shall now return to Minkowski space and employ the following relation $(-i\Gamma^{(1)} = \Gamma^{(1)}_E)$:

$$\langle j^\mu_{odd}\rangle = \frac{\delta\Gamma^{(1)}_{odd}}{\delta A_\mu}. \tag{4.60}$$

In three-dimensional QED, the Dirac matrices are given by (4.7), so that in $\slashed{\Pi}^2 = -\Pi^2 + (e/2)\sigma^{\mu\nu}F_{\mu\nu}$,

$$\sigma^{\mu\nu} = \frac{i}{2}[\gamma^\mu,\gamma^\nu],$$

$$\text{with}\quad \sigma^{12} = \epsilon^{123}\gamma^0,\quad \sigma^{13} = -\epsilon^{132}\gamma^2,\quad \sigma^{23} = -\epsilon^{231}\gamma^1,$$

$$\Rightarrow\quad \sigma^{\mu\nu} = -\epsilon^{\mu\nu\alpha}\gamma_\alpha. \tag{4.61}$$

Hence, we obtain

$$\frac{1}{2}\sigma^{\mu\nu}F_{\mu\nu} = -\gamma^\lambda \, {}^\star F_\lambda. \tag{4.62}$$

This enables us to write

$$\rlap{/}\Pi^2 = -\Pi^2 - e\,\gamma^\lambda{}^\star F_\lambda \tag{4.63}$$

and

$$\langle j^\mu_{\mathrm{odd}}\rangle = -\mathrm{i}e\sqrt{\pi}\sqrt{\tau}\,\mathrm{Tr}\left[\gamma^\mu\,\mathrm{e}^{(-\Pi^2 - e\,\gamma^\lambda{}^\star F_\lambda)\tau}\right]_{\tau=0}. \tag{4.64}$$

At this point it is convenient to replace τ by $\mathrm{i}\tau$, so that

$$\langle j^\mu_{\mathrm{odd}}\rangle = -\mathrm{i}e\sqrt{\pi\tau}\,\mathrm{e}^{\mathrm{i}(\pi/4)}\,\mathrm{Tr}\left[\gamma^\mu\,\mathrm{e}^{-\mathrm{i}(\Pi^2 + e\,\gamma^\lambda{}^\star F_\lambda)\tau}\right]_{\tau=0}, \tag{4.65}$$

where the exponent reads in detail

$$-\mathrm{i}(\Pi^2 + e\,\gamma^\lambda{}^\star F_\lambda)\tau = -\mathrm{i}\tau\left[-\partial^2 + \mathrm{i}e(\partial\cdot A + A\cdot\partial) + e^2 A^2 + e\,\gamma^\lambda{}^\star F_\lambda\right]$$
$$= -\mathrm{i}\tau\left(-\partial^2 + Q\right). \tag{4.66}$$

Introducing

$$U(\tau) = \mathrm{e}^{-\mathrm{i}\tau(-\partial^2 + Q)}, \qquad\qquad U_0(\tau) = \mathrm{e}^{\mathrm{i}\tau\partial^2}, \tag{4.67}$$

we obtain

$$\langle j^\mu_{\mathrm{odd}}\rangle = -\mathrm{i}e\sqrt{\pi\tau}\,\mathrm{e}^{\mathrm{i}(\pi/4)}\,\mathrm{Tr}\left\{\gamma^\mu\left[U(\tau) - U_0(\tau)\right]\right\}_{\tau=0}, \tag{4.68}$$

where we have regularized $\langle j^\mu_{\mathrm{odd}}\rangle$ by subtracting the zero-field part.

For $\tau \to 0$, we may expand the right-hand side of (4.68) [148]:

$$\mathrm{Tr}\left\{\gamma^\mu\left[U(\tau) - U_0(\tau)\right]\right\}$$

$$= \mathrm{i}\tau\,\mathrm{Tr}\left[\gamma^\mu\,U_0(\tau)\,Q\right] + \mathrm{i}^2\int_0^\tau \mathrm{d}t\,(\tau - t)\,\mathrm{Tr}\left[\gamma^\mu\,U_0(\tau-t)\,Q\,U_0(t)\,Q\right] + \ldots$$

$$= \mathrm{i}\tau\,\mathrm{Tr}\left[\gamma^\mu\,U_0(\tau)\,Q\right] + \mathcal{O}[\tau^2, U_0(\tau)]. \tag{4.69}$$

After substituting (4.69) into (4.68), we obtain in Minkowski space,

$$\langle j^\mu_{\mathrm{odd}}\rangle = -\mathrm{i}e\sqrt{\pi\tau}\,\mathrm{e}^{\mathrm{i}(\pi/4)}\,\mathrm{i}\tau$$
$$\times\mathrm{Tr}\left\{\gamma^\mu\,\mathrm{e}^{\mathrm{i}\tau\partial^2}\left[\mathrm{i}e(\partial\cdot A + A\cdot\partial) + e^2 A^2 + e\,\gamma^\lambda{}^\star F_\lambda\right]\right\}_{\tau=0}$$
$$= -\mathrm{i}^2 e\sqrt{\pi}\tau^{3/2}\mathrm{e}^{\mathrm{i}(\pi/4)}\,\mathrm{tr}_x\,\mathrm{tr}_\gamma\left[\gamma^\mu\gamma^\lambda\,\mathrm{e}^{\mathrm{i}\tau\partial^2}\,e\,{}^\star F_\lambda\right]_{\tau=0}$$
$$= -2e^2\sqrt{\pi}\tau^{3/2}\,\mathrm{e}^{\mathrm{i}(\pi/4)}\,\mathrm{tr}_x\left[\mathrm{e}^{\mathrm{i}\tau\partial^2}\,{}^\star F^\mu\right]_{\tau=0}$$
$$= -\frac{e^2}{4\pi}\,{}^\star F^\mu, \tag{4.70}$$

where $\mathrm{tr}\gamma^\mu\gamma^\lambda = -2g^{\mu\lambda}$. Here we have made use of the fact that the fields are assumed to be constant, and that $\mathrm{tr}_x\,\mathrm{e}^{\mathrm{i}\tau\partial^2} = (1/8\pi^{3/2})(1/\tau^{3/2})\,\mathrm{e}^{-\mathrm{i}(\pi/4)}$.

The effective action that contributes to the parity-odd part is given by $(F_{\mu\nu} = \text{const.})$

$$\Gamma^{(1)}_{\text{odd}}[A] = -\frac{e^2}{8\pi} \int d^3x \, A_\mu \, {}^\star F^\mu. \tag{4.71}$$

The relation to (4.70) via (4.60) can easily be checked, after a straightforward differentiation.

Introducing the Chern–Simons term

$$W[A] = \frac{e^2}{16\pi^2} \int d^3x \, A_\mu \, {}^\star F^\mu = \frac{e^2}{16\pi^2} \int d^3x \, \epsilon^{\mu\rho\sigma} A_\mu \partial_\rho A_\sigma, \tag{4.72}$$

we can rewrite our findings for the effective action in massless QED$_{2+1}$:

$$\Gamma^{(1)}[A] = -2\pi \, W[A] - \frac{1}{\pi^2} \zeta\left(\frac{3}{2}\right) \int d^3x \left(\frac{e|{}^\star F|}{2}\right)^{3/2}. \tag{4.73}$$

Comparison with (4.2) proves our statement that the topological photon mass term can be generated perturbatively from the parity-breaking fermion mass even in the zero-mass limit (for the fermion). To be precise, we find a topological photon mass of $|\mu| = \alpha/\pi$.

So far, we have neglected the mass of the electron. In the following, we want to find out how far a nonvanishing fermion mass modifies our result of (4.73).

Hence, let us start with

$$\Gamma^{(1)}_{\text{E,odd}} = i\frac{\pi}{2} \, \eta_{m+\slashed{H}} = -i\Gamma^{(1)}_{\text{odd}} = i \, \text{Tr} \arctan \frac{i\slashed{H}}{m}.$$

We want to perform the following calculations in Minkowski space so that we can write

$$\langle j^\mu_{\text{odd}}(x) \rangle = \frac{\delta \Gamma^{(1)}_{\text{odd}}}{\delta A_\mu(x)} = -\text{Tr} \frac{\delta}{\delta A_\mu(x)} \arctan \frac{i\slashed{H}(x')}{m}$$

$$= -\text{Tr} \frac{m}{m^2 + (i\slashed{H})^2}(-ie\gamma^\mu)\Big|_{x' \to x}$$

$$= iem \, \text{Tr} \frac{\gamma^\mu}{m^2 - \slashed{H}^2}\Big|_{x' \to x}. \tag{4.74}$$

With the aid of Schwinger's proper-time technique, we rewrite (4.74) in the form

$$\langle j^\mu_{\text{odd}}(x) \rangle = -em \int\limits_0^\infty ds \, e^{-im^2 s} \, \text{Tr} \left[\gamma^\mu \, e^{is\slashed{H}^2}\right]\Big|_{x' \to x}. \tag{4.75}$$

As before (see Sect. 2.1, (2.13)), we introduce the pseudo-Hamiltonian

$$H = -\slashed{H}^2 = \Pi^2 + e\gamma^\lambda \, {}^\star F_\lambda, \tag{4.76}$$

which enters the proper-time evolution operator $U(s) = e^{-iHs} = e^{i\not{\Pi}^2 s}$. The transition amplitude $\langle x_1, s | x_2, 0 \rangle \equiv \langle x_1 | U(s) | x_2 \rangle$ satisfies the "Schrödinger equation"

$$i\partial_s \langle x_1 | U(s) | x_2 \rangle = \langle x_1 | U(s) H | x_2 \rangle, \tag{4.77}$$

with the boundary condition $\langle x_1, s | x_2, 0 \rangle \big|_{s \to 0} = \delta(x_1 - x_2)$. The proper-time dynamical equations can be found in (2.15); for constant fields, they are simply given by

$$\frac{\mathrm{d}x^\mu}{\mathrm{d}s} = 2\,\Pi^\mu(s), \tag{4.78}$$

$$\frac{\mathrm{d}\Pi^\mu}{\mathrm{d}s} = 2e\,F^{\mu\nu}\Pi_\nu. \tag{4.79}$$

Equation (4.78) can easily be integrated:

$$\Pi^\mu(s) = \left(e^{2eFs}\right)^{\mu\lambda}\Pi_\lambda(0). \tag{4.80}$$

When this expression is substituted into (4.80), we arrive at the standard result:

$$x^\mu(s) - x^\mu(0) = \left[\frac{1}{eF}\left(e^{2eFs} - 1\right)\right]^{\mu\lambda}\Pi_\lambda(0)$$

$$= 2\left[\frac{1}{eF}\,e^{eFs}\sinh(eFs)\right]^{\mu\lambda}\Pi_\lambda(0). \tag{4.81}$$

Equation (4.81) can be used to express $\Pi^\mu(0)$ in terms of $x^\mu(s) - x^\mu(0)$:

$$\Pi^\mu(0) = \frac{1}{2}\left[eF\,e^{-eFs}\sinh^{-1}(eFs)\right]^{\mu\lambda}[x(s) - x(0)]_\lambda. \tag{4.82}$$

This equation, when inserted into (4.80), yields

$$\Pi^\mu(s) = \frac{1}{2}\left[eF\,e^{eFs}\sinh^{-1}(eFs)\right]^{\mu\lambda}[x(s) - x(0)]_\lambda. \tag{4.83}$$

For the Hamiltonian, we need the square of (4.83) (note that $F = -F^T$):

$$\Pi^2(s) = [x(s) - x(0)]_\lambda\frac{1}{4}\left[(eF)^2\sinh^{-2}eFs\right]^{\lambda\sigma}[x(s) - x(0)]_\sigma$$

$$= [x(s) - x(0)]_\lambda\,K^{\lambda\sigma}\,[x(s) - x(0)]_\sigma. \tag{4.84}$$

At different proper times, we have

$$[x^\mu(s), x^\nu(0)] \overset{(4.81)}{=} -2i\left[(eF)^{-1}\,e^{eFs}\sinh eFs\right]^{\mu\nu}. \tag{4.85}$$

This result is useful for deriving the result that $(x \equiv x(0))$

$$\Pi^2(s) = x(s)Kx(s) - 2x(s)Kx + xKx - \frac{i}{2}\,\mathrm{tr}_L\left[(eF)\coth eFs\right], \tag{4.86}$$

where we have also employed the fact that the field strength tensor is traceless. Now the Schrödinger equation (4.77) becomes

$$i\partial_s \langle x_1|U(s)|x_2\rangle = \Big(x_1 K x_1 - 2x_1 K x_2 + x_2 K x_2$$

$$-\frac{i}{2}\,\mathrm{tr_L}\left[(eF)\coth eFs\right] + e\gamma^{\lambda\star}F_\lambda \Big)\langle x_1|U(s)|x_2\rangle.$$

The solution of this equation is given by

$$\langle x_1|U(s)|x_2\rangle = \frac{C(x_1,x_2)}{s^{3/2}}\,\exp\left[\frac{i}{4}(x_1-x_2)eF\coth eFs(x_1-x_2)\right.$$

$$\left. -\frac{1}{2}\mathrm{tr}\,\ln\left(\frac{\sinh eFs}{eFs}\right) - ie\,\gamma^{\lambda\star}F_\lambda s\right],\quad (4.87)$$

which can be checked by a straightforward computation. The factor $C(x_1,x_2)$ carries the complete gauge dependence of the transition amplitude and is known as the "counter-gauge factor":

$$C(x_1,x_2) = C\,\exp\left[ie\int_{x_2}^{x_1}d\xi_\mu\,A^\mu(\xi)\right].\tag{4.88}$$

To determine the constant C, we recall the boundary condition,

$$\lim_{s\to 0}\int d^3x_1\langle x_1|U(s)|x_2\rangle = 1.\tag{4.89}$$

Hence, we need the value of the following integral:

$$\lim_{s\to 0}\int d^3x_1\frac{1}{s^{3/2}}\exp\left[\frac{i}{4}(x_1-x_2)eF\coth eFs(x_1-x_2)\right.$$

$$\left. -\frac{1}{2}\mathrm{tr}\,\ln\left(\frac{\sinh eFs}{eFs}\right) - ie\,\gamma^{\lambda\star}F_\lambda s\right]$$

$$= \lim_{s\to 0}\int d^3x_1\frac{1}{s^{3/2}}\exp\left[\frac{i}{4}(x_1-x_2)eF\coth eFs(x_1-x_2)\right]$$

$$= \lim_{s\to 0}\frac{1}{s^{3/2}}(i\pi)^{3/2}\left(\det\frac{1}{4s}eFs\coth eFs\right)^{-1/2}$$

$$= \left(\frac{i\pi}{s}\right)^{3/2}e^{-(1/2)\mathrm{tr}\,\ln(1/4s)[1+\mathcal{O}(s^2)]} = \left(\frac{i\pi}{s}\right)^{3/2}e^{(3/2)\ln 4s} = \left(4\pi i\right)^{3/2}$$

$$= 8\pi^{3/2}\,e^{i(\pi/4)}.$$

Therefore, to satisfy (4.89), we have to choose

$$C = \frac{1}{8\pi^{3/2}}\,e^{-i(\pi/4)} = -\frac{1}{8\pi^{3/2}}\,e^{(3\pi/4)i}.\tag{4.90}$$

So far, we found the following for the transition amplitude:

$$\langle x_1|U(s)|x_2\rangle = -\frac{e^{(3\pi i/4)}}{(4\pi s)^{3/2}} e^{ie \int_{x_2}^{x_1} d\xi_\mu A^\mu(\xi)} \tag{4.91}$$

$$\times \exp\left[\frac{i}{4}(x_1 - x_2)eF \coth eFs(x_1 - x_2)\right.$$

$$\left. -\frac{1}{2}\mathrm{tr}\,\ln\left(\frac{\sinh eFs}{eFs}\right) - ie\,\gamma^{\lambda\star}F_\lambda s\right].$$

Incidentally, differences from the four-dimensional version of $\langle x_1|U(s)|x_2\rangle \equiv \langle x_1, s|x_2\rangle$ can be identified in the Dirac structure, in the power dependence of the proper time and in the numerical prefactors.

Finally, regarding (4.75), we have to evaluate the following trace:

$$\mathrm{Tr}\left[\gamma^\mu U(s)\right]_{x'\to x} = \mathrm{Tr}\left[\gamma^\mu e^{-is(\Pi^2 + e\gamma^{\lambda\star}F_\lambda)}\right]_{x'\to x}$$

$$= -\frac{e^{(3\pi i/4)}}{(4\pi s)^{3/2}}\mathrm{tr}_\gamma\left\{\gamma^\mu \exp\left[-\frac{1}{2}\mathrm{tr}\,\ln\left(\frac{\sinh eFs}{eFs}\right) - ie\,\gamma^{\lambda\star}F_\lambda s\right]\right\}. \tag{4.92}$$

For the first term in the exponential, we need the spectrum of $F^{\mu\nu}$. For this, first note that

$$F^{\mu\nu}F_{\nu\alpha}F^{\alpha\beta} = -\frac{1}{2}F^2\,F^{\mu\beta} \equiv -{}^\star F^2\,F^{\mu\beta}. \tag{4.93}$$

To prove the last identity, it is sufficient to recognize that the left-hand side is antisymmetric and, therefore, must be proportional to $F^{\mu\beta}$ – the only available antisymmetric tensor. Now, iterating the eigenvalue equation

$$F^{\mu\nu}\psi_\nu = \lambda\psi^\mu, \quad \Rightarrow \quad F^{\mu\nu}F_{\nu\alpha}F^{\alpha\beta}\psi_\beta = \lambda^3\psi^\mu \overset{(4.93)}{=} -{}^\star F^2\,\lambda\psi^\mu,$$

we can identify the eigenvalues:

$$\lambda = 0, \pm i|{}^\star F|, \quad \text{where} \quad |{}^\star F| = \sqrt{B^2 - E^2}. \tag{4.94}$$

Employing these eigenvalues, we obtain

$$\exp\left[-\frac{1}{2}\mathrm{tr}\,\ln\left(\frac{\sinh eFs}{eFs}\right)\right] = \exp\left(\ln\det\sqrt{\frac{eFs}{\sinh eFs}}\right)$$

$$= \left(\frac{ie|{}^\star F|s}{\sinh ie|{}^\star F|s}\right)^{1/2}\left(\frac{-ie|{}^\star F|s}{\sinh -ie|{}^\star F|s}\right)^{1/2}$$

$$= \frac{e|{}^\star F|s}{\sin e|{}^\star F|s}. \tag{4.95}$$

The second exponential in (4.92) is given by

$$e^{-ie\gamma^{\lambda\star}F_\lambda s} = \cos\left(es|{}^\star F|\right) - i\gamma^\lambda\frac{{}^\star F}{|{}^\star F|}\sin\left(es|{}^\star F|\right). \tag{4.96}$$

Now we can continue with (4.92), and find

$$\text{Tr}\left[\gamma^\mu U(s)\right]_{x'\to x} = \frac{e^{-i(\pi/4)}}{(4\pi s)^{3/2}}\text{tr}_\gamma\left[\gamma^\mu e|{}^\star F|s\left(\cot\left(es|{}^\star F|\right) - i\gamma^\lambda\frac{{}^\star F}{|{}^\star F|}\right)\right]$$

$$= i\frac{e^{-i(\pi/4)}}{4\pi^{3/2}}\frac{1}{s^{1/2}}\, e^\star F^\mu. \tag{4.97}$$

This brings us to the expression

$$\langle j_{\text{odd}}^\mu(x)\rangle = -em\int_0^\infty ds\, e^{-im^2 s}\, i\,\frac{e^{-i(\pi/4)}}{4\pi^{3/2}}\frac{1}{s^{1/2}}\, e^\star F^\mu$$

$$= 2ie^2 m\,\frac{e^{3\pi i/4}}{8\pi^{3/2}}\int_0^\infty \frac{ds}{s^{1/2}}\, e^{-im^2 s}\, {}^\star F^\mu$$

$$= -\frac{m}{|m|}\frac{e^2}{4\pi}\, {}^\star F^\mu, \tag{4.98}$$

where we have made use of the standard proper-time integral [91],

$$\int_0^\infty \frac{ds}{s^{1/2}}\, e^{-im^2 s} = -ie^{i(\pi/4)}\frac{\sqrt{\pi}}{|m|}.$$

Equation (4.98) is in perfect agreement with, for example, Redlich's result [137].[4] The same result can be found in [44, 131]. Note that the modulus of the generated topological photon mass is independent of the value of the fermion mass.

This concludes our investigation of the symmetry-breaking patterns of 2+1-dimensional quantum electrodynamics.

4.3.1 QED$_{2+1}$ for Nonconstant Fields

We now turn to a discussion of external *inhomogeneous* electromagnetic fields interacting with massive fermions in the two-component formulation of 2+1-dimensional QED. In the following, we shall omit the discussion of parity-violating contributions and concentrate on the parity-even part of the effective action. The aim is to determine the next-to-leading-order invariants involving derivatives of the field strength tensor. The variation of the electromagnetic field is considered to be small. Therefore, the scale is set by the fermion mass m.

We shall employ the heat-kernel method, which takes advantage of the similarities between the Schrödinger equation and the diffusion equation (e.g. [70]). In the present context, we are dealing with a Schrödinger equation (e.g. (4.77)) that describes an evolution in the proper time. An analytic continuation of this proper-time variable, therefore, leads to a kind of diffusion

---

[4] The discrepancy of an overall minus sign stems from a different definition of the effective action.

equation. Again, the Hamiltonian is equal to the squared Dirac operator $\not{D}^2$, and acts in a fictitious quantum mechanical Hilbert space whose coordinate representation is given by

$$\langle x|H|y\rangle = (\not{D}_x^2)\delta(x-y), \qquad \not{D}_x = \gamma^\mu[\partial_x^\mu - \mathrm{i}eA^\mu(x)]. \qquad (4.99)$$

We work in Euclidean space and the γ matrices are the 2×2 matrices $\gamma_{2\times 2}$ mentioned earlier (just before (4.39)). The time development in this three-dimensional space is governed by the operator

$$K(t) = \mathrm{e}^{-Ht}, \quad t > 0, \qquad (4.100)$$

where t denotes the Wick-rotated proper time. The well-known relation between the kernel $K(t)$ and the Green's function Δ is stated in (2.27):

$$\Delta = \int_0^\infty \mathrm{d}t\, \mathrm{e}^{-m^2 t}\, K(t) = \int_0^\infty \mathrm{d}t\, \mathrm{e}^{-(H+m^2)t} = \frac{1}{\not{D}^2 + m^2}. \qquad (4.101)$$

As usual, we shall be interested in the matrix element

$$K(x,y;t) = \langle x,t|y,0\rangle = \langle x|\mathrm{e}^{-Ht}|y\rangle, \qquad (4.102)$$

which satisfies the diffusion equation

$$\left(\frac{\partial}{\partial t} + \not{D}_x^2\right) K(x,y;t) = 0, \qquad (4.103)$$

with the boundary condition $K(x,y;0) = \langle x|y\rangle = \delta(x-y)$. Introducing the kinetic momentum $\Pi_\mu = p_\mu - eA_\mu$ and $p_\mu = -\mathrm{i}\partial_\mu$, we can rewrite the Hamiltonian in the following form ($\sigma_{\mu\nu} = (\mathrm{i}/2)[\gamma_\mu,\gamma_\nu]$):

$$H = \not{D}^2 = -(\gamma\Pi)^2 = \Pi_\mu\Pi_\mu - \frac{e}{2}\sigma_{\mu\nu}F_{\mu\nu},$$

$$\text{or}\quad H = -\partial_\mu\partial_\mu + \mathrm{i}e\partial_\mu A_\mu + \mathrm{i}eA_\mu\partial_\mu + e^2 A_\mu A_\mu - \frac{e}{2}\sigma_{\mu\nu}F_{\mu\nu}.$$

When this Hamiltonian is substituted into (4.103), we find for the diffusion equation in QED$_3$

$$\left(-\partial^2 + 2\mathrm{i}eA_\mu\partial_\mu + X + \frac{\partial}{\partial t}\right) K(x,y;t) = 0, \qquad (4.104)$$

$$X = \mathrm{i}e\partial_\mu A_\mu + e^2 A_\mu A_\mu - \frac{e}{2}\sigma_{\mu\nu}F_{\mu\nu}.$$

For vanishing fields, (4.104) can easily be solved by

$$K_0(x,y;t) = \frac{1}{(4\pi t)^{3/2}}\mathrm{e}^{-(x-y)^2/4t}. \qquad (4.105)$$

For nonvanishing fields we try to solve (4.104) by the ansatz

$$K(x,y;t) = \frac{1}{(4\pi t)^{3/2}}\mathrm{e}^{-(x-y)^2/4t}\sum_{k=0}^\infty a_k(x,y)\, t^k. \qquad (4.106)$$

This expression, when inserted in (4.104), yields the following recursion relation for the coefficients a_k:

$$k = 0: \quad (x - y)_\mu D_\mu a_0(x, y) = 0, \tag{4.107}$$

$$k \geq 0: \quad \left(D^2 + \frac{e}{2}\sigma F\right) a_k(x, y) = (k + 1)a_{k+1}(x, y)$$
$$+ (x - y)_\mu(\partial_\mu - ieA_\mu)a_{k+1}(x, y).$$

These recursion relations are formally the same as the ones that one encounters in QED$_4$ [97]. Furthermore, from our experience with constant-field configurations, we decompose $K(x, y; t)$ into a gauge-independent factor and the counter-gauge factor

$$\Phi(x, y) = \exp\left(ie \int_y^x d\xi_\mu\, A_\mu(\xi)\right),$$

which allows us to separate off the gauge dependence from the coefficients a_k by writing

$$a_k(x, y) = \Phi(x, y)\, f_k(x, y). \tag{4.108}$$

Now we take over the known results from QED$_4$ [97] and obtain, using the convenient definition $[f_k](x) = \lim_{y \to x} f_k(x, y)$,

$$[f_0](x) = 1, \quad [f_1](x) = \frac{e}{2}\sigma F, \quad [f_2](x) = \frac{e^2}{12}F^2 + \frac{e}{12}\sigma F^{,\mu\mu} + \frac{1}{2}\left(\frac{e}{2}\sigma F\right)^2,$$

$$[f_3](x) = \frac{1}{3!}\left(\frac{e}{2}\sigma F\right)^3 + \frac{e}{2}\sigma F\left(\frac{e}{12}\sigma F^{,\mu\mu} + \frac{e^2}{12}F^2\right) + \frac{e^2}{48}\sigma F^{,\mu}\sigma F^{,\mu} \tag{4.109}$$

$$+ \frac{e}{120}\sigma F^{,\mu\mu\nu\nu} - \frac{e^2}{30}F^{\alpha\beta}F^{\alpha\beta,\mu\mu} - \frac{e^2}{45}(F^{,\mu})^2 - \frac{e^2}{180}F^{\alpha\mu,\mu}F^{\alpha\nu,\nu},$$

where

$$\sigma F = \sigma^{\alpha\beta}F^{\alpha\beta}, \qquad F^2 = F^{\alpha\beta}F^{\beta\alpha},$$

$$(F^{,\mu})^2 = F^{\alpha\beta,\mu}F^{\alpha\beta,\mu}, \qquad F^{\alpha\beta,\lambda} = \frac{\partial}{\partial x_\lambda}F^{\alpha\beta}. \tag{4.110}$$

These $[f_k](x)$ contain contributions with and without space derivatives. Hence, it is only natural to write

$$\sum_{k=0}^{\infty}[f_k]\, t^k = [1 + b_1(x)\, t + b_2(x)\, t^2 + \ldots]\sum_{k=0}^{\infty}[f_k^c](x)\, t^k. \tag{4.111}$$

The coefficients $[f_k^c]$ depend on nonconstant fields in the same manner as they would on constant fields. The $b_i(x)$ represent corrections for nonconstant fields. According to (4.109) and (4.110), the terms without derivatives are given by

$$[f_1^c] = \frac{e}{2}\sigma F, \quad [f_2^c] = \frac{e^2}{12}F^2 + \frac{1}{2}\left(\frac{e}{2}\sigma F\right)^2,$$

$$[f_3^c] = \frac{1}{3!}\left(\frac{e}{2}\sigma F\right)^3 + \frac{e}{2}\sigma F\frac{e^2}{12}F^2. \tag{4.112}$$

Substituting the results of (4.112), (4.109) and (4.110) into (4.111) produces

$$b_1(x) = 0, \qquad b_2(x) = \frac{e}{12}\sigma F^{,\mu\mu} \tag{4.113}$$

$$b_3(x) = \frac{e^2}{48}\sigma F^{,\mu}\sigma F^{,\mu} + \frac{e}{120}\sigma F^{,\mu\mu\nu\nu} - \frac{e^2}{30}F^{\alpha\beta}F^{\alpha\beta,\mu\mu}$$

$$- \frac{e^2}{45}(F^{,\mu})^2 - \frac{e^2}{180}F^{\alpha\mu,\mu}F^{\alpha\nu,\nu}.$$

The series in (4.111), containing the form of the constant-field contributions, can be summed to give the well-known result [38]

$$K^c(x,y;t) = \frac{\Phi(x,y)}{(4\pi t)^{3/2}}e^{(e/2)\sigma Ft}\exp\left[-\frac{1}{2}\operatorname{tr}\ln\left(\frac{\sin(eFt)}{eFt}\right)\right]$$

$$\times \exp\left[-\frac{1}{4}(x-y)eF\cot(eFt)(x-y)\right], \tag{4.114}$$

which represents the proper-time Wick-rotated version of (4.91). In the coincidence limit $x \to y$, we then obtain for nonconstant fields

$$K(x,x;t) = \frac{1}{(4\pi t)^{3/2}}\exp\left[\frac{e}{2}\sigma Ft - \frac{1}{2}\operatorname{tr}\ln\left(\frac{\sin(eFt)}{eFt}\right)\right] \tag{4.115}$$

$$\times\left[1 + \frac{e}{12}\sigma F^{,\mu\mu}t^2 + \left(\frac{e^2}{48}\sigma F^{,\mu}\sigma F^{,\mu} + \frac{e}{120}\sigma F^{,\mu\mu\nu\nu}\right.\right.$$

$$\left.\left. - \frac{e^2}{30}F^{\alpha\beta}F^{\alpha\beta,\mu\mu} - \frac{e^2}{45}(F^{,\mu})^2 - \frac{e^2}{180}F^{\alpha\nu,\mu}F^{\alpha\nu,\nu}\right)t^3 + \cdots\right].$$

Finally we arrive at the (Euclidean) one-loop effective action for nonconstant-field configurations ($^\star F_\mu = (1/2)\epsilon_{\mu\nu\rho}F_{\nu\rho}$):

$$\Gamma^{(1)} = \int d^3x\,\mathcal{L}^{(1)}(x) = \frac{1}{2}\int d^3x\int_0^\infty\frac{dt}{t}e^{-m^2t}\operatorname{Tr}[K(x,x;t) + \text{c.t.}], \tag{4.116}$$

where

$$\operatorname{Tr}[K(x,x;t) + \text{c.t.}] \tag{4.117}$$

$$= \frac{1}{(4\pi t)^{3/2}}\operatorname{tr}_\gamma\left\{\left(e^\star Ft\coth(e^\star Ft) + \frac{e}{2}\sigma Ft\right)\right.$$

$$\times\left[1 + \frac{e}{12}\sigma F^{,\mu\mu}t^2 + \left(\frac{e^2}{48}\sigma F^{,\mu}\sigma F^{,\mu} + \frac{e}{120}\sigma F^{,\mu\mu\nu\nu}\right.\right.$$

$$-\frac{e^2}{30}F^{\alpha\beta}F^{\alpha\beta,\mu\mu} - \frac{e^2}{45}(F^{,\mu})^2 - \frac{e^2}{180}F^{\alpha\mu,\mu}F^{\alpha\nu,\nu}\bigg)t^3\bigg]\bigg\}.$$

Deviating from QED$_4$, we now need the following trace identities valid for QED$_3$ only:

$$\mathrm{tr}_\gamma\, 1 = 2, \quad \mathrm{tr}\,\sigma^{\mu\nu} = 0, \quad \mathrm{tr}\,\sigma^{\mu\nu}\sigma^{\lambda\sigma} = 2(\delta^{\mu\lambda}\delta^{\nu\sigma} - \delta^{\mu\sigma}\delta^{\nu\lambda}),$$
$$\mathrm{tr}\,\sigma^{\mu\nu}\sigma^{\lambda\sigma}\sigma^{\rho\kappa} = 2\mathrm{i}(\epsilon^{\mu\nu\rho}\epsilon^{\lambda\sigma\kappa} - \epsilon^{\mu\nu\kappa}\epsilon^{\lambda\sigma\rho}).$$

Performing the various trace operations in (4.117), we finally end up with

$$\Gamma^{(1)} = \frac{1}{16\pi^{3/2}}\int\limits_0^\infty \frac{dt}{t^{5/2}}e^{-m^2t}\int d^3x\,\bigg\{2[e^\star F t\coth(e^\star F t) - 1]$$

$$+\frac{e^2}{15}e^\star F t\coth e^\star F t\left(\frac{7}{12}F^{\alpha\beta,\mu}F^{\alpha\beta,\mu} - F^{\alpha\beta}F^{\alpha\beta,\mu\mu}\right. \qquad (4.118)$$

$$\left.-\frac{1}{6}F^{\alpha\mu,\mu}F^{\alpha\nu,\nu}\right)t^3 + \frac{e^2}{6}F^{\lambda\sigma}F^{\lambda\sigma,\mu\mu}t^3 + \frac{e^2}{60}F^{\lambda\sigma}F^{\lambda\sigma,\mu\mu\nu\nu}t^4\bigg\}.$$

Equation (4.118) represents the central result of this section. Note that all terms on the right-hand side yield finite contributions. It should be mentioned that the last term is not the only one proportional to t^4 and involving four derivatives; for a complete list, we have to evaluate the fourth-order coefficient $[f_4](x)$ in (4.109). Hence, the last term in (4.118) should be regarded as only an example of a fourth-order term.

In the following, we shall mostly be interested in nonconstant (static) external magnetic fields,

$$F^{12} = B, \qquad {}^\star F = B > 0. \qquad (4.119)$$

Then the first term in the curly brackets of (4.118) yields the well-known form for the effective Lagrangian in QED$_3$ for constant B fields (cf. (4.50)):

$$\mathcal{L}^{(1)} = \frac{\sqrt{2}}{2\pi}(eB)^{3/2}\,\zeta\left(-\frac{1}{2}, \frac{m^2}{2eB}\right) - \frac{meB}{4\pi}. \qquad (4.120)$$

The correction terms for a nonconstant B field that follow from (4.118) are contained in three contributions (labeled by A, B and C). The first is

$$\mathcal{L}_A^{(1)} = \frac{1}{16\pi^{3/2}}\frac{e^2}{15}Q\int\limits_0^\infty dt\,\sqrt{t}\,eBt\coth(eBt)\,e^{-m^2t}, \qquad (4.121)$$

where

$$Q = \frac{7}{12}F^{\alpha\beta,\mu}F^{\alpha\beta,\mu} - F^{\alpha\beta}F^{\alpha\beta,\mu\mu} - \frac{1}{6}F^{\alpha\mu,\mu}F^{\alpha\nu,\nu}$$
$$= \frac{7}{6}\partial^\mu B\partial^\mu B - B\partial^2 B - \frac{1}{6}\partial^j B\partial^j B, \qquad j = 1,2. \qquad (4.122)$$

The value of the integral occurring in (4.121) turns out to be

$$\frac{1}{(eB)^{3/2}} \frac{3\sqrt{\pi}}{4} \left[\frac{1}{2\sqrt{2}} \zeta\left(\frac{5}{2}, \frac{m^2}{2eB}\right) - \left(\frac{eB}{m^2}\right)^{5/2} \right].$$

Hence, we obtain

$$\mathcal{L}_A^{(1)} = \frac{e^2}{640\pi\sqrt{2}} \zeta\left(\frac{5}{2}, 1 + \frac{m^2}{2eB}\right) \frac{Q}{(eB)^{3/2}} + \frac{e^2}{320\pi} \frac{eB}{m^2} \frac{Q}{m^3}, \qquad (4.123)$$

where, in the limit of a static inhomogeneous magnetic field, Q reduces to

$$Q = \partial^j B \partial^j B - 2B \partial^2 B. \qquad (4.124)$$

The last two terms in (4.118) contribute

$$\mathcal{L}_B^{(1)} = \frac{1}{16\pi^{3/2}} \frac{e^2}{6} F^{\lambda\sigma} F^{\lambda\sigma,\mu\mu} \int_0^\infty dt\, \sqrt{t}\, e^{-m^2 t} = \frac{e^2}{96\pi} \frac{B \partial^2 B}{|m|^3}, \qquad (4.125)$$

$$\mathcal{L}_C^{(1)} = \frac{1}{16\pi^{3/2}} \frac{e^2}{60} F^{\lambda\sigma} F^{\lambda\sigma,\mu\mu\nu\nu} \int_0^\infty dt\, t^{3/2} e^{-m^2 t} = \frac{e^2}{640\pi} \frac{B \partial^2 \partial^2 B}{|m|^5}. \quad (4.126)$$

Equations (4.123)–(4.126) represent the next-to-leading-order contributions to the derivative expansion of the (parity-even) effective action. To be precise, this expansion is valid as long as the variation of the field is small compared with the fermion mass.[5] Recently, second-order derivative expansions have been established with the aid of the elegant worldline method, which are capable of analyzing the $m \to 0$ limit as well [38, 95]. In this limit, it becomes apparent that the presence of an inhomogeneous magnetic field may lower the energy of the vacuum configuration in QED$_{2+1}$; it has therefore been speculated that the QED$_{2+1}$ ground state might resemble the QCD vacuum state [38]. A lower bound was placed on the fermionic determinant of Euclidean QED$_3$ in [80].

Finally, let us mention that exact effective one-loop actions for special nonconstant field configurations have been found in 2+1 as well as 3+1 dimensions [39,69]; from the viewpoint of a derivative expansion, these solutions represent "all-order" results.

For the remainder of this section, we consider the weak-field limit of the effective action (4.118). Then, (4.118) reduces to

$$\Gamma^{(1)} = \frac{1}{16\pi^{3/2}} \int_0^\infty \frac{dt}{t^{5/2}} e^{-m^2 t} \int d^3x \left[\frac{e^2}{3} F^{\mu\nu} F^{\mu\nu} t^2 + \frac{e^2}{6} F^{\lambda\sigma} F^{\lambda\sigma,\mu\mu} t^3 \right. \qquad (4.127)$$

$$\left. + \frac{e^2}{15} \left(\frac{7}{12} F^{\alpha\beta,\mu} F^{\alpha\beta,\mu} - F^{\alpha\beta} F^{\alpha\beta,\mu\mu} - \frac{1}{6} F^{\alpha\mu,\mu} F^{\alpha\nu,\nu} \right) t^3 \right].$$

---

[5] We take the opportunity to point out that this expansion is not valid in the limit $m \to 0$, which was erroneously investigated in the original paper [59].

Partial integration leads us to

$$\Gamma^{(1)} = \frac{1}{16\pi^{3/2}} \int\limits_0^\infty \frac{dt}{t^{5/2}} e^{-m^2 t} \tag{4.128}$$

$$\int d^3 x \left[\frac{e^2}{3} F^{\mu\nu} F^{\mu\nu} t^2 - \frac{e^2}{15} \left(\frac{11}{12} F^{\alpha\beta,\mu} F^{\alpha\beta,\mu} + \frac{1}{6} F^{\alpha\mu,\mu} F^{\alpha\nu,\nu} \right) t^3 \right].$$

With the aid of the Bianchi identity, we find

$$\int d^3 x \, F^{\alpha\beta,\mu} F^{\alpha\beta,\mu} = - \int d^3 x \, F^{\alpha\beta} F^{\alpha\beta,\mu\mu}$$

$$= - \int d^3 x \, F^{\alpha\beta} \left(F^{\mu\beta,\alpha\mu} + F^{\alpha\mu,\beta\mu} \right)$$

$$= \int d^3 x \left(F^{\alpha\beta,\alpha} F^{\mu\beta,\mu} + F^{\alpha\beta,\beta} F^{\alpha\mu,\mu} \right)$$

$$= 2 \int d^3 x \, F^{\alpha\mu,\mu} F^{\alpha\nu,\nu}. \tag{4.129}$$

Inserting (4.129) into (4.128), we arrive at

$$\Gamma^{(1)} = \frac{1}{16\pi^{3/2}} \int\limits_0^\infty \frac{dt}{t^{5/2}} e^{-m^2 t} \int d^3 x \left(\frac{e^2}{3} F^{\mu\nu} F^{\mu\nu} t^2 - \frac{2e^2}{15} F^{\alpha\mu,\mu} F^{\alpha\nu,\nu} t^3 \right). \tag{4.130}$$

Adding the classical Maxwell part, the complete action becomes

$$\Gamma^{(1)} = \int d^3 x \left[\frac{1}{4} F^{\mu\nu} F^{\mu\nu} \left(1 + \frac{e^2}{12\pi^{3/2}} \int\limits_0^\infty \frac{dt}{\sqrt{t}} e^{-m^2 t} \right) \right.$$

$$\left. - \frac{e^2}{120\pi^{3/2}} F^{\alpha\mu,\mu} F^{\alpha\nu,\nu} \int\limits_0^\infty dt \sqrt{t} \, e^{-m^2 t} \right]$$

$$= \int d^3 x \left[\frac{1}{4} F^{\mu\nu} F^{\mu\nu} \left(1 + \frac{1}{12\pi} \frac{e^2}{|m|} \right) - \frac{1}{240\pi} \frac{e^2}{|m|} F^{\alpha\mu,\mu} F^{\alpha\nu,\nu} \right]. \tag{4.131}$$

We can easily derive the modified Maxwell equations by variation with respect to A^μ:

$$\partial^\alpha F^{\alpha\mu} \left(1 + \frac{\alpha_D}{3} \right) = - \frac{\alpha_D}{30 m^2} \partial^2 \partial^\alpha F^{\alpha\mu}, \tag{4.132}$$

where we have introduced the dimensionless coupling constant $\alpha_D = e^2/4\pi|m|$. Equation (4.132) precisely describes the 2+1-dimensional analogue to the Uehling vacuum polarization effect in the absence of any external sources. Note that the coefficient is half as large as in the 3+1-dimensional theory.

    This concludes our investigation of the influence of field inhomogeneities on the effective action in QED$_{2+1}$.

5. Scattering of Light by Light

The application of the Heisenberg–Euler Lagrangian can be used to describe the interaction of low-energy electromagnetic fields with each other. The nonlocal expression for the fundamental box graph of QED with two incoming and two outgoing photons is thereby reduced to the local Heisenberg–Euler Lagrangian.

However, there is also the other extremal situation, namely high-energy photon–photon scattering. Here the asymptotic form of the elastic scattering amplitude with one or two intermediate charged electron–positron pairs can be obtained using the impact factor or eikonal approximation.

5.1 Photon–Photon Scattering at Low Energies

Let the incoming photons have momenta k_a and k_b, while the scattered photons have momenta k_c and k_d. As the interaction Lagrangian, we take

$$\mathcal{L}^1(x) = \frac{2}{45} \frac{\alpha^2}{m^4} \left[(\boldsymbol{E}^2 - \boldsymbol{B}^2)^2 + 7(\boldsymbol{E} \cdot \boldsymbol{B})^2 \right]$$

$$= \frac{2}{45} \frac{\alpha^2}{m^4} \left(4\mathcal{F}^2 + 7\mathcal{G}^2 \right), \tag{5.1}$$

so that the effective Heisenberg–Euler action is given by

$$W^1 = \frac{2}{45} \frac{\alpha^2}{m^4} \int \mathrm{d}^4x \left(4\mathcal{F}^2 + 7\mathcal{G}^2 \right). \tag{5.2}$$

The transition matrix connecting the initial and final states is

$$W_{\mathrm{fi}} = \langle \mathrm{f} | \int \mathrm{d}^4x \left(4\mathcal{F}^2 + 7\mathcal{G}^2 \right) | \mathrm{i} \rangle$$

$$= \int \mathrm{d}^4x \, \langle a_{\lambda_c}(k_c) \, a_{\lambda_d}(k_d) | \left(4\mathcal{F}^2 + 7\mathcal{G}^2 \right) | a_{\lambda_a}^\dagger(k_a) a_{\lambda_b}^\dagger(k_b) \rangle. \tag{5.3}$$

The operators $\mathcal{F}^2$ and $\mathcal{G}^2$ can be obtained from the mode expansion

$$A_\mu(x) = \frac{1}{\sqrt{V}} \sum_k \frac{1}{\sqrt{2\omega}} \sum_{\lambda=1}^4 \left[a_\lambda(k) \, \epsilon_\mu^\lambda(k) \, \mathrm{e}^{\mathrm{i}kx} + a_\lambda^\dagger(k) \, \epsilon_\mu^{\lambda *}(k) \, \mathrm{e}^{-\mathrm{i}kx} \right].$$

Upon using the relation

$$F_{\mu\nu}(x) = \frac{\partial A_\mu}{\partial x^\nu} - \frac{\partial A_\nu}{\partial x^\mu},$$

we obtain the field intensity operator,

$$F_{\mu\nu}(x) = \frac{i}{\sqrt{V}} \sum_k \frac{1}{\sqrt{2\omega}} \sum_{\lambda=1}^4 \left[a_\lambda(k) f_{\mu\nu}^\lambda(k) e^{ikx} - a_\lambda^\dagger(k) f_{\mu\nu}^{\lambda*}(k) e^{-ikx} \right],$$

where $f_{\mu\nu}^\lambda(k) = \epsilon_\mu^\lambda(k) k_\nu - \epsilon_\nu^\lambda(k) k_\mu$. Using $\mathcal{F} = (1/4)F_{\mu\nu}F^{\mu\nu}$ and $\mathcal{G} = (1/4)F_{\mu\nu}{}^*F^{\mu\nu} = (1/8)F_{\mu\nu}\epsilon^{\mu\nu\rho\sigma}F_{\rho\sigma}$, the following two integrals can be evaluated:

$$\int d^4x \, \langle f|\mathcal{F}^2|i\rangle$$
$$= \frac{(2\pi)^4}{2V^2} \frac{\delta(k_a + k_b - k_c - k_d)}{\sqrt{16\omega_a\omega_b\omega_c\omega_d}} \tag{5.4}$$
$$\times \left(f_{\mu\nu}^a f^{b\mu\nu} f_{\alpha\beta}^c f^{d\alpha\beta} + f_{\mu\nu}^a f^{c\mu\nu} f_{\alpha\beta}^b f^{d\alpha\beta} + f_{\mu\nu}^a f^{d\mu\nu} f_{\alpha\beta}^b f^{c\alpha\beta} \right),$$

$$\int d^4x \, \langle f|\mathcal{G}^2|i\rangle$$
$$= \frac{(2\pi)^4}{V^2} \frac{\delta(k_a + k_b - k_c - k_d)}{\sqrt{16\omega_a\omega_b\omega_c\omega_d}}$$
$$\times \left(-f_{\rho\sigma}^a f^{b\rho\sigma} f_{\lambda\tau}^c f^{d\lambda\tau} - f_{\rho\sigma}^a f^{c\rho\sigma} f_{\lambda\tau}^b f^{d\lambda\tau} - f_{\rho\sigma}^a f^{d\rho\sigma} f_{\lambda\tau}^b f^{c\lambda\tau} \right.$$
$$+ f_{\rho\lambda}^a f^{b\lambda\tau} f_{\tau\sigma}^c f^{d\sigma\rho} + f_{\rho\lambda}^a f^{b\lambda\tau} f_{\tau\sigma}^d f^{c\sigma\rho} + f_{\rho\lambda}^a f^{c\lambda\tau} f_{\tau\sigma}^d f^{b\sigma\rho}$$
$$\left. + f_{\rho\lambda}^a f^{d\lambda\tau} f_{\tau\sigma}^c f^{b\sigma\rho} + 2f_{\rho\sigma}^a f^{d\sigma\tau} f_{\tau\lambda}^b f^{c\lambda\rho} \right). \tag{5.5}$$

Using (5.4) and (5.5), the transition matrix becomes

$$M_{fi} = \frac{2}{45} \frac{\alpha^2}{m^4} \frac{(2\pi)^4}{V^2} \frac{\delta(k_a + k_b - k_c - k_d)}{\sqrt{16\omega_a\omega_b\omega_c\omega_d}} \tag{5.6}$$
$$\times \left(-5f_{\mu\nu}^a f^{b\mu\nu} f_{\alpha\beta}^c f^{d\alpha\beta} - 5f_{\mu\nu}^a f^{c\mu\nu} f_{\alpha\beta}^b f^{d\alpha\beta} - 5f_{\mu\nu}^a f^{d\mu\nu} f_{\alpha\beta}^b f^{c\alpha\beta} \right.$$
$$\left. + 14f_{\mu\nu}^a f^{b\nu\alpha} f_{\alpha\beta}^c f^{d\beta\mu} + 14f_{\mu\nu}^a f^{c\nu\alpha} f_{\alpha\beta}^b f^{d\beta\mu} + 14f_{\mu\nu}^a f^{d\nu\alpha} f_{\alpha\beta}^b f^{c\beta\mu} \right),$$

or

$$M_{fi} = \frac{2}{45} \frac{\alpha^2}{m^4} \frac{(2\pi)^4}{V^2} \frac{\delta(k_a + k_b - k_c - k_d)}{\sqrt{16\omega_a\omega_b\omega_c\omega_d}} \, f, \tag{5.7}$$

where f is the quantity in the last pair of parentheses in (5.6).

If we observe the scattered photon with momentum k_c, the kinematics of the problem can be described as shown in Fig. 5.1.

A detailed calculation for this kinematics yields the differential cross section for scattering of a photon with energy $k_c^0 = \omega_c$ within the solid angle $d\Omega$:

$$\frac{d\sigma}{d\Omega} = \left(\frac{\alpha^2}{4\pi \times 45m^4} \right)^2 \left(\frac{\omega_c}{\omega_a\omega_b} \right)^2 \frac{1}{(1 - \cos\delta)^2} \, |f|^2. \tag{5.8}$$

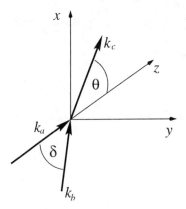

Fig. 5.1. Kinematics for γ–γ scattering, $(k_a = \omega_a \hat{z},\ k_b = \omega_b \sin\delta\,\hat{x} + \omega_b \cos\delta\hat{z},$
$k_c = \omega_c \sin\theta\cos\phi\,\hat{x} + \omega_c \sin\theta\sin\phi\,\hat{y} + \omega_c \cos\theta\,\hat{z})$

Since we are not interested in observing the polarization states of the photons, we must average over the polarization of the incident photons (a,b) and sum over the polarization of the scattered photons (c,d). Hence the differential cross section is given by

$$\frac{d\sigma}{d\Omega} = \left(\frac{\alpha^2}{4\pi \times 45m^4}\right)^2 \left(\frac{\omega_c}{\omega_a\omega_b}\right)^2 \frac{1}{(1-\cos\delta)^2}\frac{1}{4}\sum_{\text{pol.a,b,c,d}} |f|^2. \tag{5.9}$$

After a lengthy calculation, the summation over the polarization turns out to give

$$\sum_{\text{pol.a,b,c,d}} |f|^2 = 1112\Big[(k_a\cdot k_b)^4 + (k_a\cdot k_b)^4 + \{(k_a\cdot k_b)-(k_a\cdot k_c)\}^4\Big]. \tag{5.10}$$

Here, then, is our final answer for the differential cross section for $\gamma\gamma$ scattering:

$$\frac{d\sigma}{d\Omega} = r_0^2 \left(\frac{\alpha}{2\pi}\right)^2 \frac{278}{8100}\left(\frac{\omega_c}{m}\right)^2 \frac{1}{(1-\cos\delta)^2} \tag{5.11}$$

$$\times \left\{\left(\frac{\omega_a\omega_b}{m^2}\right)^2 (1-\cos\delta)^4 + \left(\frac{\omega_a}{\omega_b}\right)^2\left(\frac{\omega_c}{m}\right)^4 (1-\cos\theta)^4\right.$$

$$\left. + \left(\frac{\omega_a}{\omega_b}\right)^2\left[\frac{\omega_b}{m}(\cos\delta-1) - \frac{\omega_c}{m}(\cos\theta-1)\right]^4\right\},$$

where $r_0 = e^2/(4\pi m) = \alpha/m \simeq 2.8\times10^{-13}$ cm is the classical electron radius. In the CM system, where

$$(k_a^0 =)\,\omega_a = \omega_b = \omega_c = \omega_d \equiv \omega$$

and $\delta = \pi$, we find for the differential cross section [76, 106, 112]

$$\frac{d\sigma}{d\Omega} = \frac{139}{8100}\left(\frac{\alpha}{2\pi}\right)^2 r_0^2 \left(\frac{\omega}{m}\right)^6 (3 + \cos^2\theta)^2 \quad \text{(CM)}, \quad \text{for} \quad \omega \ll m. \quad (5.12)$$

The integrated total elastic cross section is then given by

$$\sigma = \frac{973}{10125}\frac{\alpha^2}{\pi} r_0^2 \left(\frac{\omega}{m}\right)^6. \tag{5.13}$$

We may add that, at very high energies, one finds (for $\omega \gg m$) a logarithmic variation in the forward direction:

$$\frac{d\sigma}{d\Omega}(\omega) \sim \left(\frac{\alpha}{\pi}\right)^2 r_0^2 \left(\frac{m}{\omega}\right)^2 \left(\ln\frac{\omega}{m}\right)^4. \tag{5.14}$$

5.2 High-Energy Photon Scattering
Close to the Forward Direction

In the previous sections, we have treated the photon field mainly as an external slowly varying c-number field. There, we presented several applications using low-energy Heisenberg–Euler-type Lagrangians (see Fig. 5.2). It is the purpose of the present section to switch to the other extreme, high-energy photons; i.e. we want to study processes in which two real photons undergo an elastic scattering with one or two intermediate charged electron–positron pairs.

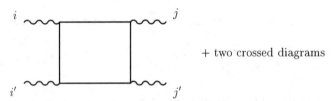

+ two crossed diagrams

Fig. 5.2. The fourth-order Feynman diagrams for γ–γ scattering

Needless to say, these processes are hard to deal with for arbitrary photon energies. Hence we are going to resort to the impact factor, or eikonal, approximation, which allows one to calculate closed-form expressions for the high-energy photon–photon scattering amplitude. Our treatment follows the papers by Cheng and Wu [41] and Dittrich [53,62].

Let us begin with the well-known lowest-order process possible in the photon–photon collision, $\gamma + \gamma \rightarrow e^+ + e^-$. The single-pair-creation cross section for this process is given as (e.g. [150])

$$\sigma_a(s) = 2\pi\frac{\alpha^2}{s}\kappa\left[(3 - \kappa^4)\frac{1}{\kappa}\ln\frac{1+\kappa}{1-\kappa} - 4 + 2\kappa^2\right]$$

$$= \frac{4\pi\alpha^2}{s}\left\{\left(2 + \frac{8m^2}{s} - \frac{16m^4}{s^2}\right)\ln\left[\left(\frac{s}{4m^2}\right)^{1/2} + \left(\frac{s}{4m^2} - 1\right)^{1/2}\right]\right.$$

$$-\left(1+\frac{4m^2}{3}\right)\left(1-\frac{4m^2}{3}\right)^{1/2}\Bigg\},\quad (5.15)$$

where $\kappa = \left(1-4m^2/s\right)^{1/2}$. The limiting value of (5.15) for $s \to \infty$ is given by

$$\sigma_{\rm b}(s) \simeq \frac{4\pi\alpha^2}{s}\left(\ln\frac{s}{m^2}-1\right), \quad \frac{s}{m^2}\to\infty, \quad (5.16)$$

which vanishes asymptotically as $s \to \infty$. Following [41], we want to show that the eighth-order two-pair-creation process (see Fig. 5.3), $\gamma + \gamma \to e^+ + e^- + e^+ + e^-$, has a cross section whose asymptotic value is given by a constant

$$\lim_{s\to\infty}\sigma_{\rm c}(s) \simeq \frac{\alpha^4}{36\pi m^2}\left[175\zeta(3)-38\right] \simeq 6.45\,\mu\text{b}. \quad (5.17)$$

Thus, although $\sigma_{\rm b}$ becomes of lower order in α, $\sigma_{\rm c}$ becomes larger than $\sigma_{\rm b}$ as the photon energy becomes sufficiently high (see Fig. 5.4). The cross section (5.17) dominates over (5.16) when $s \geq 1\,\text{GeV}^2$, where s is the square of the total energy in the CM system.

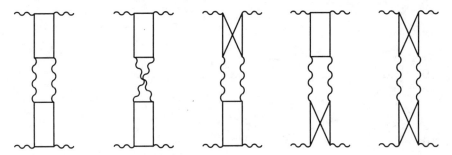

Fig. 5.3. The eighth-order γ–γ scattering that gives a finite cross section at infinite energy

For the problem under discussion, where the incoming and outgoing photon energy s is very large, while the momentum transfer $\boldsymbol{\Delta}$ is very small (essentially zero) compared with the rest energy of the electron, the impact factor representation [41] can be applied:

$$M_0^{(\gamma\gamma)} \sim \text{i}s\frac{1}{(2\pi)^2}\int d^2q_\perp\,\frac{1}{|\boldsymbol{q}_\perp|^2}\,J_{ij}^\gamma(\boldsymbol{\Delta}=0,\boldsymbol{q}_\perp)\,J_{i'j'}^\gamma(\boldsymbol{\Delta}=0,\boldsymbol{q}_\perp). \quad (5.18)$$

Here, i and i', and j and j' are the polarization indices for the incoming and outgoing photons, respectively, and the photon impact factor at $\boldsymbol{\Delta} = 0$ is given by [41]

$$J_{ij}^\gamma(0,\boldsymbol{q}_\perp) = -8\alpha^2\int_0^1 d\beta\int_0^1 dx\,\big[\boldsymbol{q}_\perp^2\,x(1-x)+m^2\big]^{-1}$$

$$\times \left\{ 2\beta(1-\beta)x(1-x)q_{\perp i}q_{\perp j} \right.$$

$$\left. -q_\perp^2 \delta_{ij}\left[\frac{1}{4}-2\beta(1-\beta)\left(x-\frac{1}{2}\right)^2\right]\right\}$$

$$= \frac{2}{3}\alpha^2 \int\limits_0^1 dx \left[q_\perp^2 x(1-x)+m^2\right]^{-1}$$

$$\times \left\{q_\perp^2 \delta_{ij}[3-(1-2x)^2]-4x(1-x)q_{\perp i}q_{\perp j}\right\}. \quad (5.19)$$

Substituting (5.19) into (5.18) and carrying out the various integrations, we end up with

$$M_0^{\gamma\gamma} \sim \delta_{ij}\delta_{i'j'}\,A + \frac{1}{2}\left(\delta_{ii'}\delta_{jj'}+\delta_{ij'}\delta_{i'j}-\delta_{ij}\delta_{i'j'}\right)A_{\mathrm{ex}},$$

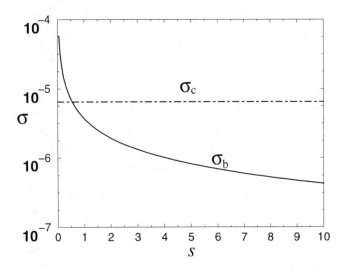

Fig. 5.4. Asymptotic cross sections σ_{b} and σ_{c} in units of barns as a function of the CM energy s in units of GeV^2

where

$$A = \mathrm{i}s\alpha^4(9\pi m^2)^{-1}\int\limits_0^1 dy\,dy'\,\frac{(5-y^2)(5-y'^2)}{y'^2-y^2}\ln\frac{1-y^2}{1-y'^2}, \quad (5.20)$$

and

$$A_{\mathrm{ex}} = \mathrm{i} s \alpha^4 (9\pi m^2)^{-1} \int\limits_0^1 \mathrm{d}y \, \mathrm{d}y' \frac{(1-y^2)(1-y'^2)}{y'^2 - y^2} \ln \frac{1-y^2}{1-y'^2}. \tag{5.21}$$

Numerical integration yields

$$A = \mathrm{i} s \alpha^4 (9\pi m^2)^{-1} \times 43.0899,$$
$$A_{\mathrm{ex}} = \mathrm{i} s \alpha^4 (9\pi m^2)^{-1} \times 6.03599 \times 10^{-1}, \tag{5.22}$$

which is equivalent to the analytical result given in [41]:

$$A = \mathrm{i} s \alpha^4 (36\pi m^2)^{-1} \big[175\zeta(3) - 38 \big],$$
$$A_{\mathrm{ex}} = \mathrm{i} s \alpha^4 (36\pi m^2)^{-1} \big[6\zeta(3) - 6 \big], \tag{5.23}$$

where $\zeta(3) \simeq 1.2020569$, the Riemann zeta function for an argument of 3.

The total cross section σ for light–light scattering is related to $M_0^{\gamma\gamma}$ in the forward direction by the optical theorem,

$$\sigma(s) = \frac{1}{s} \operatorname{Im} M_0^{\gamma\gamma}. \tag{5.24}$$

Thus we obtain, up to the fourth order in α,

$$\lim_{s \to \infty} \sigma_{\mathrm{c}}(s) \sim \frac{\alpha^4}{36\pi m^2} \big[175\zeta(3) - 38 \big] \sim 6.5\mu\mathrm{b}, \tag{5.25}$$

which is independent of s as well as the helicities of the incoming photons. Equation (5.25) is the value of the total cross section for $\gamma\gamma \to$ two electron–positron pairs, which dominates $\gamma\gamma \to$ one electron–positron pair at high energies. This is what we wanted to show.

Although the absorptive cross sections $\sigma_{\mathrm{a,b,c}}(s)$ are extremely small at high energies, these processes may well have important cosmological implications. Therefore we now turn to the absorption mechanism of high-energy cosmic γ rays.

It has been known for quite some time that cosmic γ rays of sufficiently large energy ω are absorbed, owing to the presence of the universal blackbody radiation. For $\omega \geq 10^{14}$ one has to take into account the absorption cross section for single-pair creation, σ_{b}. Since this cross section decreases with increasing ω, one could conclude that photons with energies comparable with the most energetic protons observed (10^{22}–10^{23} eV) may have a mean free path as large as the universe. However, for increasing energies, the mean free path for cosmic γ rays in fact eventually levels off to a constant. Because of the large constant cross section $\sigma_{\mathrm{c}}(s)$ for double-pair production, we shall never observe extragalactic photons from the edge of the universe. Just when the single-pair cross section becomes small enough to see out to 10^{26} cm, the double-pair-creation cross section adds a constant (in s) absorption probability per unit path length $\sim 10^{-26}/\mathrm{cm}$.

In order to prove these statements, we follow the calculations of [36, 90, 110], and begin with the absorption probability per unit length,

$$\frac{d\tau}{dx} = \frac{1}{4\pi} \int d\epsilon \, d\Omega \, \sigma(s) \, n(\epsilon) \, (1 - \cos\theta). \tag{5.26}$$

Here, $\sigma(s)$ is the cross section for real photons with CM energy $\sqrt{s}$, and $n(\epsilon)$ is the energy spectrum of the target, i.e. the number of photons per unit volume and energy. For an incoming photon with energy ω, we have

$$s = 2\epsilon\omega(1 - \cos\theta), \tag{5.27}$$

where θ is the angle between the incident and target photons. A change of variables with the aid of (5.27) yields

$$\frac{d\tau}{dx} = \frac{1}{8\omega^2} \int\limits_{s_0/4\omega}^{\infty} d\epsilon \, \frac{n(\epsilon)}{\epsilon^2} \int\limits_{s_0}^{4\epsilon\omega} ds \, s \, \sigma(s), \tag{5.28}$$

where

$$s_0 = \begin{cases} 16m^2 & \text{for} & \sigma_c \\ 4m^2 & \text{for} & \sigma_a \end{cases},$$

and we use in (5.28) [36]

$$\sigma_c(s) \simeq \left(1 - \frac{16m^2}{s}\right)^6 \sigma_c(\infty) = \left(1 - \frac{16m^2}{s}\right)^6 \times 6.45\mu b \tag{5.29}$$

and $\sigma_a(s)$ as is given in (5.15).

An integration by parts in (5.28) produces

$$\frac{d\tau}{dx} = \frac{1}{8\omega^2} \int\limits_{s_0}^{\infty} ds \, s \, \sigma(s) \, V\left(\frac{s}{4\omega}\right), \quad \text{where} \quad V(\epsilon) = \int\limits_{\epsilon}^{\infty} d\epsilon' \, \frac{n(\epsilon')}{\epsilon'^2}. \tag{5.30}$$

Since black-body radiation is the most important absorber for γ rays, we use

$$n(\epsilon) = \frac{1}{\pi^2} \frac{\epsilon^2}{e^{\epsilon/T} - 1}; \tag{5.31}$$

hence,

$$V(\epsilon) = -\frac{T}{\pi} \ln\left(1 - e^{-\epsilon/T}\right). \tag{5.32}$$

Now we can compute $d\tau/dx$ in (5.28) as a function of the γ ray energy ω. For single- and double-pair creation, we then obtain, with $T = 3$ K, the results shown in Fig. 5.5 for the absorption probability per unit length.

Therefore, we conclude that high-energy ($\geq 10^{21}$eV) γ rays, while undergoing little absorption through single-pair production, are almost entirely absorbed by double-pair creation when they traverse more than $\sim 10^{26}$ cm of black-body radiation. If such black-body radiation extends throughout the universe, then we cannot expect to see any further than 10^{26} cm.

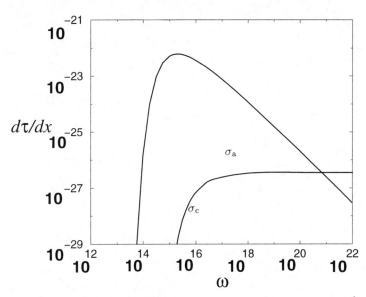

Fig. 5.5. Absorption probability per unit length (in units of cm^{-1}) as a function of the γ ray energy ω (in units of eV)

Appendices

A Units, Metric and the Gamma Matrices

We use exclusively the Heaviside–Lorentz system with "God-given" units, i.e. we set $\hbar = c = 1$. Then, the fine structure constant is $\alpha = e^2/(4\pi)$.

In this system, the SI units of the field of human experience are related in the following way to the energy unit of the electron Volt:

$$1\,\mathrm{m} \simeq 5.1 \times 10^6 \, \mathrm{eV}^{-1}, \tag{A.1}$$

$$1\,\mathrm{s} \simeq 1.5 \times 10^{15} \, \mathrm{eV}^{-1}, \tag{A.2}$$

$$1\,\mathrm{kg} \simeq 5.6 \times 10^{35} \, \mathrm{eV}, \tag{A.3}$$

$$1\,\mathrm{A} \simeq 1244 \, \mathrm{eV}, \tag{A.4}$$

$$1\,\mathrm{N/C} \simeq 6.5 \times 10^{-7} \, \mathrm{eV}^2, \tag{A.5}$$

$$1\,\mathrm{T} \simeq 195.5 \, \mathrm{eV}^2, \tag{A.6}$$

$$1\,\mathrm{K} \simeq 8.61735 \times 10^{-5} \, \mathrm{eV}. \tag{A.7}$$

We have as the metric $g = \mathrm{diag}(-,+,+,+)$, so that the field strength tensor can be written as

$$(F^{\mu\nu}) = \begin{pmatrix} 0 & E_1 & E_2 & E_3 \\ -E_1 & 0 & B_3 & -B_2 \\ -E_2 & -B_3 & 0 & B_1 \\ -E_3 & B_2 & -B_1 & 0 \end{pmatrix}. \tag{A.8}$$

Lowering the indices of the field strength tensor corresponds to replacing E_i by $-E_i$. It is, furthermore, useful to note that

$$F^{0i} = E_i = -F_{0i}, \tag{A.9}$$

$$F^{ij} = \epsilon_{ijk} B_k = F_{ij}, \tag{A.10}$$

$${}^\star F^{0i} = B_i = -{}^\star F_{0i}, \tag{A.11}$$

$${}^\star F^{ij} = -\epsilon_{ijk} E_k = {}^\star F_{ij}, \tag{A.12}$$

where the dual of the field strength tensor, ${}^\star F^{\mu\nu}$, is defined by

$${}^\star F^{\mu\nu} = \frac{1}{2} \epsilon^{\mu\nu\alpha\beta} F_{\alpha\beta}, \tag{A.13}$$

and the convention $\epsilon^{0123} = 1$ is understood.

For the product of two ϵ tensors, we find

$$\epsilon^{\alpha\nu\lambda\kappa}\epsilon_{\alpha\mu\rho\sigma} = -\Big(\delta^\nu_\mu\delta^\lambda_\rho\delta^\kappa_\sigma + \delta^\nu_\rho\delta^\lambda_\sigma\delta^\kappa_\mu + \delta^\nu_\sigma\delta^\lambda_\mu\delta^\kappa_\rho$$
$$-\delta^\nu_\rho\delta^\lambda_\mu\delta^\kappa_\sigma - \delta^\nu_\mu\delta^\lambda_\sigma\delta^\kappa_\rho - \delta^\nu_\sigma\delta^\lambda_\rho\delta^\kappa_\mu\Big), \tag{A.14}$$

$$\epsilon^{\alpha\beta\lambda\kappa}\epsilon_{\alpha\beta\rho\sigma} = -2\Big(\delta^\lambda_\rho\delta^\kappa_\sigma - \delta^\lambda_\sigma\delta^\kappa_\rho\Big), \tag{A.15}$$

$$\epsilon^{\alpha\beta\gamma\kappa}\epsilon_{\alpha\beta\gamma\sigma} = -6\,\delta^\kappa_\sigma, \tag{A.16}$$

$$\epsilon^{\alpha\beta\gamma\delta}\epsilon_{\alpha\beta\gamma\delta} = -24. \tag{A.17}$$

The Dirac γ matrices satisfy, by definition, the anticommutation relations

$$\{\gamma^\mu, \gamma^\nu\} = -2\,g^{\mu\nu}. \tag{A.18}$$

We employ the standard representation [29]:

$$\gamma^0 = \begin{pmatrix} 1 & 0 \\ 0 & -1 \end{pmatrix}, \qquad \{\gamma^i\} \equiv \gamma = \begin{pmatrix} 0 & \sigma \\ -\sigma & 0 \end{pmatrix},$$
$$\gamma_5 = i\gamma^0\gamma^1\gamma^2\gamma^3 = \begin{pmatrix} 0 & 1 \\ 1 & 0 \end{pmatrix}, \tag{A.19}$$

where σ denotes the set of Pauli matrices. The Dirac matrices commute with γ_5:

$$\{\gamma_5, \gamma^\mu\} = 0. \tag{A.20}$$

While the trace over an odd number of Dirac matrices vanishes, we find the following identities for an even number of γ^μ's:

$$\text{tr}\,\{1\} = 4,$$
$$\text{tr}\,\{\gamma_\mu\gamma_\nu\} = -4g_{\mu\nu},$$
$$\text{tr}\,\{\gamma_\mu\gamma_\nu\gamma_\alpha\gamma_\beta\} = 4(g_{\mu\nu}g_{\alpha\beta} - g_{\mu\alpha}g_{\nu\beta} + g_{\mu\beta}g_{\nu\alpha}), \tag{A.21}$$
$$\text{tr}\,\{\gamma_5\} = 0,$$
$$\text{tr}\,\{\gamma_5\gamma^\mu\gamma^\nu\} = 0,$$
$$\text{tr}\,\{\gamma_5\gamma^\mu\gamma^\nu\gamma^\kappa\gamma^\lambda\} = -4i\epsilon^{\mu\nu\kappa\lambda}.$$

Contractions of γ matrices with each other simplify to

$$\gamma^\mu\gamma_\mu = -4,$$
$$\gamma^\mu\gamma^\alpha\gamma_\mu = 2\gamma^\alpha. \tag{A.22}$$

Of special interest (cf. Appendix C) is the commutator of two γ matrices:

$$\sigma^{\mu\nu} = \frac{i}{2}[\gamma^\mu, \gamma^\nu]. \tag{A.23}$$

One of its properties is that is commutes with γ_5:

$$[\gamma_5, \sigma^{\mu\nu}] = 0, \tag{A.24}$$

which can be easily proved by employing the definition (A.23) and (A.20).

B Invariants and Spectral Representation of the Maxwell Field Strength Tensor

In this appendix, we review the various different convenient choices of sets of invariants of the Maxwell field strength tensor and present its spectral representation. The latter enables us to conveniently and economically deal with functions of the field strength tensor without making reference to a special Lorentz frame or choice of gauge.

B.1 Invariants

The gauge-invariant building blocks for the construction of invariants are the field strength tensor $F^{\mu\nu}$ and its dual $^\star F^{\mu\nu}$ defined in (A.13). The standard choice of invariant scalars (and pseudoscalars) reads

$$\mathcal{F} = \frac{1}{4} F_{\mu\nu} F^{\mu\nu} = \frac{1}{2}(\boldsymbol{B}^2 - \boldsymbol{E}^2), \tag{B.1}$$

$$\mathcal{G} = \frac{1}{4} F_{\mu\nu} {}^\star F^{\mu\nu} = -\boldsymbol{E} \cdot \boldsymbol{B}. \tag{B.2}$$

Further important relations between the field strength tensors and the invariants are represented by the following fundamental algebraic identities, which can easily be proved by straightforward matrix multiplication:

$$F^{\mu\alpha} F^\nu{}_\alpha - {}^\star F^{\mu\alpha} {}^\star F^\nu{}_\alpha = 2\,\mathcal{F}\,g^{\mu\nu}, \tag{B.3}$$

$$F^{\mu\alpha} {}^\star F^\nu{}_\alpha = {}^\star F^{\mu\alpha} F^\nu{}_\alpha = \mathcal{G}\,g^{\mu\nu}. \tag{B.4}$$

A different but convenient choice of a set of linearly independent gauge and Lorentz invariants is given by the secular invariants

$$a = \left(\sqrt{\mathcal{F}^2 + \mathcal{G}^2} + \mathcal{F}\right)^{1/2}, \quad b = \left(\sqrt{\mathcal{F}^2 + \mathcal{G}^2} - \mathcal{F}\right)^{1/2}; \tag{B.5}$$

the inverse relations between the two sets of invariants can be written as

$$|\mathcal{G}| = ab, \quad \mathcal{F} = \frac{1}{2}(a^2 - b^2). \tag{B.6}$$

Since the product of the secular invariants is equal not to $\mathcal{G}$ but to its absolute value, we confine ourselves to the case of $\mathcal{G} > 0$ (or $\boldsymbol{E} \cdot \boldsymbol{B} < 0$) in this work; otherwise we would have to perform the whole formalism with a minus sign at the appropriate places. This subtlety has sometimes caused confusion in the literature.

In a Lorentz frame where the field vectors $\boldsymbol{E}$ and $\boldsymbol{B}$ are antiparallel, the invariants a and b correspond to

$$a = |\boldsymbol{B}|, \quad b = |\boldsymbol{E}|, \quad \text{for} \quad \boldsymbol{B} \parallel -\boldsymbol{E}. \tag{B.7}$$

At this stage, it is necessary to remark that these sets of invariants are complete only for the case of a constant electromagnetic field in an isotropic space–time. Otherwise, the number of invariants will not remain so small; for example, if $F^{\mu\nu}$ is a C^∞ function with nonvanishing derivatives there will be infinitely many invariants which can be constructed from $F^{\mu\nu}$, $^*F^{\mu\nu}$ and ∂^μ. For the case of finite temperature discussed in Sect. 3.5, we encounter two more invariants, which are constructed from an additional vector n^μ that can be interpreted as the four-velocity of the heat bath.

However, for the case of a purely constant electromagnetic field modifying the vacuum, the field configuration space is spanned by either one of the above-mentioned sets of two invariants.

In Sect. 3.2, we need the Laplacian differential operator acting on functions of the field space coordinates $\mathcal{F}$ and $\mathcal{G}$:

$$\nabla^2\phi \equiv (\partial_{\mathcal{F}}^2 + \partial_{\mathcal{G}}^2)\phi. \tag{B.8}$$

To perform a transition to the secular invariants, we first obtain the gradient in these new field space coordinates using the standard formula

$$\boldsymbol{\nabla}\phi = \left(A\frac{\partial\phi}{\partial a}\right)\hat{e}_a + \left(B\frac{\partial\phi}{\partial b}\right)\hat{e}_b, \tag{B.9}$$

where the quantities A and B are given by

$$A = \left|\frac{\partial}{\partial a}\left(\frac{\mathcal{F}}{\mathcal{G}}\right)\right|^{-1} = \frac{1}{\sqrt{a^2 + b^2}} = \left|\frac{\partial}{\partial b}\left(\frac{\mathcal{F}}{\mathcal{G}}\right)\right|^{-1} = B. \tag{B.10}$$

Hence, we obtain for the gradient in (B.7)

$$\boldsymbol{\nabla}\phi = \frac{1}{\sqrt{a^2 + b^2}}(\partial_a\phi\,\hat{e}_a + \partial_b\phi\,\hat{e}_b). \tag{B.11}$$

For the desired Laplacian in terms of the secular invariants, we immediately find

$$\nabla^2\phi = AB\left[\partial_a\left(\frac{(\boldsymbol{\nabla}\phi)_a}{B}\right) + \partial_b\left(\frac{(\boldsymbol{\nabla}\phi)_b}{A}\right)\right]$$

$$= \frac{1}{a^2 + b^2}(\partial_a^2 + \partial_b^2)\phi. \tag{B.12}$$

The invariants a and b are called *secular* invariants because they are related to the solutions of the secular equation for the field strength tensor, i.e. the eigenvalues of the field strength tensor $F^{\mu\nu}$. As we shall shortly prove, these are given by

$$f_{1,2} = \pm b, \qquad f_{3,4} = \pm ia. \tag{B.13}$$

The eigenvalues of the dual of the field strength tensor $i^*F^{\mu\nu}$ (the factor of i is inserted for reasons of convenience) are similarly found to be

$$\bar{f}_{1,2} = \mp ia, \qquad \bar{f}_{3,4} = \mp b. \tag{B.14}$$

B.2 Spectral Representation of $F^{\mu\nu}$

Let us define a set of four matrices using the eigenvalues given in (B.13) and (B.14):

$$A_{j\,\mu\nu} = \frac{1}{2(f_j^2 - \bar{f}_j^2)}(-\bar{f}_j^2\, g_{\mu\nu} + f_j\, F_{\mu\nu} + F_\mu{}^\lambda F_{\lambda\nu} - i\bar{f}_j\,{}^\star F_{\mu\nu}). \qquad (B.15)$$

In the following, we shall prove that the $A_{j\,\mu\nu}$'s $(j = 1, 2, 3, 4)$ are the projection matrices of the field strength tensor (and its dual) which project on the eigenspaces with eigenvalue f_j (or $\bar{f}_j$). The main properties of the $A_{j\,\mu\nu}$'s are summarized at the end of this section. These projectors were originally introduced by Batalin and Shabad [22] (using a different metric). Since

$$g_{\mu\nu} F^{\nu\alpha} = F_{\mu\nu} g^{\nu\alpha},$$

$$(F_\mu{}^\lambda F_{\lambda\nu}) F^{\nu\alpha} \overset{\nu\leftrightarrow\lambda}{=} F_{\mu\nu}(F^\nu{}_\lambda F^{\lambda\alpha})$$

and $\quad {}^\star F_{\mu\nu} F^{\nu\alpha} \overset{(B.4)}{=} F_{\mu\nu}{}^\star F^{\nu\alpha},$

it is obvious that the matrices A_j and F commute:

$$[A_j, F] = 0, \qquad \text{or} \qquad A_{j\,\mu\nu} F^{\nu\alpha} = F_{\mu\nu} A_j^{\nu\alpha}. \qquad (B.16)$$

With the aid of the fundamental algebraic identities for the field strength tensors (B.3) and (B.4), we find for the product

$$A_{j\,\mu}{}^\nu F_{\nu\alpha} = \frac{1}{2(f_j^2 - \bar{f}_j^2)}(-\bar{f}_j^2\, F_{\mu\alpha} + f_j\, F_\mu{}^\nu F_{\nu\alpha}$$

$$+ F_\mu{}^\lambda \underbrace{F_\lambda{}^\nu F_{\nu\alpha}}_{\overset{(B.3)}{=}-2\mathcal{F} g_{\lambda\alpha} + {}^\star F_\lambda{}^\nu{}^\star F_{\nu\alpha}} - i\bar{f}_j \underbrace{{}^\star F_\mu{}^\nu F_{\nu\alpha}}_{\overset{(B.4)}{=}-\mathcal{G} g_{\mu\alpha}})$$

$$= \frac{f_j}{2(f_j^2 - \bar{f}_j^2)}\left(i\frac{\bar{f}_j}{f_j}\mathcal{G} g_{\mu\alpha} - \frac{2\mathcal{F} + \bar{f}_j^2}{f_j} F_{\mu\alpha} + F_\mu{}^\nu F_{\nu\alpha} - \frac{\mathcal{G}}{f_j}{}^\star F_{\mu\alpha}\right)$$

$$= f_j\, A_{j\,\mu\alpha}. \qquad (B.17)$$

For the last equality, we have employed the relations

$$- \bar{f}_j\, f_j = i\mathcal{G},$$
$$f_j^2 + \bar{f}_j^2 = -2\,\mathcal{F}. \qquad (B.18)$$

The validity of (B.18) can be easily proved with the aid of the definitions (B.5), (B.13) and (B.14). Equation (B.17) confirms the claim that the f_j's are the eigenvalues of the field strength tensor. A similar calculation can be performed for the dual objects:

$$A_j{}^{\mu\nu}(i{}^\star F_{\nu\alpha}) = \frac{i}{2(f_j^2 - \bar{f}_j^2)}(-\bar{f}_j^2\,{}^\star F^\mu{}_\alpha + f_j \underbrace{F^{\mu\nu\,\star}F_{\nu\alpha}}_{=-\mathcal{G} g^\mu{}_\alpha}$$

$$+ F^{\mu\lambda} \underbrace{F_\lambda{}^\nu {}^\star F_{\nu\alpha}}_{\overset{(B.4)}{=} -\mathcal{G}\, g_{\lambda\alpha}} - i \bar{f}_j \underbrace{{}^\star F^{\mu\nu} {}^\star F_{\nu\alpha}}_{\overset{(B.3)}{=} F^{\mu\nu} F_{\nu\alpha} + 2\mathcal{F}\, g^\mu{}_\alpha} \Big)$$

$$= \frac{-i\bar{f}_j}{2(f_j^2 - \bar{f}_j^2)}$$

$$\times \left(\frac{\mathcal{G} f_j + 2i\mathcal{F} \bar{f}_j}{\bar{f}_j}\, g^\mu{}_\alpha + \frac{\mathcal{G}}{\bar{f}_j}\, F^\mu{}_\alpha + i F^{\mu\nu} F_{\nu\alpha} + \bar{f}_j\, {}^\star F_{\mu\alpha} \right)$$

$$= \bar{f}_j\, A_j{}^\mu{}_\alpha. \tag{B.19}$$

Finally, we have to prove the projector properties of the matrices A_j, for which the following identities are helpful:

$$\sum_{j=1}^4 \frac{\bar{f}_j^2}{f_j^2 - \bar{f}_j^2} = -2, \qquad\qquad \sum_{j=1}^4 \frac{1}{f_j^2 - \bar{f}_j^2} = 0,$$

$$\sum_{j=1}^4 \frac{\bar{f}_j}{f_j^2 - \bar{f}_j^2} = 0, \qquad\qquad \sum_{j=1}^4 \frac{f_j}{f_j^2 - \bar{f}_j^2} = 0. \tag{B.20}$$

The sum of the projectors should yield the following identity:

$$\sum_{j=1}^4 A_j{}^{\mu\nu} = \frac{1}{2} \left[g^{\mu\nu} \left(-\sum_{j=1}^4 \frac{\bar{f}_j^2}{f_j^2 - \bar{f}_j^2} \right) + F^{\mu\nu} \left(\sum_{j=1}^4 \frac{f_j}{f_j^2 - \bar{f}_j^2} \right) \right.$$

$$\left. + F^{\mu\lambda} F_\lambda{}^\nu \left(\sum_{j=1}^4 \frac{1}{f_j^2 - \bar{f}_j^2} \right) - i {}^\star F^{\mu\nu} \left(\sum_{j=1}^4 \frac{\bar{f}_j}{f_j^2 - \bar{f}_j^2} \right) \right]$$

$$\overset{(B.20)}{=} g^{\mu\nu}. \tag{B.21}$$

Since the spectrum is generally nondegenerate, the trace of each projector should be equal to 1:

$$A_j{}^\mu{}_\mu = \frac{1}{2(f_j^2 - \bar{f}_j^2)} (-4\bar{f}_j^2 - 4\mathcal{F}) \overset{(B.13),(B.14)}{=} 1. \tag{B.22}$$

The orthogonality and idempotency properties of the projectors can be demonstrated by considering the product

$$A_k{}^{\mu\nu} A_{j\,\nu\alpha} = \frac{1}{2(f_j^2 - \bar{f}_j^2)} (-\bar{f}_j^2 A_k{}^\mu{}_\alpha + f_k f_j A_k{}^\mu{}_\alpha$$

$$+ f_k^2 A_k{}^\mu{}_\alpha - i\bar{f}_j \underbrace{A_k{}^{\mu\nu} {}^\star F_{\nu\alpha}}_{\overset{(B.19)}{=} -i\bar{f}_k A_k{}^\mu{}_\alpha})$$

$$= \frac{1}{2(f_j^2 - \bar{f}_j^2)} (-\bar{f}_j^2 + f_k^2 + f_k f_j - \bar{f}_j \bar{f}_k)\, A_k{}^\mu{}_\alpha$$

$$= \begin{cases} A_k{}^\mu{}_\alpha & \text{for} \quad k = j \\ 0 & \text{otherwise} \end{cases}. \tag{B.23}$$

This concludes the investigation of the projector algebra of the A_j matrices.

Finally, let us summarize the recipe for the use of the spectral representation of the field strength tensors $F^{\mu\nu}$ and $^*F^{\mu\nu}$:

$$A_{j\,\mu\nu} = \frac{1}{2(f_j^2 - \bar{f}_j^2)}$$
$$\times(-\bar{f}_j^2\, g_{\mu\nu} + f_j\, F_{\mu\nu} + F_\mu{}^\lambda F_{\lambda\nu} - i\bar{f}_j\, {}^*F_{\mu\nu}), \quad \text{(B.24)}$$
$$F^\mu{}_\nu\, A_j{}^{\nu\alpha} = f_j\, A_j{}^{\mu\alpha}, \quad \text{(B.25)}$$
$$i^*F^\mu{}_\nu\, A_j{}^{\nu\alpha} = \bar{f}_j\, A_j{}^{\mu\alpha}, \quad \text{(B.26)}$$
$$A_k{}^{\mu\nu} A_{j\,\nu\alpha} = \begin{cases} A_k{}^\mu{}_\alpha & \text{for} \quad k = j, \\ 0 & \text{otherwise} \end{cases} \quad \text{(B.27)}$$
$$\sum_{j=1}^4 A_j{}^{\mu\nu} = g^{\mu\nu}, \quad \text{(B.28)}$$
$$A_j{}^\mu{}_\mu = 1. \quad \text{(B.29)}$$

By employing (B.24)–(B.29), we can represent any function h of the field strength tensor as a sum over the projectors times a spectral weight that is given by the function h of the eigenvalues:

$$h(\mathsf{F})^{\mu\nu} \equiv \sum_{j=1}^4 A_j{}^{\mu\nu}\, h(f_j). \quad \text{(B.30)}$$

B.3 Applications

As a simple application, we derive the explicit representation of the function $e^{-iY(is)}$ defined in (2.35):

$$Y(is) = \frac{1}{2}\mathrm{tr}\, \ln[\cos(ie\mathsf{F}s)]. \quad \text{(B.31)}$$

With the aid of the spectral representation, we find

$$Y(is) \quad = \quad \exp\left(-\frac{1}{2}\mathrm{tr}_\mathrm{L}\sum_j A_j^{\mu\nu} \ln\cos(ief_js)\right)$$
$$\overset{(B.29)}{=} \quad \exp\left(-\frac{1}{2}\sum_j \ln\cos(ief_js)\right) = \left(\prod_j \cos(ief_js)\right)^{-1/2}$$
$$\overset{(B.13),(B.14)}{=} \quad \frac{1}{\cos eas\,\cosh ebs}. \quad \text{(B.32)}$$

This representation is required, for example, in Sect. 3.5. Further examples of the power of the spectral representation are given there.

C More about the Commutator $\sigma^{\mu\nu}$ of Dirac Matrices

The properties of the commutator of the Dirac matrices, $\sigma^{\mu\nu} = \mathrm{i}/2\,[\gamma^{\mu}, \gamma^{\nu}]$, play a central role in the list of calculational techniques for quantum electrodynamics with external fields. We dedicate this Appendix to deriving and enumerating various identities, which in part are necessary for the main text and in part are of general interest in their own right.

Let us begin with two basic trace identities which can immediately be read off from (A.21):

$$\mathrm{tr}\,\{\sigma^{\mu\nu}\} = 0,$$
$$\mathrm{tr}\,\{\gamma_5\,\sigma^{\mu\nu}\} = 0. \tag{C.1}$$

For the study of further identities, we introduce the anticommutator of two σ's:

$$\frac{1}{2}\,\{\sigma^{\mu\nu}, \sigma^{\alpha\beta}\} = g^{\mu\alpha}g^{\nu\beta} - g^{\mu\beta}g^{\nu\alpha} + \mathrm{i}\,\gamma_5\,\epsilon^{\mu\nu\alpha\beta}. \tag{C.2}$$

Equation (C.2) can be derived rudimentarily on the level of the underlying γ algebra or understood by symmetry arguments, while the coefficients of an appropriate ansatz can be determined by using the trace identities given above and in Appendix A. For reasons of completeness, we also mention an equation for the commutator,

$$[\sigma^{\mu\nu}, \sigma^{\alpha\beta}] = -2\mathrm{i}\big(g^{\nu\alpha}\sigma^{\mu\beta} - g^{\mu\alpha}\sigma^{\nu\beta} - g^{\nu\beta}\sigma^{\mu\alpha} + g^{\mu\beta}\sigma^{\nu\alpha}\big), \tag{C.3}$$

which can also be proved by means of the γ algebra. Incidentally, the commutation rule (C.3) is identical to the defining commutator of the Lorentz algebra, which reflects the fact that the σ's generate a representation of the Lorentz group.

With the aid of (C.2), we can easily find the following trace identities:

$$\mathrm{tr}\,\{\sigma^{\mu\nu}\sigma^{\alpha\beta}\} = 4\big(g^{\mu\alpha}g^{\nu\beta} - g^{\mu\beta}g^{\nu\alpha}\big), \tag{C.4}$$
$$\mathrm{tr}\,\{\gamma_5\sigma^{\mu\nu}\sigma^{\alpha\beta}\} = 4\mathrm{i}\,\epsilon^{\mu\nu\alpha\beta}. \tag{C.5}$$

In the second identity, we have employed the fact that γ_5 commutes with $\sigma^{\mu\nu}$ (see (A.24)).

In most cases, Lorentz scalars are constructed from $\sigma^{\mu\nu}$ only by contraction with the field strength tensor $F_{\mu\nu}$. For example, the trace identities in (C.4) and (C.5) then become ($\sigma F = \sigma^{\mu\nu}F_{\mu\nu}$)

$$\mathrm{tr}\,\Big\{(\sigma F)^2\Big\} = 8\,F_{\mu\nu}F^{\mu\nu} \equiv 32\,\mathcal{F}, \tag{C.6}$$

$$\mathrm{tr}\,\Big\{\gamma_5\,(\sigma F)^2\Big\} = 8\mathrm{i}\,F_{\mu\nu}\,{}^\star F^{\mu\nu} \equiv 32\mathrm{i}\,\mathcal{G}. \tag{C.7}$$

These results could have been expected, since there are no other scalars or pseudoscalars constructable from $F^{\mu\nu}$ and $\star F^{\mu\nu}$. The same type of argument can be used to prove that

$$\text{tr}\left\{(\sigma F)^{2n+1}\right\} = \text{tr}\left\{\gamma_5 (\sigma F)^{2n+1}\right\} = 0, \quad \text{for} \quad n \in \mathbb{N}_0, \tag{C.8}$$

since there are no invariant combinations of an odd number of field strength tensors.

In the proper-time formalism, the interaction of the fermion's spin with an electromagnetic field is characterized by the function $\exp\left[(ie/2)s\,\sigma F\right]$, which is responsible for the nontrivial Dirac structure of the proper-time evolution operator. In the following, we shall project this quantity onto the fundamental elements of the Dirac algebra, which allows a practical treatment during calculations. An appropriate ansatz that is complete in the sense of a Fierz transformation is given by

$$e^{i(e/2)s\,\sigma F} = S + \frac{ie}{2}T\,\sigma F - iP\gamma_5 + \frac{e}{2}T^\star\,\gamma_5\sigma F. \tag{C.9}$$

The quantities S and T are scalar functions and P and $T^\star$ are pseudoscalar functions of the field invariants and the proper-time variable s, and have to be determined term by term. According to the identities given above, the trace of (C.9) reduces to

$$\text{tr}\left\{e^{i(e/2)s\,\sigma F}\right\} = 4S. \tag{C.10}$$

Hence, the eigenvalues of σF are required, for which we employ the anticommutator relation (C.2) to obtain

$$(\sigma F)^2 = \frac{1}{2}F_{\mu\nu}F_{\alpha\beta}\left\{\sigma^{\mu\nu}, \sigma^{\alpha\beta}\right\} = 8\left(\mathcal{F} + i\gamma_5\,\mathcal{G}\right). \tag{C.11}$$

Since the eigenvalues of γ_5 are ± 1, we find for the eigenvalues of σF

$$\frac{1}{2}(\sigma F)' = \pm\left[2(\mathcal{F} \pm i\mathcal{G})\right]^{1/2} = \pm(a \pm ib), \tag{C.12}$$

where in the last step we have used the secular invariants defined in (B.5). Inserting these eigenvalues into the left-hand side of (C.10), we obtain

$$S = \cos esa\,\cosh esb. \tag{C.13}$$

The basic trick for projecting out the other functions P, T and $T^\star$ by tracing goes as follows:

$$\text{tr}\left\{(\sigma F)\,e^{i(e/2)s\,\sigma F}\right\} = \frac{\partial}{\partial i(e/2)s}\,\text{tr}\left\{e^{i(e/2)s\,\sigma F}\right\}$$
$$= 8i(a\sin esa\,\cosh esb - b\cos esa\,\sinh esb). \tag{C.14}$$

On the other hand, we may insert the ansatz (C.9) into the left-hand side of (C.14), which yields, after employing the trace identities given above,

$$\text{tr}\left\{(\sigma F)\,e^{i(e/2)s\,\sigma F}\right\} = 16ie\left(\mathcal{F}T + \mathcal{G}T^\star\right). \tag{C.15}$$

Equating (C.14) and (C.15), we obtain

$$\mathcal{F}T + \mathcal{G}T^\star = \frac{1}{2e}\left(a\sin esa\,\cosh esb - b\cos esa\,\sinh esb\right). \tag{C.16}$$

The same procedure can be applied to the exponential function in (C.9) for a second time:

$$\mathrm{tr}\left\{(\sigma F)^2\,\mathrm{e}^{\mathrm{i}(e/2)s\,\sigma F}\right\} = \frac{\partial^2}{(\partial\mathrm{i}(e/2)s)^2}\,\mathrm{tr}\left\{\mathrm{e}^{\mathrm{i}(e/2)s\,\sigma F}\right\}$$

$$= 32\left[\frac{1}{2}(a^2 - b^2)\cos esa\,\cosh esb\right.$$

$$\left. + ab\sin esa\,\sinh esb\right]$$

$$= 32\,\mathcal{F}S + 32\,\mathcal{G}\sin esa\,\sinh esb. \tag{C.17}$$

Inserting the ansatz (C.9) into the left-hand side of (C.17), on the other hand, leads to

$$\mathrm{tr}\left\{(\sigma F)^2\,\mathrm{e}^{\mathrm{i}(e/2)s\,\sigma F}\right\} = 32\,\mathcal{F}S + 32\,\mathcal{G}P, \tag{C.18}$$

where we have made use of (C.6)–(C.8). Combining the last two equations clearly results in the following final expression for P:

$$P = \sin esa\,\sinh esb. \tag{C.19}$$

Instead of proceeding with the direct differentiation of the exponential function, we first state the trace identity

$$\mathrm{tr}\left\{\gamma_5\,\mathrm{e}^{\mathrm{i}(e/2)s\,\sigma F}\right\} = -4\mathrm{i}\,P, \tag{C.20}$$

which can be read off from the ansatz (C.9). Applying our basic trick to (C.20), we first obtain

$$\mathrm{tr}\left\{\gamma_5\,(\sigma F)\,\mathrm{e}^{\mathrm{i}(e/2)s\,\sigma F}\right\} = \frac{\partial}{\partial\mathrm{i}(e/2)s}\,\mathrm{tr}\left\{\gamma_5\,\mathrm{e}^{\mathrm{i}(e/2)s\,\sigma F}\right\} \tag{C.21}$$

$$= -8\left(b\sin esa\,\cosh esb + a\cos esa\,\sinh esb\right),$$

whereas the insertion of the ansatz (C.9) into the left-hand side leads to

$$\mathrm{tr}\left\{\gamma_5\,(\sigma F)\,\mathrm{e}^{\mathrm{i}(e/2)s\,\sigma F}\right\} = 16es\left(\mathcal{F}T^\star - \mathcal{G}T\right). \tag{C.22}$$

The last two equations imply

$$\mathcal{F}T^\star - \mathcal{G}T = -\frac{1}{2e}\left(b\sin esa\,\cosh esb + a\cos esa\,\sinh esb\right). \tag{C.23}$$

Finally, (C.16) and (C.23) can be solved for T and $T^\star$.

To summarize, we have found the following representation of the exponential function $\exp\left[\mathrm{i}(e/2)s\,\sigma F\right]$ in terms of the fundamental elements of the Dirac algebra and the invariants of the electromagnetic field:

$$e^{i(e/2)s\,\sigma F} = S + \frac{ie}{2}\,T\,\sigma F - i\,P\,\gamma_5 + \frac{e}{2}\,T^\star\,\gamma_5\sigma F, \tag{C.24}$$

$$S = \cos esa\,\cosh esb, \tag{C.25}$$

$$P = \sin esa\,\sinh esb, \tag{C.26}$$

$$T = \frac{1}{e(a^2 + b^2)}$$
$$\times (a \sin esa\,\cosh esb + b \cos esa\,\sinh esb), \tag{C.27}$$

$$T^\star = \frac{1}{e(a^2 + b^2)}$$
$$\times (b \sin esa\,\cosh esb - a \cos esa\,\sinh esb), \tag{C.28}$$

At this point, we want to stress that we have obtained this representation without referring to any particular Lorentz system or any special field configuration. Relativistic invariance is completely maintained.[1] Of course, no assumption about the profile of the field strength was required either.

This representation of the exponential function in (C.24) serves as the starting point for most of the Dirac trace calculations in external-field electrodynamics. Only in this form does a systematic calculation become manageable. Without going into details, we want to give two examples which appear in basic Feynman graph evaluations with external fields; we begin with

$$\mathrm{tr}\left\{\gamma_\mu\,e^{i(e/2)s\,\sigma F}\,\gamma_\nu\,e^{i(e/2)s'\,\sigma F}\right\}$$
$$= 4\left\{\left[-(SS' + PP') + 2e^2\,\mathcal{F}(TT' + T^\star T^{\star\prime})\right]g_{\mu\nu}\right.$$
$$+ e\left[(ST' - TS') + (PT^{\star\prime} - T^\star P')\right]F_{\mu\nu}$$
$$+ e\left[(ST^{\star\prime} - T^\star S') + (TP' - PT')\right]{}^\star F_{\mu\nu}$$
$$\left. + 2e^2(TT' + T^\star T^{\star\prime})\,F_{\mu\nu}^2\right\}, \tag{C.29}$$

where primed functions depend on the proper time s', and $F_{\mu\nu}^2 = F_\mu{}^\alpha F_{\alpha\nu}$. It is interesting to note that this second-rank tensor is indeed constructed from a maximum set of linearly independent tensors, all of which can be derived from the field strength tensors $F_{\mu\nu}$ and ${}^\star F_{\mu\nu}$.

For the second trace identity, we introduce a list of convenient definitions of standardized Lorentz tensors of fourth rank involving the field strength tensors and the metric:

---

[1] In a system where $\boldsymbol{E}$ and $\boldsymbol{B}$ are (anti)parallel, (C.24)–(C.28) reduce to the findings of Ritus [141].

$$K_{\mu\alpha\nu\beta} = g_{\mu\alpha}g_{\nu\beta} - g_{\mu\nu}g_{\alpha\beta} + g_{\mu\beta}g_{\nu\alpha},$$
$$S_{\mu\alpha\nu\beta} = \left(g_{\mu\alpha}F_{\nu\beta} - g_{\mu\nu}F_{\alpha\beta} + g_{\mu\beta}F_{\nu\alpha}\right) + \left(F_{\mu\alpha}g_{\nu\beta} - F_{\mu\nu}g_{\alpha\beta} + F_{\mu\beta}g_{\nu\alpha}\right)$$
$$S^{\star}_{\mu\alpha\nu\beta} = \left(g_{\mu\alpha}{}^{\star}F_{\nu\beta} - g_{\mu\nu}{}^{\star}F_{\alpha\beta} + g_{\mu\beta}{}^{\star}F_{\nu\alpha}\right)$$
$$+\left({}^{\star}F_{\mu\alpha}g_{\nu\beta} - {}^{\star}F_{\mu\nu}g_{\alpha\beta} + {}^{\star}F_{\mu\beta}g_{\nu\alpha}\right),$$
$$U_{\mu\alpha\nu\beta} = \mathcal{F}\,K_{\mu\alpha\nu\beta} + \left(-F_{\mu\alpha}F_{\nu\beta} - F_{\mu\nu}F_{\alpha\beta} + F_{\mu\beta}F_{\alpha\nu}\right) \tag{C.30}$$
$$+\left(-F^2_{\mu\nu}g_{\alpha\beta} + F^2_{\mu\beta}g_{\alpha\nu} - g_{\mu\nu}F^2_{\alpha\beta} + g_{\mu\beta}F^2_{\alpha\nu}\right),$$
$$V_{\mu\alpha\nu\beta} = \mathcal{G}\,K_{\mu\alpha\beta\nu} + \left({}^{\star}F_{\mu\beta}F_{\alpha\nu} + F_{\mu\beta}{}^{\star}F_{\alpha\nu}\right)$$
$$-\left({}^{\star}F_{\mu\nu}F_{\alpha\beta} + F_{\mu\nu}{}^{\star}F_{\alpha\beta}\right) - \mathcal{F}\,\epsilon_{\mu\alpha\nu\beta}.$$

With the aid of these definitions, we present the trace identity for the following object:

$$\text{tr}\left\{\gamma_\mu\gamma_\alpha\,\mathrm{e}^{\mathrm{i}(e/2)s\,\sigma F}\,\gamma_\nu\gamma_\beta\,\mathrm{e}^{\mathrm{i}(e/2)s'\,\sigma F}\right\}$$

$$= 4\Big[\left(SS' - PP'\right)K_{\mu\alpha\nu\beta} - \left(SP' + PS'\right)\epsilon_{\mu\alpha\nu\beta}$$
$$+e\left(PT^{\star\prime} - ST'\right)S_{\mu\alpha\nu\beta} + e\left(T^{\star}P' - TS'\right)S_{\nu\beta\mu\alpha}$$
$$-e\left(ST^{\star\prime} + PT'\right)S^{\star}_{\mu\alpha\nu\beta} - e\left(T^{\star}S' + TP'\right)S^{\star}_{\nu\beta\mu\alpha}$$
$$+2e^2\left(T^{\star}T^{\star\prime} - TT'\right)U_{\mu\alpha\nu\beta} - 2e^2\left(TT^{\star\prime} + T^{\star}T'\right)V_{\mu\alpha\nu\beta}\Big]. \tag{C.31}$$

In view of the considerable extent of this expression, it appears doubtful to us that the formulation of spinor electrodynamics employing the usual Dirac algebra is appropriate for higher-order external-field calculations. Even computer-aided algebraic-manipulation programs seem to be of limited use.

For the remainder of this Appendix, we concentrate on the case of a purely magnetic background field, i.e. we take the limits $b \to 0$ and $a \to |\boldsymbol{B}| \equiv B$. Assuming that the magnetic field points along the z direction, the only nonvanishing field strength components are

$$F^{12} = -F^{21} = B. \tag{C.32}$$

It is useful to leave the Lorentz-invariant notation and take a look at the explicit representation of $\sigma^{\mu\nu}$:

$$\sigma^{ij} = \begin{pmatrix} \sigma_k & 0 \\ 0 & \sigma_k \end{pmatrix}, \quad \text{for} \quad i,j,k = 1,2,3 \quad \text{and cyclic permutations,}$$

$$\sigma^{0i} = \mathrm{i}\begin{pmatrix} 0 & \sigma_i \\ \sigma_i & 0 \end{pmatrix}, \quad \text{for} \quad i = 1,2,3. \tag{C.33}$$

The σ_i's obviously denote the set of Pauli matrices.

Equations (C.33) imply that we may write for σF

$$\sigma F = 2\,B\,\sigma_3, \tag{C.34}$$

where it is understood that the (2×2) Pauli matrix σ_3 is blockwise doubled in (4×4) Dirac space.

Specializing the general results given above to the case of a purely magnetic field in the z direction, we immediately obtain for (C.24)

$$e^{i(e/2)s\,\sigma F} = e^{ie\,BS\,\sigma_3} = \cos eBs + i\sin eBs\,\sigma_3. \tag{C.35}$$

With regard to Sect. 2.3, we are also interested in

$$\mathrm{tr}\left\{\sigma_{\mu\nu}\,e^{i(e/2)s\,\sigma F}\right\} \overset{(C.35)}{=} i\sin eBs\,\mathrm{tr}\left\{\sigma_{\mu\nu}\sigma_{12}\right\}$$

$$\overset{(C.4)}{=} 4i\sin eBs\left(g_{\mu 1}g_{\nu 2} - g_{\mu 2}g_{\nu 1}\right)$$

$$= -4\sin eBs\left(\sigma_{t2}\right)_{\mu\nu}, \tag{C.36}$$

where the subscript "t" denotes the transverse subspace of Lorentz space, $\mu, \nu = 1, 2$, and the nonvanishing components of $(\sigma_{t2})_{\mu\nu}$ therefore read $(\sigma_{t2})_{12} = -i = -(\sigma_{t2})_{21}$ (see Sect. 2.3).

The last desired trace identity appears in (2.188) in Sect. 2.3 during the discussion of the interaction of a plane wave $f_{\mu\nu}$ with an external magnetic field:

$$\mathrm{tr}\left\{\left(f^*_{\|,\perp}\sigma\right)e^{i(e/2)(s-t)\sigma F}\left(f_{\|,\perp}\sigma\right)e^{i(e/2)t\sigma F}\right\}. \tag{C.37}$$

This resembles the trace identity given in (C.31). Instead of employing the latter, we demonstrate how the special assumption of a purely magnetic field reduces the calculational effort.

Adopting the notation of Sect. 2.3, we find, with the aid of (C.33),

$$\frac{1}{i\omega}\left(f_{\|}\sigma\right) = -2\begin{pmatrix} \sigma_2 & i\sigma_3 \\ i\sigma_3 & \sigma_2 \end{pmatrix}e^{i\omega\xi} = -2\,\Sigma_{\|}\,e^{i\omega\xi},$$

$$\frac{1}{i\omega}\left(f_{\perp}\sigma\right) = 2\begin{pmatrix} \sigma_3 & -i\sigma_2 \\ -i\sigma_2 & \sigma_3 \end{pmatrix}e^{i\omega\xi} = 2\,\Sigma_{\perp}\,e^{i\omega\xi}, \tag{C.38}$$

where we have defined the matrices $\Sigma_{\|,\perp}$, which satisfy the nilpotency property:

$$\Sigma^2_{\|,\perp} = 0. \tag{C.39}$$

After insertion of (C.35) into the trace (C.37), this property greatly simplifies the desired expression:

$$\mathrm{tr}\left[\left(f^*_{\|,\perp}\sigma\right)e^{i(e/2)(s-t)\sigma F}\left(f_{\|,\perp}\sigma\right)e^{i(e/2)t\sigma F}\right]$$

$$= 4\omega^2\sin eB(s-t)\,\sin eBt\,\mathrm{tr}\left\{\Sigma_{\|,\perp}\,\sigma_3\,\Sigma_{\|,\perp}\,\sigma_3\right\}$$

$$= \left(\mp 32\right)_{\|,\perp}\omega^2\sin eB(s-t)\,\sin eBt$$

$$= \left(\mp 16\right)_{\|,\perp}\omega^2\left(\cos eBvs - \cos eBs\right), \tag{C.40}$$

where $v = 2t/s - 1$, and the trace on the right-hand side of the first line can be evaluated by using the explicit representation of the Dirac algebra quantities in terms of Pauli matrices. This concludes the proof of (2.188).

D Elementary Calculations

This Appendix presents some techniques necessary for performing elementary calculations which are required in various chapters of the main text.

D.1 ϵ expansions

The following list of ϵ expansions is a useful tool for the evaluation of a finite sum of infinite integrals (C denotes Euler's constant: $C = 0.57721\ldots$):

$$a^{-\epsilon} = 1 - \epsilon \ln a + \frac{1}{2}(\ln a)^2 \epsilon^2 + \ldots,$$

$$\Gamma(1+\epsilon) = 1 - C\epsilon + \left(\frac{1}{12}\pi^2 + \frac{1}{2}C^2\right)\epsilon^2 + \ldots,$$

$$\Gamma(\epsilon) = \frac{1}{\epsilon} - C + \left(\frac{1}{12}\pi^2 + \frac{1}{2}C^2\right)\epsilon^2 + \ldots,$$

$$\Gamma(\epsilon - 1) = -\frac{1}{\epsilon} + (C - 1) + \left(-\frac{\pi^2}{12} - \frac{C^2}{2} - 1 + C\right)\epsilon + \ldots,$$

$$\Gamma(\epsilon - 2) = \frac{1}{2\epsilon} + \left(\frac{3}{4} + \frac{C}{2}\right) + \left(\frac{\pi^2}{24} + \frac{C^2}{4} + \frac{7}{8} - \frac{3}{4}C\right)\epsilon + \ldots,$$

$$\zeta(\epsilon, h) = \frac{1}{2} - h + \epsilon\,\zeta'(0, h) + \frac{\epsilon^2}{2}\,\zeta''(0, h) + \ldots,$$

$$\zeta(\epsilon - 1, h) = -\frac{1}{2}h^2 + \frac{1}{2}h - \frac{1}{12} + \epsilon\,\zeta'(-1, h) + \frac{\epsilon^2}{2}\,\zeta''(-1, h) + \ldots,$$

$$\zeta(1 + \epsilon, h) = \frac{1}{\epsilon} - \psi(h) + \mathcal{O}(\epsilon). \tag{D.1}$$

The last equation follows from [91]:

$$\lim_{\epsilon \to 0} \frac{\zeta(1 + \epsilon, h)}{\Gamma(-\epsilon)} = -1 \quad \text{and} \quad \lim_{\epsilon \to 0}\left(\zeta(1 + \epsilon, h) - \frac{1}{\epsilon}\right) = -\psi(h). \tag{D.2}$$

D.2 Identities for Special Functions

Stirling's formula:

$$\ln \Gamma(z) = z \ln z - z - \frac{1}{2}\ln z + \frac{1}{2}\ln 2\pi + \mathcal{O}(1/z). \tag{D.3}$$

Γ–ζ relations:

$$\ln \Gamma(z) = \zeta'(0, h) + \frac{1}{2}\ln 2\pi. \tag{D.4}$$

D.3 Nonstandard Integrals

The following integrals are the basic tools of the ϵ-integration technique; they can be found in [91] (the formula numbers used in [91] are given):

$$(3.551.1) \quad \int_0^\infty dx\, x^\mu\, e^{-2hx}\, \sinh 2x \tag{D.5}$$

$$= \frac{1}{2}\Gamma(1+\mu)\Big[(2h-2)^{-1-\mu} - (2h+2)^{-1-\mu}\Big], \quad \mu > -2,$$

$$(3.551.2) \quad \int_0^\infty dx\, x^\mu\, e^{-2hx}\, \cosh x \tag{D.6}$$

$$= \frac{1}{2}\Gamma(1+\mu)\Big[(2h-2)^{-1-\mu} + (2h+2)^{-1-\mu}\Big], \quad \mu > -1,$$

$$(3.551.3) \quad \int_0^\infty dx\, x^\mu\, e^{-2hx}\, \coth x$$

$$= \Gamma(1+\mu)\Big[2^{-\mu}\, \zeta(1+\mu, h) - (2h)^{-1-\mu}\Big], \tag{D.7}$$

$$(3.381.4) \quad \int_0^\infty dx\, x^{\mu-1}\, e^{-2hx} = (2h)^{-\mu}\, \Gamma(\mu), \quad \mu > 0. \tag{D.8}$$

D.4 ϵ-Integration Techniques

One often encounters finite, well-defined integrals in analytical proper-time calculations, where the integrand consists of a sum over various functions, (e.g. $\int (A + B) < \infty$). After interchanging summation and integration, one obtains a finite sum of infinite integrals (e.g. $\int A + \int B,\ |\int A| \to \infty$ and $|\int B| \to \infty$), which in combination is finite. The origin of such finite combinations, of course, lies in the procedure of renormalization. These integrals can be evaluated by introducing a convergence-enforcing factor that depends on an appropriately chosen parameter ϵ and goes to 1 in the limit $\epsilon \to 0$.

In the present work, we need the following integrals:

$$\int_0^\infty dz\, e^{-2ihz}\left(\cot z - \frac{1}{z}\right) = \int_0^\infty dx\, e^{-2hx}\left(\coth x - \frac{1}{x}\right)$$

$$= \lim_{\epsilon \to 0}\left(\int_0^\infty dx\, x^\epsilon e^{-2hx}\, \coth x - \int_0^\infty dx\, x^{\epsilon-1} e^{-2hx}\right)$$

$$\overset{(D.7),(D.8)}{=} \lim_{\epsilon \to 0}\left\{\Gamma(\epsilon+1)\Big[2^{-\epsilon}\zeta(\epsilon+1, h) - (2h)^{-1-\epsilon}\Big] - (2h)^{-\epsilon}\Gamma(\epsilon)\right\}$$

$$\overset{(D.1)}{=} \lim_{\epsilon \to 0}\left[-\psi(h) - \frac{1}{2h} + \ln h + \mathcal{O}(\epsilon)\right]$$

$$= \ln h - \psi(1 + h) + \frac{1}{2h}, \tag{D.9}$$

where we have employed the identity $\psi(1 + x) = \psi(x) + 1/x$ in the last line. From this integral, we can deduce

$$2ih \int_0^1 d\nu \int_0^\infty dz\, e^{-2ihz} \frac{\nu \sin \nu z}{\sin z} = 2ih \int_0^\infty dz\, e^{-2ihz} \left(\frac{1}{z^2} - \frac{\cot z}{z} \right)$$

$$= 2h \int_0^\infty dx\, \frac{e^{-2hx}}{x} \left(\coth x - \frac{1}{x} \right) = 4h \int_h^\infty dh' \int_0^\infty dx\, e^{-2hx} \left(\coth x - \frac{1}{x} \right)$$

$$\overset{(D.9)}{=} \int_h^\infty dh' \left[\ln h' - \psi(h') - \frac{1}{2h'} \right]$$

$$= 4h \lim_{h' \to \infty} \left(h' \ln h' - h' - \ln \Gamma(h') - \frac{1}{2} \ln h' \right)$$

$$- 4h \left(h \ln h - h - \ln \Gamma(h) - \frac{1}{2} \ln h \right)$$

$$\overset{(D.3),(D.4)}{=} 4h\zeta'(0, h) + \left(2h - 4h^2 \right) \ln h + 4h^2. \tag{D.10}$$

In addition, we need

$$\int_0^\infty \frac{dz}{z} e^{-2ihz} \left[\left(\frac{2}{z} + 2ih \right) \cot z + \frac{2}{3} - \frac{2ih}{z} - \frac{2}{z^2} \right]$$

$$= \int_0^\infty \frac{dx}{x} e^{-2hx} \left[\left(-\frac{2}{x} - 2h \right) \coth x + \frac{2}{3} + \frac{2h}{x} + \frac{2}{x^2} \right]$$

$$= \lim_{\epsilon \to 0} \left(-2 \int_0^\infty dx\, x^{\epsilon-2} e^{-2hx} \coth x - 2h \int_0^\infty dx\, x^{\epsilon-1} e^{-2hx} \coth x \right.$$

$$\left. + \frac{2}{3} \int_0^\infty dx\, x^{\epsilon-1} e^{-2hx} + 2h \int_0^\infty dx\, x^{\epsilon-2} e^{-2hx} + 2 \int_0^\infty dx\, x^{\epsilon-3} e^{-2hx} \right)$$

$$\overset{(D.7),(D.8)}{=} \lim_{\epsilon \to 0} \left\{ -2\Gamma(\epsilon - 1) \left[2^{2-\epsilon} \zeta(\epsilon - 1, h) - (2h)^{1-\epsilon} \right] \right.$$

$$- 2h\Gamma(\epsilon) \left[2^{1-\epsilon} \zeta(\epsilon, h) - (2h)^{-\epsilon} \right]$$

$$\left. + \frac{2}{3} (2h)^{-\epsilon} \Gamma(\epsilon) + 2h(2h)^{1-\epsilon} \Gamma(\epsilon - 1) + 2(2h)^{2-\epsilon} \Gamma(\epsilon - 2) \right\}$$

$$\overset{(D.1)}{=} -2h^2 + 8\zeta'(-1, h) - 4h\zeta'(0, h) + 2h \ln h - \frac{2}{3} \ln h - \frac{2}{3}. \tag{D.11}$$

The ϵ-integration technique can also be applied to infinite integrals. First, by introducing an appropriate ϵ, the integral is made finite; secondly, taking the limit $\epsilon \to 0$ separates the finite terms from the divergences in a well-defined way.

The following integrals will be useful (terms of order ϵ are neglected):

$$\int_0^\infty dz\, e^{-2hz} \coth z \;\to\; \int_0^\infty dz\, z^\epsilon\, e^{-2hz} \coth z$$

$$\stackrel{(D.7)}{=} \Gamma(1+\epsilon)\big[2^{-\epsilon}\,\zeta(1+\epsilon,h) - (2h)^{-1-\epsilon}\big]$$

$$\stackrel{(D.1)}{=} \frac{1}{\epsilon} - \psi(h) - C - \ln 2 - \frac{1}{2h}, \tag{D.12}$$

$$\int_0^\infty dz\, e^{-2hz} \sinh 2z \stackrel{(D.5)}{=} \frac{1}{2}\big[(2h-2)^{-1} - (2h+2)^{-1}\big] = \frac{1}{2}\frac{1}{h^2-1}, \tag{D.13}$$

$$\int_0^\infty \frac{dz}{z}\, e^{-2hz} \;\to\; \int_0^\infty dz\, z^{\epsilon-1}\, e^{-2hz}$$

$$= (2h)^{-\epsilon}\Gamma(\epsilon) \stackrel{(D.1)}{=} \frac{1}{\epsilon} - \ln 2h - C, \tag{D.14}$$

$$\int_0^\infty dz\, e^{-2hz} \frac{\cosh 2z}{z} \;\to\; \int_0^\infty dz\, z^{\epsilon-1}\, e^{-2hz} \cosh 2z$$

$$\stackrel{(D.6)}{=} \frac{1}{2}\Gamma(\epsilon)\big[(2h-2)^{-\epsilon} + (2h+2)^{-\epsilon}\big]$$

$$\stackrel{(D.1)}{=} \frac{1}{\epsilon} - C - \frac{1}{2}\ln(4h^2-4), \tag{D.15}$$

$$\int_0^\infty \frac{dz}{z^2}\, e^{-2hz} \;\to\; \int_0^\infty dz\, z^{\epsilon-2}\, e^{-2hz} = (2h)^{1-\epsilon}\,\Gamma(\epsilon-1)$$

$$= -\frac{2h}{\epsilon} + 2h\ln 2h + 2h(C-1), \tag{D.16}$$

$$\int_0^\infty dz\, e^{-2hz} \frac{\cosh 2z}{z^2} \;\to\; \int_0^\infty dz\, z^{\epsilon-2}\, e^{-2hz} \cosh 2z$$

$$\stackrel{(D.6)}{=} \frac{1}{2}\Gamma(\epsilon-1)\big[(2h-2)^{1-\epsilon} + (2h+2)^{1-\epsilon}\big]$$

$$\stackrel{(D.1)}{=} -\frac{2h}{\epsilon} + 2h(C-1) + (h-1)\ln(2h-2)$$

$$+(h+1)\ln(2h+2) \tag{D.17}$$

$$= -\frac{2h}{\epsilon} + 2h(C-1) + h\ln(4h^2-4) + \ln\frac{h+1}{h-1},$$

$$\int_0^\infty dz\, e^{-2hz}\frac{\sinh 2z}{z} \;\to\; \int_0^\infty dz\, z^{\epsilon-1}\, e^{-2hz}\sinh 2z$$

$$\overset{(D.5)}{=} \frac{1}{2}\Gamma(\epsilon)\big[(2h-2)^{-\epsilon} - (2h+2)^{-\epsilon}\big]$$

$$\overset{(D.1)}{=} \frac{1}{2}\ln\frac{h+1}{h-1}, \tag{D.18}$$

$$\int_0^\infty dz\, e^{-2hz}\frac{\coth z}{z} \;\to\; \int_0^\infty dz\, z^{\epsilon-1}\, e^{-2hz}\coth z$$

$$\overset{(D.7)}{=} \Gamma(\epsilon)\big[2^{1-\epsilon}\,\zeta(\epsilon,h) - (2h)^{-\epsilon}\big] \tag{D.19}$$

$$\overset{(D.1)}{=} -\frac{2h}{\epsilon} + 2Ch + 2h\ln 2 + 2\ln\Gamma(h) - \ln 2\pi + \ln h,$$

$$\int_0^\infty dz\, e^{-2hz}\cosh 2z \overset{(D.6)}{=} \frac{1}{2}\big[(2h-2)^{-1} + (2h+2)^{-1}\big] = \frac{1}{2}\frac{h}{h^2-1}, \tag{D.20}$$

$$\int_0^\infty dz\, e^{-2hz}\frac{\coth z}{z^2} \;\to\; \int_0^\infty dz\, z^{\epsilon-2}\, e^{-2hz}\coth z$$

$$\overset{(D.7)}{=} \Gamma(\epsilon-1)\big[2^{2-\epsilon}\,\zeta(\epsilon-1,h) - (2h)^{1-\epsilon}\big]$$

$$\overset{(D.1)}{=} \left(2h^2 + \frac{1}{3}\right)\frac{1}{\epsilon} + (1-C)\left(2h^2 + \frac{1}{3}\right) \tag{D.21}$$

$$- \left(2h^2 + \frac{1}{3}\right)\ln 2 - 4\zeta'(-1,h) - 2h\ln h,$$

$$\int_0^\infty dz\, e^{-2hz}\frac{1}{z\sinh^2 z} \;\to\; -\frac{\partial}{\partial\alpha}\int_0^\infty dz\, z^{\epsilon-2}\, e^{-2hz}\coth\alpha z\bigg|_{\alpha=1}$$

$$\overset{z'\equiv\alpha z}{=} -\frac{\partial}{\partial\alpha}\alpha^{1-\epsilon}\int_0^\infty dz'\, z'^{\epsilon-2}\, e^{-2(h/\alpha)z'}\coth z'\bigg|_{\alpha=1}$$

$$= (\epsilon-1)\int_0^\infty dz'\, z'^{\epsilon-2}\, e^{-2hz'}\coth z' - 2h\int_0^\infty dz'\, z'^{\epsilon-1}\, e^{-2hz'}\coth z'$$

$$\overset{(D.21),(D.19)}{=\!=} (\epsilon - 1)\left[\left(2h^2 + \frac{1}{3}\right)\frac{1}{\epsilon} + (1 - C)\left(2h^2 + \frac{1}{3}\right)\right.$$
$$\left. - \left(2h^2 + \frac{1}{3}\right)\ln 2 - 4\zeta'(-1, h) - 2h \ln h\right]$$
$$- 2h\left[-\frac{2h}{\epsilon} + 2C h + 2h \ln 2 + 2 \ln \Gamma(h) - \ln 2\pi + \ln h\right]$$
$$= \left(2h^2 - \frac{1}{3}\right)\frac{1}{\epsilon} - C\left(2h^2 - \frac{1}{3}\right) - \left(2h^2 - \frac{1}{3}\right)\ln 2$$
$$+ 4\zeta'(-1, h) - 4h \ln \Gamma(h) + 2h \ln 2\pi. \tag{D.22}$$

D.5 $\Pi^{\mu\nu}$ for Strong Magnetic Fields

The equivalence between (2.94) and (2.95) will be proved in the following. For reasons of convenience, we substitute $z = eBs$ and $h = B_{cr}/(2B)$. Equation (2.94) then reads

$$\eta_{\parallel,\perp} = -\int\limits_0^\infty \frac{dz}{z} \int\limits_{-1}^1 \frac{d\nu}{2} e^{-2ihz} N_{\parallel,\perp}, \tag{D.23}$$

where the functions $N_{\parallel,\perp}$ are found in (2.92):

$$N_\parallel = \frac{z \cos \nu z}{\sin z} - z \cot z \left(1 - \nu^2 + \nu \frac{\sin \nu z}{\sin z}\right), \tag{D.24}$$

$$N_\perp = -\frac{z \cos \nu z}{\sin z} + \frac{\nu z \sin \nu z \cot z}{\sin z} + \frac{2z(\cos \nu z - \cos z)}{\sin^3 z}.$$

In various terms of these integrands, we observe second-order poles, which can be eliminated by partial integration. The main technique is to isolate $1/\sin^2 z$ terms, which yield $(-\cot z)$ upon partial integration. In particular, we study the following integral identities:

$$-\int\limits_0^\infty dz\, e^{-2ihz} \frac{\nu \sin \nu z \cot z}{\sin z} = -\int\limits_0^\infty dz \left(e^{-2ihz}\, \nu \sin \nu z \cos z\right)\left(\frac{1}{\sin^2 z}\right)$$

$$\overset{\text{i.b.p.}}{=\!=} \left[\left(e^{-2ihz}\, \nu \sin \nu z \cos z\right)\cot z\right]_0^\infty$$
$$-\int\limits_0^\infty dz \cot z \frac{d}{dz}\left(e^{-2ihz}\, \nu \sin \nu z \cos z\right)$$

$$= -\nu^2 + \int\limits_0^\infty dz\, \frac{e^{-2ihz}}{\sin z}\left(-\nu^2 \cos \nu z \underbrace{\cos^2 z}_{=1-\sin^2 z}\right.$$
$$\left. + 2ih\nu \sin \nu z \underbrace{\cos^2 z}_{=1-\sin^2 z} + \nu \sin \nu z \sin z \cos z\right)$$

$$= -\nu^2 + \int_0^\infty dz \, \frac{e^{-2ihz}}{\sin z} \left(-\nu^2 \cos \nu z + 2ih\nu \sin \nu z \right)$$

$$+ \int_0^\infty dz \, e^{-2ihz} \left(\nu^2 \cos \nu z \sin z - 2ih\nu \sin \nu z \sin z + \nu \sin \nu z \cos z \right)$$

$$\underbrace{\qquad\qquad\qquad\qquad\qquad\qquad\qquad\qquad\qquad\qquad\qquad\qquad\qquad\qquad}$$

$$= \nu \int_0^\infty dz \frac{d}{dz} \left(e^{-2ihz} \sin \nu z \sin z \right) = 0$$

$$= -\nu^2 + \int_0^\infty dz \, \frac{e^{-2ihz}}{\sin z} \left(-\nu^2 \cos \nu z + 2ih\nu \sin \nu z \right), \qquad (D.25)$$

$$2 \int_0^\infty dz \, \frac{e^{-2ihz}}{\sin^3 z} \left(\cos \nu z - \cos z \right) = 2 \int_0^\infty dz \left[\frac{e^{-2ihz}}{\sin z} \left(\cos \nu z - \cos z \right) \right] \left(\frac{1}{\sin^2 z} \right)$$

$$\stackrel{\text{i.b.p.}}{=} 2 \left[\frac{e^{-2ihz}}{\sin^3 z} \left(\cos \nu z - \cos z \right)(- \cot z) \right]_0^\infty$$

$$+ 2 \int_0^\infty dz \, \cot z \frac{d}{dz} \left[\frac{e^{-2ihz}}{\sin^3 z} \left(\cos \nu z - \cos z \right) \right]$$

$$= 2 \lim_{z \to 0} \frac{\cos \nu z - \cos z}{\sin^2 z} - 2 \int_0^\infty dz \, e^{-2ihz} \cot z$$

$$\times \left[\frac{\cos z (\cos \nu z - \cos z)}{\sin^2 z} + 2ihz \frac{\cos \nu z - \cos z}{\sin z} + \frac{\nu \sin \nu z - \sin z}{\sin z} \right]$$

$$= 1 - \nu^2 - 2 \int_0^\infty dz \, e^{-2ihz} \frac{\cos^2 z (\cos \nu z - \cos z)}{\sin^3 z}$$

$$+ 2 \int_0^\infty dz \, \frac{e^{-2ihz}}{\sin z} \left[-2ihz \cot z (\cos \nu z - \cos z) - \nu \cot z \sin \nu z + \cos z \right]$$

$$= 1 - \nu^2 + 2 \int_0^\infty dz \, \frac{e^{-2ihz}}{\sin z} (\cos \nu z - \cos z) \underbrace{- 2 \int_0^\infty dz \, \frac{e^{-2ihz}}{\sin^3 z} (\cos \nu z - \cos z)}_{= \text{left-hand side}}$$

$$+ 2 \int_0^\infty dz \, \frac{e^{-2ihz}}{\sin z} \left[-2ihz \cot z (\cos \nu z - \cos z) - \nu \cot z \sin \nu z + \cos z \right]$$

$$\Rightarrow \quad 2 \int_0^\infty dz \, \frac{e^{-2ihz}}{\sin^3 z} (\cos \nu z - \cos z)$$

$$= \frac{1}{2}(1 - \nu^2)$$

$$+ \int_0^\infty dz \frac{e^{-2ihz}}{\sin z} \left[-\nu \cot z \sin \nu z + \cos \nu z - 2ihz \cot z (\cos \nu z - \cos z) \right],$$

(D.26)

$$\int_0^\infty dz\, e^{-2ihz} \frac{\cot z}{\sin z} (\cos \nu z - \cos z) = \int_0^\infty \frac{dz}{\sin^2 z} \left[e^{-2ihz} \cos z (\cos \nu z - \cos z) \right]$$

$$\stackrel{\text{i.b.p.}}{=} \left[e^{-2ihz} \cos z (\cos \nu z - \cos z)(-\cot z) \right]_0^\infty$$

$$+ \int_0^\infty dz \cot z \frac{d}{dz} \left[e^{-2ihz} \cos z (\cos \nu z - \cos z) \right]$$

$$= 0 + \int_0^\infty dz \cot z e^{-2ihz} \left[-2ih \cos z (\cos \nu z - \cos z) \right.$$

$$\left. - \sin z (\cos \nu z - \cos z) - \cos z (\nu \sin \nu z - \sin z) \right]$$

$$= \int_0^\infty dz\, e^{-2ihz} \left[-2ih \frac{1 - \sin^2 z}{\sin z} (\cos \nu z - \cos z) \right.$$

$$\left. - \cos z (\cos \nu z - \cos z) - \frac{1 - \sin^2 z}{\sin z} (\nu \sin \nu z - \sin z) \right]$$

$$= \int_0^\infty dz\, e^{-2ihz} \left[-2ih \frac{\cos \nu z - \cos z}{\sin z} + 2ihz \sin z (\cos \nu z - \cos z) \right.$$

$$\left. - \cos z (\cos \nu z - \cos z) - \frac{\nu \sin \nu z - \sin z}{\sin z} + \sin z (\nu \sin \nu z - \sin z) \right]$$

$$= \int_0^\infty dz\, e^{-2ihz} \left(1 - \frac{\nu \sin \nu z}{\sin z} - 2ih \frac{\cos \nu z - \cos z}{\sin z} \right)$$

(D.27)

$$+ \int_0^\infty dz\, e^{-2ihz} \left[2ih \sin z (\cos \nu z - \cos z) \right.$$

$$\left. \underbrace{- \cos z (\cos \nu z - \cos z) + \sin z (\nu \sin \nu z - \sin z)}_{= -\frac{d}{dz} \left[e^{-2ihz} \sin z (\cos \nu z - \cos z) \right] \to 0} \right].$$

With the aid of these identities, (D.25)–(D.27), we are able to rewrite $\eta_{\parallel,\perp}$ as follows:

$$\eta_\| = \int_0^\infty \frac{dz}{z}\, e^{-2ihz} \int_0^1 d\nu \left[z \cot z \left(1 - \nu^2 + \frac{\nu \sin \nu z}{\sin z} \right) - \frac{z \cos \nu z}{\sin z} \right]$$

$$= \int_0^\infty dz\, e^{-2ihz} \cot z \int_0^1 d\nu (1 - \nu^2) + \int_0^1 d\nu \underbrace{\int_0^\infty dz\, e^{-2ihz} \frac{\nu \sin \nu z \cot z}{\sin z}}_{}$$

$$\qquad\qquad\qquad\qquad\qquad\qquad\qquad\qquad\qquad\qquad (D.25)$$

$$- \int_0^\infty dz\, e^{-2ihz} \frac{1}{\sin z} \int_0^1 d\nu\, \cos \nu z$$

$$= \frac{2}{3} \int_0^\infty dz\, e^{-2ihz} \cot z + \int_0^1 d\nu\, \nu^2$$

$$- \int_0^1 d\nu \int_0^\infty dz\, \frac{e^{-2ihz}}{\sin z} \left(-\nu^2 \cos \nu z + 2ih\nu \sin \nu z \right) - \int_0^\infty dz\, e^{-2ihz} \frac{1}{z}$$

$$= \frac{1}{3} + \frac{2}{3} \int_0^\infty dz\, e^{-2ihz} \left(\cot z - \frac{1}{z} \right) - \frac{1}{3} \int_0^\infty dz\, e^{-2ihz} \frac{1}{z}$$

$$+ \int_0^\infty dz\, \frac{e^{-2ihz}}{\sin z} \left[\frac{2}{z^2} \cos z + \left(\frac{1}{z} - \frac{2}{z^3} \right) \sin z - 2ih \frac{\sin z}{z^2} + 2ih \frac{\cos z}{z} \right]$$

$$= \frac{1}{3} + \frac{2}{3} \underbrace{\int_0^\infty dz\, e^{-2ihz} \left(\cot z - \frac{1}{z} \right)}_{(D.9)}$$

$$+ \underbrace{\int_0^\infty \frac{dz}{z}\, e^{-2ihz} \left[\left(\frac{2}{z} + 2ih \right) \cot z + \frac{2}{3} - \frac{2ih}{z} - \frac{2}{z^2} \right]}_{(D.11)}$$

$$= \frac{1}{3} + \frac{2}{3} \left[\ln h - \psi(1+h) + \frac{1}{2h} \right] - 2h^2 + 8\zeta'(-1,h) - 4h\zeta'(0,h)$$

$$+ 2h \ln h - \frac{2}{3} \ln h - \frac{2}{3}. \qquad\qquad (D.28)$$

Substituting $h = B_{\mathrm{cr}}/(2B)$ into (D.28) leads us exactly to the expression for $\eta_\|(B)$ cited in (2.95).

The function $\eta_\perp$ can be evaluated similarly; according to (D.23) and (D.24), $\eta_\perp$ reads

$$\eta_\perp = \int\limits_0^\infty dz\, e^{-2ihz} \int\limits_0^1 d\nu \left[\frac{\cos \nu z}{\sin z} - \frac{\nu \cot z \sin \nu z}{\sin z} - \underbrace{\frac{2(\cos \nu z - \cos z)}{\sin^3 z}}_{\to (D.26)} \right]$$

$$= \int\limits_0^\infty dz\, e^{-2ihz} \int\limits_0^1 d\nu \left(\frac{\cos \nu z}{\sin z} - \frac{\nu \cot z \sin \nu z}{\sin z} \right) - \int\limits_0^1 d\nu \frac{1}{2}(1 - \nu^2)$$

$$- \int\limits_0^1 d\nu \int\limits_0^\infty dz\, \frac{e^{-2ihz}}{\sin z}$$

$$\times \left[-\nu \sin \nu z \cot z + \cos \nu z - 2ih \cot z (\cos \nu z - \cos z) \right]$$

$$= -\frac{1}{3} + 2ih \int\limits_0^1 d\nu \int\limits_0^\infty dz\, e^{-2ihz} \underbrace{\frac{\cot z}{\sin z} (\cos \nu z - \cos z)}_{(D.27)}$$

$$= -\frac{1}{3} + 2ih \int\limits_0^1 d\nu \int\limits_0^\infty dz\, e^{-2ihz} \left(1 - \frac{\nu \sin \nu z}{\sin z} - 2ih \frac{\cos \nu z - \cos z}{\sin z} \right)$$

$$= -\frac{1}{3} + 2ih \int\limits_0^1 d\nu \frac{1}{2ih} - 2ih \int\limits_0^1 d\nu \int\limits_0^\infty dz\, e^{-2ihz} \frac{\nu \sin \nu z}{\sin z}$$

$$+ 4h^2 \int\limits_0^\infty dz\, e^{-2ihz} \left(\int\limits_0^1 d\nu \frac{\cos \nu z}{\sin z} - \int\limits_0^1 d\nu \frac{\cos z}{\sin z} \right)$$

$$= \frac{2}{3} - \underbrace{2ih \int\limits_0^1 d\nu \int\limits_0^\infty dz\, e^{-2ihz} \frac{\nu \sin \nu z}{\sin z}}_{(D.10)} + \underbrace{4h^2 \int\limits_0^\infty dz\, e^{-2ihz} \left(\frac{1}{z} - \cot z \right)}_{(D.9)}$$

$$= -4h\zeta'(0, h) + 4h^2\psi(1 + h) - 2h \ln h - 2h - 4h^2 + \frac{2}{3}. \qquad (D.29)$$

Substituting $h = B_{\mathrm{cr}}/(2B)$ into (D.29) finally leads us to the expression for $\eta_\perp(B)$ cited in (2.95). This completes the proof of (2.95).

D.6 Generalized Gamma Function and Hurwitz Zeta Function

During the preceding calculation, we encountered the integral given in (D.11):

$$I = \int\limits_0^\infty \frac{dz}{z} e^{-2ihz} \left[\left(\frac{2}{z} + 2ih \right) \cot z + \frac{2}{3} - \frac{2ih}{z} - \frac{2}{z^2} \right]$$

$$= -2h^2 + 8\zeta'(-1, h) - 4h\zeta'(0, h) + 2h \ln h - \frac{2}{3} \ln h - \frac{2}{3}. \tag{D.30}$$

Employing a slightly different decomposition of the function $\eta_{\parallel}$, Tsai and Erber [160] evaluated the same integral in two separated steps:

$$I = -\mathcal{J}_2 - \mathcal{J}_3, \tag{D.31}$$

where

$$\mathcal{J}_2 = 2ih \int\limits_0^\infty dz \, e^{-2ihz} \int\limits_0^1 d\nu \frac{\nu \sin \nu z}{\sin z}, \tag{D.32}$$

$$\mathcal{J}_3 = \int\limits_0^\infty dz \, e^{-2ihz} \int\limits_0^1 d\nu \, (1 - \nu^2) \left(\frac{\cos \nu z}{\sin z} - \frac{1}{z} \right). \tag{D.33}$$

Note, incidentally, that $\mathcal{J}_2$ corresponds to the integral given in (D.10). Following the computation of [160], one finds

$$\mathcal{J}_2 + \mathcal{J}_3 = 2h^2 + 2h \ln h + \frac{2}{3} \ln h - 2h \ln 2\pi + 4h \ln \Gamma(1 + h)$$
$$-8 \ln \Gamma_1(1 + h) + 8L_1. \tag{D.34}$$

Here we discover the generalized Γ function of the first kind [26], which in this case appears by virtue of the relation

$$\ln \Gamma_1(x) = \int\limits_0^x dt \, \ln \Gamma(t) + \frac{x}{2}(x - 1) - \frac{x}{2} \ln 2\pi. \tag{D.35}$$

This function satisfies the functional equation

$$\Gamma_1(1 + x) = x^x \, \Gamma_1(x), \tag{D.36}$$

with the constraints $\Gamma_1(0) = \Gamma_1(1) = \Gamma_1(2) = 1$. The resemblance to the usual Γ function (of the zeroth kind) becomes obvious on observing that Γ_1 for integer values of the argument reduces to

$$\Gamma_1(1 + n) = 1^1 \times 2^2 \times \cdots \times n^n, \quad n > 0. \tag{D.37}$$

The constant L_1 that appears in (D.34) can be obtained from the Raabe integral:

$$L_1 = \frac{1}{3} + \int\limits_0^1 dx \, \ln \Gamma_1(1 + x) \simeq 0.248\,754\,477. \tag{D.38}$$

Comparing this representation of the integral I with ours, we obtain the following identity:

$$4h\zeta'(0, h) - 8\zeta'(-1, h) = 4h \ln h - 2h \ln 2\pi + 4h \ln h\Gamma(h)$$
$$-8 \ln \Gamma_1(1 + h) + 8L_1 - \frac{2}{3}. \tag{D.39}$$

With the aid of the Γ-ζ relation (D.4), we finally find a direct connection between the generalized Γ function of the first kind and a derivative of the Hurwitz ζ function:

$$\zeta'(-1, h) = \ln \Gamma_1(h) - L_1 + \frac{1}{12}. \tag{D.40}$$

An important special case occurs at $h = 1$,

$$\zeta'(-1, 1) \equiv \zeta'(-1) = \frac{1}{12} - L_1, \tag{D.41}$$

since the exact value of $\zeta'(-1)$ plays an important role in the effective-action approach to QED [55]. Comparing (D.40) with (D.4), it should be interesting to investigate whether a general connection exists between the generalized Γ functions of the nth kind and the first derivative of the Hurwitz ζ function.

D.7 Q Factor for Strong Magnetic Fields

Next, we prove the equivalence between (3.47) and (3.48). For reasons of convenience, we substitute $h = B_{cr}/(2B)$. The integral part of (3.47) then reads

$$I_Q = \int_0^{i\infty} \frac{dz}{z} e^{-2hz} \left(\frac{z \coth z - 1}{\sinh^2 z} - \frac{1}{3} z \coth z \right). \tag{D.42}$$

Rotating the integration path in the complex plane removes the factor of i from the upper integral boundary. The first term can be integrated by parts:

$$\int_0^\infty \frac{dz}{z} e^{-2hz} \frac{z \coth z - 1}{\sinh^2 z} \overset{\text{i.b.p.}}{=} \left[(-\coth z) \frac{e^{-2hz}}{z} (z \coth z - 1) \right]_0^\infty$$

$$+ \int_0^\infty dz \coth z\, e^{-2hz} \left[-2h \left(\coth z - \frac{1}{z} \right) - \frac{1}{\sinh^2 z} + \frac{1}{z^2} \right]$$

$$= \frac{1}{3} + \int_0^\infty dz\, e^{-2hz} \left(2h \frac{\coth z}{z} - 2h \coth^2 z + \frac{\coth z}{z^2} \right)$$

$$\left. - \int_0^\infty \frac{dz}{z} e^{-2hz} \frac{z \coth z - 1}{\sinh^2 z} \right\} \triangleq \text{left} - \text{hand side}$$

$$\Rightarrow \int_0^\infty \frac{dz}{z} e^{-2hz} \frac{z \coth z - 1}{\sinh^2 z} \tag{D.43}$$

$$= \frac{1}{6} + \int_0^\infty dz\, e^{-2hz} \left(\frac{h}{z} \coth z - h \coth^2 z + \frac{\coth z}{2z^2} - \frac{1}{2z \sinh^2 z} \right).$$

Hence, I_Q can be written as

$$I_Q = \frac{1}{6} + \int_0^\infty dz\, e^{-2hz} \tag{D.44}$$

$$\times \left(\frac{h}{z} \coth z - h \coth^2 z + \frac{\coth z}{2z^2} - \frac{1}{2z \sinh^2 z} - \frac{1}{3} \coth z \right).$$

While the last three terms can be integrated by ϵ techniques (see (D.12), (D.21) and (D.22)), the integration of the $\coth^2 z$ term is much more involved. For this, consider the combination of the first two terms in I_Q, which is finite:

$$\int_0^\infty dz\, e^{-2hz} \coth z \left(\frac{1}{z} - \coth z \right) = \int_0^\infty dz\, \frac{e^{-2hz}}{\sinh^2 z} \left(\frac{\cosh z \sinh z}{z} - \cosh^2 z \right)$$

$$= \left[-\coth z \left(\frac{\sinh 2z}{2z} - \cosh^2 z \right) e^{-2hz} \right]_0^\infty \quad \Big\} \quad \to 0$$

$$+ \int_0^\infty dz\, \coth z\, e^{-2hz} \left[-2h \left(\frac{\cosh z \sinh z}{z} - \cosh^2 z \right) \right.$$

$$\left. - \frac{\cosh z \sinh z}{z^2} + \frac{\sinh^2 z}{z} + \frac{\cosh^2 z}{z} - 2 \cosh z \sinh z \right]$$

$$= \int_0^\infty dz\, e^{-2hz} \left(2h \coth z \underbrace{\cosh^2 z}_{=1+\sinh^2 z} - 2h \frac{\cosh^2 z}{z} - \frac{\cosh^2 z}{z^2} \right.$$

$$\left. + \frac{\sinh z \cosh z}{z} + \frac{1}{z} \coth z \underbrace{\cosh^2 z}_{=1+\sinh^2 z} - \cosh^2 z \right)$$

$$= \int_0^\infty dz\, e^{-2hz} \left(2h \coth z + 2h \sinh z \cosh z - 2h \frac{\cosh^2 z}{z} - \frac{\cosh^2 z}{z^2} \right.$$

$$\left. + \frac{\sinh z \cosh z}{z} + \frac{\coth z}{z} + \frac{\sinh z \cosh z}{z} - 2 \cosh^2 z \right)$$

$$= \int_0^\infty dz\, e^{-2hz} \left(2h \coth z + h \sinh 2z - \frac{h}{z} - h \frac{\cosh 2z}{z} - \frac{1}{2z^2} - \frac{\cosh 2z}{2z^2} \right.$$

$$\left. + \frac{\sinh 2z}{z} + \frac{\coth z}{z} - 1 - \cosh 2z \right). \tag{D.45}$$

In the last step, we have employed the identities $2 \sinh z \cosh z = \sinh 2z$ and $\cosh^2 z = (1/2)(\cosh 2z + 1)$. Inserting (D.45) into (D.44), we finally have to evaluate the following integral:

$$I_Q = \frac{1}{6} + \int_0^\infty dz\, e^{-2hz} \left[\left(2h^2 - \frac{1}{3} \right) \coth z + h^2 \sinh 2z - \frac{h^2}{z} - h^2 \frac{\cosh 2z}{z} \right.$$

$$-h\frac{1}{2z^2} - h\frac{\cosh 2z}{2z^2} + h\frac{\sinh 2z}{z} + h\frac{\coth z}{z}$$

$$\left. -h - h\cosh 2z + \frac{\coth z}{2z^2} - \frac{1}{2z\sinh^2 z} \right]. \quad (D.46)$$

The required integrations are listed in (D.12)-(D.22), and we can easily collect the resulting terms:

$$I_Q = \frac{1}{6} + \left(2h^2 - \frac{1}{3} \right) \left[\frac{1}{\epsilon} - \psi(h) - C - \ln 2 - \frac{1}{2h} \right] + h^2 \left(\frac{1}{2}\frac{1}{h^2-1} \right)$$

$$-h^2 \left(\frac{1}{\epsilon} - \ln 2h - C \right) - h^2 \left[\frac{1}{\epsilon} - C - \frac{1}{2}\ln(4h^2-4) \right]$$

$$-\frac{h}{2} \left[-\frac{2h}{\epsilon} + 2h\ln 2h + 2h(C-1) \right]$$

$$-\frac{h}{2} \left[-\frac{2h}{\epsilon} + 2h(C-1) + h\ln(4h^2-4) + \ln\frac{h+1}{h-1} \right]$$

$$+h \left(\frac{1}{2}\ln\frac{h+1}{h-1} \right) - h\frac{1}{2h} - h \left(\frac{1}{2}\frac{h}{h^2-1} \right)$$

$$+h \left[-\frac{2h}{\epsilon} + 2Ch + 2h\ln 2 + 2\ln \Gamma(h) - \ln 2\pi + \ln h \right]$$

$$+\frac{1}{2} \left[\left(2h^2 + \frac{1}{3} \right) \frac{1}{\epsilon} + (1-C) \left(2h^2 + \frac{1}{3} \right) \right.$$

$$\left. - \left(2h^2 + \frac{1}{3} \right) \ln 2 - 4\zeta'(-1,h) - 2h\ln h \right]$$

$$-\frac{1}{2} \left[\left(2h^2 - \frac{1}{3} \right) \frac{1}{\epsilon} - C \left(2h^2 - \frac{1}{3} \right) \right.$$

$$\left. - \left(2h^2 - \frac{1}{3} \right) \ln 2 + 4\zeta'(-1,h) - 4h\ln \Gamma(h) + 2h\ln 2\pi \right]$$

$$= - \left[\left(2h^2 - \frac{1}{3} \right) \psi(h) + h - 3h^2 \right.$$

$$\left. -4h\ln \Gamma(h) + 2h\ln 2\pi + \frac{1}{6} + 4\zeta'(-1,h) - \frac{1}{6h} \right]$$

$$= - \left[\left(2h^2 - \frac{1}{3} \right) \psi(1+h) - h - 3h^2 \right.$$

$$\left. -4h\ln \Gamma(h) + 2h\ln 2\pi + \frac{1}{6} + 4\zeta'(-1,h) + \frac{1}{6h} \right]. \quad (D.47)$$

Up to a minus sign, this corresponds exactly to the term in square brackets in (3.48), as it should; thus, the proof of (3.48) is completed.

D.8 Summation Techniques for Finite-Temperature Physics

In the following, we study the high-temperature limit of infinite sums of the form

$$S_{ij}(\lambda) = \sum_{n=1}^{\infty} (-1)^n (\lambda n)^i K_j(\lambda n), \tag{D.48}$$

where $\lambda = m/T$, and i and j denote certain integer parameters. The high-temperature limit clearly corresponds to small values of λ.

Since the appearance of Bessel functions reflects the $\mathbb{R}^3 \times S^1$ topology, which is the space of the finite-temperature field theory, the techniques described in the following are generally useful for finite-temperature applications.

The first step is to choose a representation of the modified Bessel function that shows a simple dependence on the summation index [91]:

$$K_j(\lambda n) = \int_0^{\infty} e^{-\lambda n \cosh t} \cosh jt \, dt . \tag{D.49}$$

Inserting (D.49) into (D.48), leads us to

$$S_{ij}(\lambda) = \lambda^i \int_0^{\infty} dt \cosh jt \sum_{n=1}^{\infty} n^i e^{-(i\pi + \lambda \cosh t)n} . \tag{D.50}$$

The sum appearing in (D.50) is of the form $\sum_{n=1}^{\infty} n^i q^n$, where q is a complex parameter with $|q| < 1$ and $i = 0, 1, 2, 3, \ldots$. For values of i close to zero, the result of the sum can be derived directly from the geometric series:

$$i = 0: \quad \sum_{n=1}^{\infty} q^n = \frac{q}{1-q} \quad \text{or} \quad \sum_{n=0}^{\infty} q^n = \frac{1}{1-q}. \tag{D.51}$$

Differentiation of the second form with respect to q leads us to

$$\frac{1}{(1-q)^2} = \sum_{n=1}^{\infty} n q^{n-1} = \frac{1}{1-q} + \sum_{n=0}^{\infty} n q^n,$$

$$\Rightarrow i = 1: \quad \sum_{n=0}^{\infty} n q^n \equiv \sum_{n=1}^{\infty} n q^n = \frac{q}{(1-q)^2}. \tag{D.52}$$

Further differentiation yields

$$\frac{1}{(1-q)^2} + \frac{2q}{(1-q)^3} = \frac{1+q}{(1-q)^3} \overset{(D.52)}{=} \sum_{n=1}^{\infty} n^2 q^{n-1} = \sum_{n=0}^{\infty} (n+1)^2 q^n$$

$$= \sum_{n=0}^{\infty} n^2 q^n + 2 \sum_{n=0}^{\infty} n q^n + \sum_{n=0}^{\infty} q^n$$

$$= \sum_{n=0}^{\infty} n^2 q^n + \frac{1+q}{(1-q)^2}.$$

Hence we obtain

$$i = 2: \quad \sum_{n=0}^{\infty} n^2 q^n \equiv \sum_{n=1}^{\infty} n^2 q^n = \frac{q(q+1)}{(1-q)^3}. \tag{D.53}$$

From now on, we concentrate on certain values for i and j which are of particular importance in the main text.

Sect. 3.6: Light Cone Condition at Finite Temperature. Next, we prove that (3.234) represents the high-temperature limit of (3.227), i.e. we consider the limit as $\lambda \to 0$ of $S_{11}(\lambda)$ (see (D.48)). Inserting the findings of (D.52) into (D.50), we arrive at the following ($q = -\mathrm{e}^{-\lambda \cosh t}$):

$$\begin{aligned} S_{11}(\lambda) &= -\lambda \int_0^{\infty} \mathrm{d}t \, \cosh t \, \frac{\mathrm{e}^{-\lambda \cosh t}}{\left(1 + \mathrm{e}^{-\lambda \cosh t}\right)^2} \\ &= -\int_{\lambda}^{\infty} \frac{\mathrm{d}p}{\sqrt{p^2 - \lambda^2}} \, p \, \frac{\mathrm{e}^{-p}}{\left(1 + \mathrm{e}^{-p}\right)^2} \\ &\overset{\text{i.b.p.}}{=} \int_{\lambda}^{\infty} \mathrm{d}p \, \sqrt{p^2 - \lambda^2} \, \frac{\mathrm{d}}{\mathrm{d}p} \left(\frac{\mathrm{e}^{-p}}{\left(1 + \mathrm{e}^{-p}\right)^2} \right), \end{aligned} \tag{D.54}$$

where we have employed the substitution $p = \lambda \cosh t$ again. We are interested in the Taylor expansion of $S_{11}(\lambda)$ for small values of λ:

$$S_{11}(\lambda) = S_{11}(0) + S_{11}'(0) \, \lambda + \frac{1}{2} S_{11}''(0) \, \lambda^2 + \mathcal{O}(\lambda^3). \tag{D.55}$$

For $S_{11}(0)$, we immediately find

$$S_{11}(0) = \int_0^{\infty} \mathrm{d}p \, p \frac{\mathrm{d}}{\mathrm{d}p} \left(\frac{\mathrm{e}^{-p}}{\left(1 + \mathrm{e}^{-p}\right)^2} \right) = \int_0^{\infty} \mathrm{d}p \, \frac{\mathrm{d}}{\mathrm{d}p} \left(\frac{1}{1 + \mathrm{e}^{-p}} \right) = -\frac{1}{2}. \tag{D.56}$$

The first derivative vanishes at $\lambda = 0$; we obtain for the second derivative

$$S_{11}''(\lambda) = \int_{\lambda}^{\infty} \mathrm{d}p \, \sqrt{p^2 - \lambda^2} \, \frac{\mathrm{d}}{\mathrm{d}p} \left[\frac{1}{p} \frac{\mathrm{d}}{\mathrm{d}p} \left(\frac{\mathrm{e}^{-p}}{\left(1 + \mathrm{e}^{-p}\right)^2} \right) \right] \tag{D.57}$$

$$+ \lambda^2 \int_{\lambda}^{\infty} \mathrm{d}p \, \sqrt{p^2 - \lambda^2} \, \frac{\mathrm{d}}{\mathrm{d}p} \left\{ \frac{1}{p} \frac{\mathrm{d}}{\mathrm{d}p} \left[\frac{1}{p} \frac{\mathrm{d}}{\mathrm{d}p} \left(\frac{\mathrm{e}^{-p}}{\left(1 + \mathrm{e}^{-p}\right)^2} \right) \right] \right\}.$$

In the limit $\lambda \to 0$, this yields

$$S_{11}''(0) = \int\limits_0^\infty dp\, p \frac{d}{dp}\left[\frac{1}{p}\frac{d}{dp}\left(\frac{e^{-p}}{(1+e^{-p})^2}\right)\right]$$

$$\stackrel{\text{i.b.p.}}{=} -\int\limits_0^\infty \frac{dp}{p}\frac{d}{dp}\left(\frac{e^{-p}}{(1+e^{-p})^2}\right) = \int\limits_0^\infty \frac{dp}{p}\frac{e^{-p}(1-e^{-p})}{(1+e^{-p})^3}$$

$$= k_2, \tag{D.58}$$

where $k_2 = 0.213\,139\,199\,408\,754\ldots$, a number defined by the integral given above. Finally, this leads us to

$$S_{11}(\lambda) \stackrel{(D.55)}{=} -\frac{1}{2} + \frac{1}{2}k_2\,\lambda^2 + \mathcal{O}(\lambda^4). \tag{D.59}$$

Thus, we have proved the formula which is used in (3.234).

Sect. 3.5: Debye Screening Mass. To confirm the value of the Debye screening mass in (3.171), we consider the high-temperature limit of the sum appearing in (3.170), i.e. we are interested in the limit as $\lambda \equiv m/T \to 0$ of S_{02} (see (D.48)). With the aid of (D.50) and (D.51), we may write

$$S_{02}(\lambda) = -\int\limits_0^\infty dt\, \cosh 2t \frac{e^{-\lambda\cosh t}}{1+e^{-\lambda\cosh t}}. \tag{D.60}$$

Using the decomposition $\cosh 2t = 2\cosh^2 t - 1$ and substituting $p = \lambda\cosh t$, we arrive at

$$S_{02} = -\frac{2}{\lambda^2}\int\limits_\lambda^\infty dp \frac{p^2}{\sqrt{p^2-\lambda^2}}\frac{e^{-p}}{1+e^{-p}} \int\limits_\lambda^\infty \frac{dp}{\sqrt{p^2-\lambda^2}}\frac{e^{-p}}{1+e^{-p}}. \tag{D.61}$$

The first term obviously dominates for small values of λ; to extract the leading behavior for $\lambda \to 0$, we therefore only need to know the value of the first integral for $\lambda = 0$. One easily finds [122] that

$$\int\limits_0^\infty dp\, p \frac{e^{-p}}{1+e^{-p}} = \frac{\pi^2}{12}. \tag{D.62}$$

We finally obtain

$$S_{02} = -\frac{\pi^2}{6\lambda^2} + \mathcal{O}(1) = \frac{\pi^2 T^2}{6m^2} + \mathcal{O}(1). \tag{D.63}$$

This is the formula which is required to arrive at (3.171).

E Finite-Temperature Coordinate Frame

In the following, we construct a coordinate frame that is particularly useful for finite-temperature systems with electromagnetic background fields. The coordinate frame is fixed with respect to the four-velocity of the heat bath u^μ and the (necessarily) constant field strength tensor $F^{\mu\nu}$. By a transformation into this special reference frame, the only nonvanishing components of the field strength tensors and the heat-bath velocity vector will be identical to the invariants of the system.

First, we construct the vierbein which mediates between any given system labeled by $\mu, \nu, \ldots = 0, 1, 2, 3$ and the desired system labeled by the (Lorentz) indices $A, B, \ldots = 0, 1, 2, 3$.

The simplest choice for the time-like base vector is:

$$\text{base vector 1}: \quad e_0{}^\mu = u^\mu, \qquad e_0{}^\mu e_{0\mu} = -1. \tag{E.1}$$

Owing to the antisymmetry of the field strength tensor, the first space-like base vector follows immediately:

$$\text{base vector 2}: \quad e_1{}^\mu = \frac{u^\alpha F^{\alpha\mu}}{\sqrt{\mathcal{E}}}, \qquad e_1{}^\mu e_{1\mu} = 1. \tag{E.2}$$

In (E.2), the normalization is given by

$$\mathcal{E} = \left(u_\alpha F^{\alpha\mu}\right)\left(u_\beta F^\beta{}_\mu\right), \tag{E.3}$$

which denotes a third field invariant, in addition to the usual field invariants, for fields at finite temperature. It is positive definite, since u^μ is time-like; in the rest frame of the heat bath, we find $\mathcal{E} = \boldsymbol{E}^2$.

For the construction of the third base vector, we define the auxiliary vector

$$V_\nu = u_\alpha F^{\alpha\beta} F_{\beta\nu}, \tag{E.4}$$

which obviously is linearly independent of $e_0{}^\mu$ and $e_1{}^\mu$, but not completely orthogonal. Let us first list a few properties of V_ν, which can be checked with the aid of the fundamental algebraic relations for the field strength tensor (B.3) and (B.4):

$$V^\nu e_{0\nu} = -\mathcal{E}, \tag{E.5}$$

$$V^\nu e_{1\nu} = 0, \tag{E.6}$$

$$V^\nu V_\nu = 2\mathcal{F}\mathcal{E} - \mathcal{G}^2. \tag{E.7}$$

Here, we have employed the notation $\mathcal{F}$ and $\mathcal{G}$ (see Appendix B) for the standard invariants in order to keep the formulas as transparent as possible.

The third base vector can be constructed by projecting out the $e_{0\nu}$ part of V_ν:

$$\text{base vector 3}: \quad e_2{}^\mu = \frac{1}{\sqrt{d}} V^\nu \left(\delta^\mu_\nu - \frac{e_{0\nu}\, e_0{}^\mu}{e_0^2} \right)$$

$$= \frac{1}{\sqrt{d}} V^\mu - \frac{\mathcal{E}}{\sqrt{d}} e_0{}^\mu. \tag{E.8}$$

The normalization d is given by a combination of invariants,

$$d = 2\mathcal{F}\mathcal{E} - \mathcal{G}^2 + \mathcal{E}^2 \equiv \left(|\boldsymbol{E}||\boldsymbol{B}|\right) - \left(\boldsymbol{E}\cdot\boldsymbol{B}\right)^2, \tag{E.9}$$

and is positive definite as long as $\boldsymbol{E}$ and $\boldsymbol{B}$ are not parallel.

By using the ϵ symbol, we can easily construct the fourth normalized base vector:

$$\text{base vector 4}: \quad e_3{}^\mu = \epsilon^{\alpha\beta\gamma\mu}\,e_{0\alpha}e_{1\beta}e_{2\gamma}. \tag{E.10}$$

This completes the construction of the finite-temperature coordinate frame. By definition, it satisfies the identity

$$e_{A\mu}\,e_B{}^\mu = g_{AB} \equiv \operatorname{diag}(-1,1,1,1), \tag{E.11}$$

where capital indices are used to label the four different base vectors. The metric $g_{AB} \sim g^{AB}$ can now be used to raise and lower capital indices. The base vectors $e_A{}^\mu$ form a vierbein that allows transformation into the finite-temperature coordinate frame.

For practical computations, we need, in particular, the rules for the multiplication of base vectors with the field strength tensor and its dual; the first rule is obvious from (E.2):

$$e_{0\alpha}\,F^{\alpha\beta} = \sqrt{\mathcal{E}}\,e_1{}^\beta. \tag{E.12}$$

With the help of (E.4) and (E.8), we furthermore find

$$e_{1\alpha}\,F^{\alpha\beta} = \frac{1}{\sqrt{\mathcal{E}}}\,V^\beta = \sqrt{\mathcal{E}}\,e_0{}^\beta + \sqrt{\frac{d}{\mathcal{E}}}\,e_2{}^\beta. \tag{E.13}$$

The derivation of the next multiplication rule is a little more involved; first, note from the definition (E.8) that

$$e_{2\alpha}\,F^{\alpha\beta} = \frac{1}{\sqrt{d}}\,V_\alpha F^{\alpha\beta} - \frac{\mathcal{E}}{\sqrt{d}}\,e_{0\alpha}F^{\alpha\beta}. \tag{E.14}$$

With the aid of the fundamental algebraic relations (B.3) and (B.4) for the field strength tensor, we obtain for the first term

$$V_\alpha F^{\alpha\beta} = -2\mathcal{F}\sqrt{\mathcal{E}}\,e_1{}^\beta - \mathcal{G}\,e_{0\gamma}{}^\star F^{\gamma\beta}. \tag{E.15}$$

Hence, the representation of $e_{0\gamma}{}^\star F^{\gamma\beta}$ with respect to the new coordinate frame is required; projecting this vector onto the base vectors gives

$$e_{0\gamma}{}^\star F^{\gamma\beta}\,e_{0\beta} = 0, \tag{E.16}$$

$$e_{0\gamma}{}^\star F^{\gamma\beta}\,e_{1\beta} = -\frac{\mathcal{G}}{\sqrt{\mathcal{E}}}, \tag{E.17}$$

$$e_{0\gamma}{}^\star F^{\gamma\beta}\,e_{2\beta} = 0, \tag{E.18}$$

$$e_{0\gamma}{}^\star F^{\gamma\beta}\,e_{3\beta} = \sqrt{\frac{d}{\mathcal{E}}}. \tag{E.19}$$

To arrive at these relations, we have used (B.3) and (B.4) extensively again. For (E.19), we have additionally employed the product formula (A.14) for two ϵ symbols. From (E.16)–(E.19), we learn that $e_{0\gamma}{}^{\star}F^{\gamma\beta}$ decomposes into

$$e_{0\gamma}{}^{\star}F^{\gamma\beta} = -\frac{\mathcal{G}}{\sqrt{\mathcal{E}}}\,e_1{}^{\beta} + \sqrt{\frac{d}{\mathcal{E}}}\,e_3{}^{\beta}. \tag{E.20}$$

Inserting (E.20) and (E.15) into (E.14), we obtain the desired third multiplication rule:

$$e_{2\alpha}F^{\alpha\beta} = -\sqrt{\frac{d}{\mathcal{E}}}\,e_1{}^{\beta} - \frac{\mathcal{G}}{\sqrt{\mathcal{E}}}\,e_3{}^{\beta}. \tag{E.21}$$

The last multiplication rule can be derived most easily from (E.20), which can be written as

$$e_3{}^{\beta} = \sqrt{\frac{\mathcal{E}}{d}}\,e_{0\alpha}{}^{\star}F^{\alpha\beta} + \frac{\mathcal{G}}{\sqrt{d}}\,e_1{}^{\beta}. \tag{E.22}$$

Multiplication with the field strength tensor leads us to

$$e_{3\alpha}F^{\alpha\beta} = \frac{\mathcal{G}}{\sqrt{\mathcal{E}}}\,e_2{}^{\beta}. \tag{E.23}$$

Let us summarize our findings:

$$e_{0\alpha}\,F^{\alpha\beta} = \sqrt{\mathcal{E}}\,e_1{}^{\beta}, \tag{E.24}$$

$$e_{1\alpha}\,F^{\alpha\beta} = \sqrt{\mathcal{E}}\,e_0{}^{\beta} + \sqrt{\frac{d}{\mathcal{E}}}\,e_2{}^{\beta}, \tag{E.25}$$

$$e_{2\alpha}\,F^{\alpha\beta} = -\sqrt{\frac{d}{\mathcal{E}}}\,e_1{}^{\beta} - \frac{\mathcal{G}}{\sqrt{\mathcal{E}}}\,e_3{}^{\beta}, \tag{E.26}$$

$$e_{3\alpha}\,F^{\alpha\beta} = \frac{\mathcal{G}}{\sqrt{\mathcal{E}}}\,e_2{}^{\beta}. \tag{E.27}$$

Similarly, the multiplication rules for the dual field strength tensor can be derived:

$$e_{0\alpha}\,{}^{\star}F^{\alpha\beta} = -\frac{\mathcal{G}}{\sqrt{\mathcal{E}}}\,e_1{}^{\beta} + \sqrt{\frac{d}{\mathcal{E}}}\,e_3{}^{\beta}, \tag{E.28}$$

$$e_{1\alpha}\,{}^{\star}F^{\alpha\beta} = -\frac{\mathcal{G}}{\sqrt{\mathcal{E}}}\,e_0{}^{\beta}, \tag{E.29}$$

$$e_{2\alpha}\,{}^{\star}F^{\alpha\beta} = -\sqrt{\mathcal{E}}\,e_3{}^{\beta}, \tag{E.30}$$

$$e_{3\alpha}\,{}^{\star}F^{\alpha\beta} = \sqrt{\mathcal{E}}\,e_2{}^{\beta} + \sqrt{\frac{d}{\mathcal{E}}}\,e_0{}^{\beta}. \tag{E.31}$$

The main advantage of the whole formalism becomes obvious on considering the explicit form of the field strength tensors in this finite-temperature coordinate frame; the following representations are an immediate consequence of the multiplication rules (E.24)–(E.31):

$$F_{AB} = e_{A\mu} F^{\mu\nu} e_{B\nu} = \begin{pmatrix} 0 & \sqrt{\mathcal{E}} & 0 & 0 \\ -\sqrt{\mathcal{E}} & 0 & \sqrt{d/\mathcal{E}} & 0 \\ 0 & -\sqrt{d/\mathcal{E}} & 0 & -\mathcal{G}/\sqrt{\mathcal{E}} \\ 0 & 0 & \mathcal{G}/\sqrt{\mathcal{E}} & 0 \end{pmatrix},$$

$${}^{\star}F_{AB} = e_{A\mu} {}^{\star}F^{\mu\nu} e_{B\nu} = \begin{pmatrix} 0 & -\mathcal{G}/\sqrt{\mathcal{E}} & 0 & \sqrt{d/\mathcal{E}} \\ \mathcal{G}/\sqrt{\mathcal{E}} & 0 & 0 & 0 \\ 0 & 0 & 0 & -\sqrt{\mathcal{E}} \\ -\sqrt{d/\mathcal{E}} & 0 & \sqrt{\mathcal{E}} & 0 \end{pmatrix}. \quad \text{(E.32)}$$

Note that ${}^{\star}F_{AB}$ is indeed dual to F_{AB}, which serves as an independent check of our construction.

In (E.32), the components of the field strength tensors are completely expressed in terms of the field invariants of a finite-temperature system. Hence, our construction is a direct finite-temperature analogue of the well-known (zero-temperature) reference frame in which the electric and magnetic fields are parallel and the nonvanishing field strength components are directly related to the standard invariants.

E.1 Application 1 (Sect. 3.5)

We apply the vierbein construction to the case of a matrix-valued function of the field strength tensor X which we encounter in the calculation of the QED effective action at finite temperature (see (3.152)):

$$X_{AB}(\mathrm{i}s) = \mathrm{i} \left[\frac{\tan eas}{ea} \frac{(b^2 g_{AB} - F_{AB}^2)}{a^2 + b^2} + \frac{\tanh ebs}{eb} \frac{(a^2 g_{AB} + F_{AB}^2)}{a^2 + b^2} \right]. \text{(E.33)}$$

The square of the field strength tensor, $F_{AB}^2 = F_A{}^C F_{CB}$, can be evaluated from (E.32):

$$F_{AB}^2 = \begin{pmatrix} -\mathcal{E} & 0 & \sqrt{d} & 0 \\ 0 & \mathcal{E} - d/\mathcal{E} & 0 & -\sqrt{d}\mathcal{G}/\mathcal{E} \\ \sqrt{d} & 0 & -\mathcal{G}^2/\mathcal{E} - d/\mathcal{E} & 0 \\ 0 & -\sqrt{d}\mathcal{G}/\mathcal{E} & 0 & -\mathcal{G}^2/\mathcal{E} \end{pmatrix}. \quad \text{(E.34)}$$

With the aid of this explicit representation, we can list the nonvanishing components of the symmetric tensor X_{AB}:

$$X_{00} = -\mathrm{i}\left[\langle\tan\rangle \, (b^2 - \mathcal{E}) + \langle\tanh\rangle \, (a^2 + \mathcal{E}) \right], \quad \text{(E.35)}$$

$$X_{11} = \mathrm{i}\left[\langle\tan\rangle \left(b^2 - \mathcal{E} + \frac{d}{\mathcal{E}} \right) + \langle\tanh\rangle \left(a^2 + \mathcal{E} - \frac{d}{\mathcal{E}} \right) \right], \quad \text{(E.36)}$$

$$X_{22} = i\left[\langle\tan\rangle\left(b^2 + \frac{\mathcal{G}^2}{\mathcal{E}} + \frac{d}{\mathcal{E}}\right) + \langle\tanh\rangle\left(a^2 - \frac{\mathcal{G}^2}{\mathcal{E}} - \frac{d}{\mathcal{E}}\right)\right], \qquad \text{(E.37)}$$

$$X_{33} = i\left[\langle\tan\rangle\left(b^2 + \frac{\mathcal{G}^2}{\mathcal{E}}\right) + \langle\tanh\rangle\left(a^2 - \frac{\mathcal{G}^2}{\mathcal{E}}\right)\right], \qquad \text{(E.38)}$$

$$X_{02} = i\left[\langle\tan\rangle\sqrt{d} - \langle\tanh\rangle\sqrt{d}\right], \qquad \text{(E.39)}$$

$$X_{13} = i\left[\langle\tan\rangle\frac{\mathcal{G}\sqrt{d}}{\mathcal{E}} - \langle\tanh\rangle\frac{\mathcal{G}\sqrt{d}}{\mathcal{E}}\right], \qquad \text{(E.40)}$$

where we have introduced the following two abbreviations:

$$\langle\tan\rangle = \frac{\tan eas}{ea(a^2 + b^2)} \quad \text{and} \quad \langle\tanh\rangle = \frac{\tanh ebs}{eb(a^2 + b^2)}. \qquad \text{(E.41)}$$

A further useful abbreviation will be

$$\psi = i\Big(\langle\tanh\rangle - \langle\tan\rangle\Big). \qquad \text{(E.42)}$$

With regard to Sect. 3.5, we are interested in the combinations $X_{11}X_{33} - X_{13}^2$ and $X_{00}X_{22} - X_{02}^2$. Let us first prove that these combinations are identical (up to a minus sign); for this, note that

$$X_{00} \equiv -X_{11} - \frac{d}{\mathcal{E}}\,\psi, \qquad \text{(E.43)}$$

$$X_{22} \equiv X_{33} - \frac{d}{\mathcal{E}}\,\psi, \qquad \text{(E.44)}$$

$$X_{02} \equiv \sqrt{d}\,\psi \quad \text{and} \quad X_{13} \equiv \frac{\mathcal{G}\sqrt{d}}{\mathcal{E}}\,\psi. \qquad \text{(E.45)}$$

In this language, we find

$$X_{11}X_{33} - X_{13}^2 = X_{11}X_{33} - \frac{\mathcal{G}^2 d}{\mathcal{E}^2}\,\psi^2. \qquad \text{(E.46)}$$

On the other hand, we obtain

$$-(X_{00}X_{22} - X_{02}^2) = -\left[\left(-X_{11} - \frac{d}{\mathcal{E}}\psi\right)\left(X_{33} - \frac{d}{\mathcal{E}}\psi\right) - d\psi^2\right]$$

$$= X_{11}X_{33} + \frac{d}{\mathcal{E}}\,\psi\,(X_{33} - X_{11}) + \left(d - \frac{d^2}{\mathcal{E}^2}\right)\psi^2. \quad \text{(E.47)}$$

From the explicit representations in (E.36) and (E.38), we obtain for $X_{33} - X_{11}$

$$X_{33} - X_{11} = i\langle\tanh\rangle\left(-\frac{\mathcal{G}^2}{\mathcal{E}} - \mathcal{E}\right) + i\langle\tan\rangle\left(\frac{\mathcal{G}^2}{\mathcal{E}} + \mathcal{E}\right) + \frac{d}{\mathcal{E}}\,\psi$$

$$= \left(\frac{d}{\mathcal{E}} - \frac{\mathcal{G}^2}{\mathcal{E}} - \mathcal{E}\right)\psi. \qquad \text{(E.48)}$$

Inserting (E.48) into (E.47) leads us to

$$-(X_{00}X_{22} - X_{02}^2) = X_{11}X_{33} - \frac{\mathcal{G}^2 d}{\mathcal{E}^2}\,\psi^2 \stackrel{(E.46)}{\equiv} X_{11}X_{33} - X_{13}^2, \qquad (E.49)$$

which completes the proof of the equivalence.

In order to obtain an explicit representation of $X_{11}X_{33} - X_{13}^2$, we introduce another abbreviation:

$$\phi = i\Big(b^2\,\langle\tanh\rangle + a^2\,\langle\tan\rangle\,\Big). \qquad (E.50)$$

Taking a look at (E.36) and (E.38), we find that

$$X_{11} = \phi - \Big(\frac{d}{\mathcal{E}} - \mathcal{E}\Big)\,\psi, \qquad (E.51)$$

$$X_{33} = \phi - \frac{\mathcal{G}^2}{\mathcal{E}}\,\psi. \qquad (E.52)$$

$$(E.53)$$

The desired expression therefore reads

$$X_{11}X_{33} - X_{13}^2 = \Big[\phi - \Big(\frac{d}{\mathcal{E}} - \mathcal{E}\Big)\psi\Big]\Big(\phi - \frac{\mathcal{G}^2}{\mathcal{E}}\psi\Big) - \frac{\mathcal{G}^2 d}{\mathcal{E}^2}\,\psi^2$$

$$= \phi^2 - \Big(\frac{d}{\mathcal{E}} - \mathcal{E} + \frac{\mathcal{G}^2}{\mathcal{E}}\Big)\,\phi\psi - \mathcal{G}^2\,\psi^2$$

$$= \phi^2 - 2\mathcal{F}\,\phi\psi - \mathcal{G}^2\,\psi^2. \qquad (E.54)$$

In the last step, we have employed the definition (E.9): $d = 2\mathcal{F}\mathcal{E} - \mathcal{G}^2 + \mathcal{E}^2$. Reexpressing ϕ and ψ in terms of $\langle\tan\rangle$ and $\langle\tanh\rangle$, and $\mathcal{F}$ and $\mathcal{G}$ in terms of a and b $\big(\mathcal{F} = (1/2)(a^2 - b^2),\quad \mathcal{G} = ab\big)$, we finally obtain

$$X_{11}X_{33} - X_{13}^2$$

$$= \big(\phi - \mathcal{F}\,\psi\big)^2 - \underbrace{\big(\mathcal{F}^2 + \mathcal{G}^2\big)}_{=\frac{1}{4}(a^2+b^2)^2}\,\psi^2$$

$$= -\Big[\big(a^2\langle\tanh\rangle + b^2\langle\tan\rangle\big) - \frac{1}{2}(a^2 - b^2)\big(\langle\tanh\rangle - \langle\tan\rangle\big)\Big]^2$$

$$\quad + \frac{1}{4}(a^2 + b^2)^2\big(\langle\tanh\rangle - \langle\tan\rangle\big)^2$$

$$= -\frac{1}{4}(a^2 + b^2)^2\big(\langle\tanh\rangle + \langle\tan\rangle\big) + \frac{1}{4}(a^2 + b^2)^2\big(\langle\tanh\rangle - \langle\tan\rangle\big)$$

$$\stackrel{(E.41)}{=} -\frac{\tan ebs}{eb} \frac{\tanh eas}{ea}. \qquad (E.55)$$

This representation is the desired one, which we employ in (3.155).

Finally, we need the explicit form of the function $h(s)$ defined in (3.161):

$$h(s) = \frac{iX_{22}}{X_{11}X_{33} - X_{13}^2}. \qquad (E.56)$$

For this, we rewrite the function X_{22} as given in (E.37) in the following way:

$$
\begin{aligned}
X_{22} &= i\left[\langle\tan\rangle\left(b^2 + 2\mathcal{F} + \mathcal{E}\right) + \langle\tanh\rangle\left(a^2 - 2\mathcal{F} - \mathcal{E}\right)\right] \\
&= i\left[\langle\tan\rangle\left(b^2 - \mathcal{E}\right) + \langle\tanh\rangle\left(a^2 + \mathcal{E}\right)\right] \\
&\stackrel{(E.41)}{=} i\left(\frac{a^2 + \mathcal{E}}{a^2 + b^2}\frac{\tan eas}{ea} + \frac{b^2 - \mathcal{E}}{a^2 + b^2}\frac{\tanh ebs}{eb}\right).
\end{aligned}
\tag{E.57}
$$

Here we have employed the definition of the combination of invariants d (E.9). Inserting this equation and (E.55) into (E.56), we find

$$
h(s) = \frac{b^2 - \mathcal{E}}{a^2 + b^2}\, ea \cot eas + \frac{a^2 + \mathcal{E}}{a^2 + b^2}\, eb \coth ebs.
\tag{E.58}
$$

This representation is required in (3.161).

E.2 Application 2 (Sect. 3.6)

In order to determine the form of $z_k = k_\mu F^{\mu\alpha} k_\nu F^\nu{}_\alpha \equiv -k^A F_{AC} F^C{}_B k^B$, which is required in (3.189), we employ the square of the field strength tensor as given in (E.34), and find

$$
z_k = \mathcal{E}\,(k^0)^2 - 2\sqrt{d}\, k^0 k^2 + (2\mathcal{F} + \mathcal{E})\,(k^2)^2 + \left(\frac{d}{\mathcal{E}} - \mathcal{E}\right)(k^1)^2
$$

$$
+2\frac{\sqrt{d}\mathcal{G}}{\mathcal{E}}\, k^1 k^3 + \frac{\mathcal{G}^2}{\mathcal{E}}\,(k^3)^2,
\tag{E.59}
$$

where k^0, k^1, k^2, k^3 represent the components of the rotated momentum vector $k^A = e^A{}_\mu k^\mu$.

Now we can determine the desired form of the exponent in (3.189) in terms of finite-temperature coordinates:

$$
A_z z_k + A_k k^2
$$

$$
= \left[A_k + (a^2 - b^2 + \mathcal{E})A_z\right]\left(k^2 - \frac{A_z\sqrt{d}}{A_z(2\mathcal{F} + \mathcal{E}) + A_k}\, k^0\right)^2
$$

$$
-\frac{(A_k + a^2 A_z)(A_k - b^2 A_z)}{A_k + (a^2 - b^2 + \mathcal{E})A_z}\,(k^0)^2 + \frac{(A_k + a^2 A_z)(A_k - b^2 A_z)}{A_k\frac{a^2 b^2}{\mathcal{E}} + A_k}\,(k^1)^2
$$

$$
+\left(A_z\frac{a^2 b^2}{\mathcal{E}} + A_k\right)\left(k^3 + \frac{A_z\frac{\sqrt{d}\mathcal{G}}{\mathcal{E}}}{A_z\frac{\mathcal{G}^2}{\mathcal{E}} + A_k}\, k^1\right),
\tag{E.60}
$$

where, again, k^0, k^1, k^2, k^3 represent the components of k^A.

F Two-Loop Effective Action of QED at Zero Temperature

Dedicating this appendix to the derivation of the zero-temperature two-loop Lagrangian has two reasons: first, we want to make contact with well-known results, which serve as a check of our computations; secondly, our results represent a generalization of the work of Dittrich and Reuter [55], who considered purely magnetic fields, to the case of constant electromagnetic fields.

Here, we shall give a version of the unrenormalized effective Lagrangian only, since the renormalization procedure requires technical investigations which are beyond the scope of the present work.[2] For the finite-temperature considerations, it suffices for us to know that the zero-temperature part of the action can be renormalized at the (zero-temperature) physical renormalization point given by the electron mass. Then, the parameter m appearing in the thermal contributions to the effective action automatically coincides with the physical zero-temperature electron mass.

In (3.194), we achieved a separation of the thermal and zero-temperature parts in the integral I_1; concentrating on the zero-T case, we found

$$I_1^{T=0} = \frac{1}{16\pi^2} e^{-im^2 s} \frac{1}{q_a\, q_b}, \tag{F.1}$$

where $q_a = A_k + a^2 A_z$ and $q_b = A_k - b^2 A_z$, and A_k and A_z are defined in (3.187). We also need the second integral $I_2^{T=0}$, which is related to $I_1^{T=0}$ by (3.188), leading us to

$$I_2^{T=0} = \frac{e^{-im^2 s}}{16\pi^2} \int_0^\infty ds' \left[\frac{a^2}{(s' + q_a)^2(s' + q_b)} - \frac{b^2}{(s' + q_a)(s' + q_b)^2} \right]. \tag{F.2}$$

The integration over s' can be carried out by elementary techniques:

$$I_2^{T=0} = \frac{e^{-im^2 s}}{16\pi^2} \left[a^2 \left(\frac{1}{q_a(q_b - q_a)} - \frac{1}{(q_b - q_a)^2} \ln \frac{q_b}{q_a} \right) \right.$$
$$\left. - b^2 \left(\frac{1}{q_b(q_a - q_b)} + \frac{1}{(q_b - q_a)^2} \ln \frac{q_b}{q_a} \right) \right]. \tag{F.3}$$

Inserting these $T = 0$ contributions (F.1) and (F.3) into (3.185) and reorganizing the result a little, we finally arrive at

$$\mathcal{L}^2 = -\frac{\alpha}{(4\pi)^3} \int_0^\infty \frac{ds}{s} \int_{-1}^1 \frac{d\nu}{2} e^{-im^2 s} \frac{eas\, ebs}{\sin eas \sinh ebs} \tag{F.4}$$

[2] In particular, the mass renormalization has to be treated with great care [78]; this is because the proper-time parametrization of the mass shift which is required for the mass renormalization has to coincide with the proper-time parametrization of the two-loop contribution $\mathcal{L}^2$. In the present calculation, this is not achieved immediately, since the mass shift is usually taken from the mass operator, while the construction of $\mathcal{L}^2$ relies on the polarization tensor. One method to identify the correct value of a possible finite constant in the infinite mass shift is to fix it in such a way that the imaginary part of the entire effective action reproduces the Schwinger formula to lowest order in α [140]. This, however, requires a detailed investigation of the pole and branch cut structures of the proper-time integrand, which is extremely technical in the present representation.

$$\times \left[\frac{2N_0 + \tilde{N}_2}{q_a(q_b - q_a)} + \frac{2N_0 + \tilde{N}_1}{q_b(q_a - q_b)} - \frac{\tilde{N}_2 - \tilde{N}_1}{(q_b - q_a)^2} \ln \frac{q_b}{q_a} \right],$$

where N_0, $\tilde{N}_1$ and $\tilde{N}_2$ are functions of the integration variables and of the invariants a and b, and are defined in (2.77); q_a and q_b, after insertion of (3.187), can be written as

$$q_a = \frac{is}{2} \frac{\cos \nu eas - \cos eas}{eas \sin eas}, \qquad q_b = \frac{is}{2} \frac{\cosh ebs - \cosh \nu ebs}{ebs \sinh ebs}. \tag{F.5}$$

Equation (F.4) represents our final result for the unrenormalized two-loop effective Lagrangian of QED for an arbitrary constant electromagnetic field. In the limit of vanishing electric field, we recover exactly the findings of [55]; hence our comparatively compact representation generalizes the results of [55] to an arbitrary constant electromagnetic field.

References

1. L.F. Abbott, Nucl. Phys. B **185**, 189 (1981).
2. S.L. Adler, Phys. Rev. **177**, 2426 (1969).
3. S.L. Adler and W.A. Bardeen, Phys. Rev. **182**, 1517 (1969).
4. S.L. Adler, Ann. Phys. **67**, 599 (1971).
5. S.L. Adler and C. Schubert, Phys. Rev. Lett. **77**, 1695 (1996).
6. I. Affleck, Nucl. Phys. B **265**, 409 (1986).
7. Y. Aharonov and A. Casher, Phys. Rev. A **19**, 2461 (1979).
8. E.B. Aleksandrov, A.A. Ansel'm and A.N. Moskalev, Zh. Eksp. Teor. Fiz. **89**, 1181 (1985).
9. T.W. Appelquist, M. Bowick, D. Karabali and L.C.R. Wijewardhana, Phys. Rev. D **33**, 3704 (1986).
10. T.W. Appelquist, D. Nash and L.C.R. Wijewardhana, Phys. Rev. Lett. **60**, 2575 (1988).
11. G.K. Artimovich, Sov. Phys. JETP **70**, 787 (1990).
12. M.F. Atiyah and I.M. Singer, Bull. Am. Meteorol. Soc. **69**, 422 (1963); Ann. Math. **87**, 484 (1968).
13. M.F. Atiyah, V. Patodi and I.M. Singer, Bull. London Math. Soc. **5**, 229 (1973); Proc. Cambridge Philos. Soc. **77**, 42 (1975); **78**, 405 (1975); **79**, 71 (1976).
14. R. Baier and P. Breitenlohner, Acta Phys. Austriaca **25**, 212 (1967); Nuovo Cimento **47**, 261 (1967).
15. V.N. Baier, V.M. Katkov and V.M. Strakhovenko, Zh. Eksp. Teor. Fiz. **68**, 405 (1975) (Sov. Phys. JETP **41**, 198 (1975)).
16. V.N. Baier, A.I. Milshtein and R.Z. Shaisultanov, Phys. Rev. Lett. **77**, 1691 (1996).
17. D. Bakalov et al., Nucl. Phys. B **35** (Proc. Suppl), 180 (1994).
18. D. Bakalov et al., INFN/AE-94/27 preprint, INFN Trieste (1994).
19. M.G. Baring and A.K. Harding, in *High Velocity Neutron Stars and Gamma-Ray Bursts*, Proceedings of La Jolla Workshop, American Institute of Physics, New York (1995).
20. G. Barton, Phys. Lett. B **237**, 559 (1990).
21. G. Barton and K. Scharnhorst, J. Phys A **26**, 2037 (1993).
22. I.A. Batalin and A.E. Shabad, Sov. Phys. JETP **33**, 483 (1971).
23. W. Becker and H. Mitter, J. Phys. A **8**, 1638 (1975).
24. W. Becker, Las. Part. Beams **9**, 603 (1991).
25. S. Bell and R. Jackiw, Nouvo Cimento **60**, 47 (1969).
26. L. Bendersky, Acta Mathematica (Uppsala) **61**, 263 (1933).
27. S. Ben-Menahem, Phys. Lett. B **250**, 133 (1990).
28. Z. Białynicka-Birula and I. Białynicki-Birula, Phys. Rev. D **2**, 2341 (1970).
29. J.S. Bjorken and S.D. Drell, *Relativistic Quantum Mechanics*, McGraw-Hill, New York (1964).

30. S.K. Blau, M. Visser and A. Wipf, Int. J. Mod. Phys. A **6**, 5409 (1991).
31. M. Bordag, D. Robaschik and E. Wieczorek, Ann. Phys. (New York) **165**, 192 (1985).
32. M. Bordag and K. Scharnhorst, Phys. Rev. Lett. **81**, 3815 (1998).
33. F.T. Brandt, J. Frenkel and J.C. Taylor, Phys. Rev. D **50**, 4110 (1994).
34. F.T. Brandt and J. Frenkel, Phys. Rev. Lett. **74**, 1705 (1995).
35. L.S. Brown and G.J. Maclay, Phys. Rev. **184**, 1272 (1969).
36. R.W. Brown et al., Phys. Rev. D **8**, 3083 (1973).
37. R.-G. Cai, Nucl. Phys. B **524**, 639 (1998).
38. D. Cangemi, E. D'Hoker and G. Dunne, Phys. Rev. D **51**, R2513 (1995)
39. D. Cangemi, E. D'Hoker and G. Dunne, Phys. Rev. D **52**, 3163 (1995).
40. D. Cangemi and G. Dunne, Ann. Phys. **249**, 582 (1996).
41. H. Cheng and T.T. Wu, Phys. Rev. D **1**, 3414 (1970).
42. H.T. Cho, Phys. Rev. D **56**, 6416 (1997).
43. A. Chodos, K. Everding and D.A. Owen, Phys. Rev. D **42**, 2881 (1990).
44. T.E. Clark and N. Deo, Nucl. Phys. B **291**, 535 (1987).
45. D.H. Constantinescu, Nucl. Phys. B **36**, 121 (1972).
46. M.V. Cougo-Pinto, C. Farina, F.C. Santos and A.C. Tort, J. Phys. A **32**, 4463 (1999).
47. R.A. Cover and G. Kalman, Phys. Rev. Lett. **33**, 1113 (1974).
48. P.H. Cox, W.S. Hellman and A. Yildiz, Ann. Phys. **154**, 211 (1984).
49. R.D. Daniels and G.M. Shore, Nucl. Phys. B **425**, 634 (1994).
50. R.D. Daniels and G.M. Shore, Phys. Lett. B **367**, 75 (1996).
51. J.K. Daugherty and I. Lerche, Phys. Rev. D **14**, 340 (1976).
52. S. Deser, R. Jackiw and S. Templeton, Ann. Phys. **140**, 372 (1982).
53. W. Dittrich, Fortschr. Phys. **22**, 539 (1974).
54. W. Dittrich, Phys. Rev. D **19**, 2385 (1979).
55. W. Dittrich and M. Reuter, *Effective Lagrangians in Quantum Electrodynamics*, Lecture Notes in Physics, Vol. 220, Springer, Berlin, Heidelberg (1985).
56. W. Dittrich and M. Sieber, J. Phys. A **21**, L711 (1988).
57. W. Dittrich and M. Reuter, *Classical and Quantum Dynamics*, Springer, Berlin, Heidelberg (1992).
58. W. Dittrich, *QED in Constant External Magnetic Fields*, Lecture Notes (WS 1997/98), University of Tübingen (1997); Fortschr. Phys. **26**, 289 (1978).
59. W. Dittrich and H. Gies, Phys. Lett. B **392**, 182 (1997).
60. W. Dittrich and H. Gies, Phys. Rev. D **58**, 025004 (1998).
61. W. Dittrich and H. Gies, in *The Casimir Effect 50 Years Later*, Proceedings, ed. by M. Bordag, World Scientific, Singapore (1999).
62. W. Dittrich, *QED at Extremely High Energies*, Lecture Notes (SS 1999), University of Tübingen (1999).
63. A.D. Dolgov and I.D. Novikov, Phys. Lett. B **442**, 82 (1998).
64. J.F. Donoghue and B.R. Holstein, Phys. Rev. D **28**, 340 (1983); erratum, Phys. Rev. D **29**, 3004 (1984).
65. N. Dorey and N.E. Mavromatos, Nucl. Phys. B **368**, 614 (1992).
66. I.T. Drummond and S.J. Hathrell, Phys. Rev. D **22**, 343 (1980).
67. R. Duncan and C. Thompson, Astrophysics J. (Lett.) **392**, L9 (1992); Astrophysics J. **473**, 322 (1996).
68. G. Dunne and T.M. Hall, Phys. Rev. D **53**, 2220 (1996).
69. G. Dunne and T.M. Hall, Phys. Lett. B **419**, 322 (1998); Phys. Rev. D **58** 105022 (1998).
70. D. Ebert and H. Reinhardt, Nucl. Phys. B **271**, 188 (1986).
71. P. Elmfors, D. Persson and B.-S. Skagerstam, Phys. Rev. Lett. **71**, 480 (1993); Astropart. Phys. **2**, 299 (1994).

72. P. Elmfors and B.-S. Skagerstam, Phys. Lett. B **348**, 141 (1995); erratum, Phys. Lett. B **376**, 330 (1996).
73. P. Elmfors and B.-S. Skagerstam, Phys. Lett. B **427**, 197 (1998).
74. T. Erber, Rev. Mod. Phys. **38**, 626 (1966).
75. H. Euler and B. Kockel, Naturwissenschaften **23**, 246 (1935).
76. H. Euler, Ann. Physik **26**, 398 (1936).
77. K. Farakos, G. Koutsoumbas and N.E. Mavromatos, Phys. Lett. B **431**, 147 (1998).
78. D. Fliegner, M. Reuter, M.G. Schmidt and C. Schubert, Theor. Math. Phys. **113**, 1442 (1997).
79. V. Fock, Physik. Z. Sowjetunion **12**, 404 (1937).
80. M.P. Fry, Phys. Rev. D **54**, 6444 (1996). Nucl. Phys. B **248**, 615 (1984).
81. K. Fujikawa, Phys. Rev. Lett. **42**, 1195 (1979).
82. A.K. Ganguly, P.K. Kaw and J.C. Parikh, Phys. Rev. C **51**, 2091 (1995).
83. A.K. Ganguly, preprint hep-th/9804134 (1998).
84. J.L. Gervais, *Anomalies in Ward–Takahashi Identities*, Cargèse Lectures in Physics, Vol. 5, ed. by D. Bessis, Gordon and Breach, New York (1970).
85. H. Gies, diploma thesis, Tübingen University, (1996).
86. H. Gies and W. Dittrich, Phys. Lett. B **431**, 420 (1998).
87. H. Gies, Phys. Rev. D **60**, 105002 (1999).
88. H. Gies, Phys. Rev. D **60**, 105033 (1999).
89. H. Gies, Phys. Rev. D **61**, 085021 (2000).
90. R.J. Gould and G.P. Schréder, Phys. Rev. Lett. **16**, 252 (1966).
91. I.S. Gradshteyn and I.M. Ryzhik, *Tables of Integrals, Series and Products*, Academic Press (1965).
92. P.S. Gribosky and B.R. Holstein, Z. Phys. C **47**, 205 (1990).
93. D.J. Gross and A. Neveu, Phys. Rev. D **10**, 3235 (1974).
94. V.P. Gusynin, V.A. Miransky and I.A. Shovkovy, Phys. Rev. Lett. **73**, 3499 (1994); Phys. Rev. D **52**, 4747 (1995); Nucl. Phys. B **462**, 249 (1996).
95. V.P. Gusynin and I.A. Shovkovy, J. Math. Phys. **40**, 5406 (1999); Can. J. Phys. **74**, 282 (1996).
96. J. Hallin and P. Liljenberg, Phys. Rev. D **46**, 2689 (1992).
97. J. Hauknes, Ann. Phys. (New York) **156**, 303 (1984).
98. W. Heisenberg and H. Euler, Z. Phys. **98**, 714 (1936).
99. J.S. Heyl and L. Hernquist, Phys. Rev. D **55**, 2449 (1997).
100. J.S. Heyl and L. Hernquist, J. Phys. A **30**, 6485 (1997).
101. H. Backe et al., Phys. Rev. Lett. **42**, 376 (1979); J. Schweppe et al., Phys. Rev. Lett. **51**, 2261 (1983); M. Clemente et al., Phys. Lett. B **137**, 41 (1984).
102. D.K. Hong, Phys. Rev. D **57**, 3759 (1998).
103. M. Hott and G. Metikas, Phys. Rev. D **60**, 067703 (1999).
104. E. Iacopini and E. Zavattini, Phys. Lett. B **85**, 151 (1979).
105. W. Israel, Ann. Phys. (New York) **100**, 310 (1976); Physica (Utrecht) **106A**, 204 (1981).
106. C. Itzykson and J.B. Zuber, *Quantum Field Theory*, McGraw-Hill, New York (1980).
107. R. Jackiw and K. Johnson, Phys. Rev. **182**, 1459 (1969).
108. R. Jackiw, Phys. Rev. D **29**, 2375 (1984).
109. J.D. Jackson, *Classical Electrodynamics*, Wiley, New York (1975).
110. J.V. Jelley, Phys. Rev. Lett. **16**, 479 (1966).
111. D.B. Kaplan, preprint nucl-th/9506035 (1995).
112. R. Karplus and M. Neuman, Phys. Rev. **83**, 776 (1951).
113. M.B. Kislinger and P.D. Morley, Phys. Rev. D **13**, 2765 (1976).
114. M.B. Kislinger and P.D. Morley, Phys. Rev. D **13**, 2771 (1976).

234   References

115. J.J. Klein and B.P. Nigam, Phys. Rev. **135**, B1279 (1964).
116. J.J. Klein, Rev. Mod. Phys. **40**, 523 (1968).
117. N.P. Klepikov, Zh. Eksp. Teor. Fiz. **26**, 19 (1954).
118. X. Kong and F. Ravndal, Nucl. Phys. B **526**, 627 (1998).
119. A. Kovner and B. Rosenstein, Phys. Rev. B **42**, 4748 (1990).
120. J.L. Latorre, P. Pascual and R. Tarrach, Nucl. Phys. B **437**, 60 (1995).
121. M. Loewe and J.C. Rojas, Phys. Rev. D **46**, 2689 (1992).
122. Maple V Release 3, Waterloo Maple Software, University of Waterloo (1994).
123. D.B. Melrose and R.J. Stoneham, Nuovo Cimento **32A**, 435 (1976).
124. A. Minguzzi, Nuovo Cimento **4**, 476 (1956).
125. G. Munoz, Am. J. Phys. **64**, 1285 (1996).
126. N.B. Narozhnyi, Sov. Phys. JETP **28**, 371 (1969).
127. A. Niemi and G. Semenoff, Phys. Rev. Lett. **51**, 2077 (1983).
128. Y. Ohkuwa, Prog. Theor. Phys. **65**, 1058 (1981).
129. R.R. Parwani, Phys. Lett. B **358**, 101 (1995).
130. E. Petitgirard, Z. Phys. C **54**, 673 (1992).
131. B.M. Pimentel, A.T. Suzuki and J.L. Tomazelli, Int. J. Theor. Phys. **33**, 2199 (1994).
132. R.D. Pisarski, Phys. Rev. D **29**, 2423 (1984).
133. A.P. Polychronakos, Phys. Rev. Lett. **60**, 1920 (1988).
134. S. Rao and R. Yahalom, Phys. Rev. D **34**, 1194 (1986).
135. M.E. Rassbach, Ph.D. thesis, Caltech (1971).
136. F. Ravndal, preprint hep-ph/9708449 (1997).
137. A.N. Redlich, Phys. Rev. D **29**, 2366 (1984).
138. M. Reuter, M.G. Schmidt and C. Schubert, Ann. Phys. (New York) **259**, 313 (1997); B. Körs and M.G. Schmidt, Eur. Phys. J. **C6**, 175 (1999).
139. V.I. Ritus, Ann. Phys. **69**, 555 (1972).
140. V.I. Ritus, Sov. Phys. JETP **42**, 774 (1976).
141. V.I. Ritus, *Issues in Intense-Field Quantum Electrodynamics*, Proceedings of the Lebedev Physics Institute, Vol. **168**, New York (1987).
142. D. Robaschik, K. Scharnhorst and E. Wieczorek, Ann. Phys. **174**, 401 (1987).
143. K. Scharnhorst, Phys. Lett. B **236**, 354 (1990).
144. K. Scharnhorst, talk delivered at the workshop "Superluminal(?) Velocities", Cologne, preprint hep-th/9810221 (1998); Annalen Phys. **7**, 700 (1998).
145. S. Schmidt et al., Int. J. Mod. Phys. **E7**, 709 (1998); Phys. Rev. D **59**, 094005 (1999); J.C. Bloch et al., Phys. Rev. D **60**, 116011 (1999).
146. C. Schubert, preprint hep-ph/0001288 (2000).
147. L.S. Schulman, *Techniques and Applications of Path Integration*, Wiley, New York (1981).
148. J. Schwinger, Phys. Rev. **82**, 664 (1951).
149. J. Schwinger, Phys. Rev. **128**, 2425 (1962); Phys. Rev. **125**, 397 (1962).
150. J. Schwinger, *Particles, Sources, and Fields*, Vol. 1, p. 318, Addison-Wesley, Reading, MA (1970).
151. G.W. Semenoff and L.C.R. Wijewardhana, Phys. Rev. D **45**, 1342 (1992).
152. A.E. Shabad, Ann. Phys. **90**, 166 (1975).
153. G.M. Shore, Nucl. Phys. B **460**, 379 (1996).
154. I.A. Shovkovy, Phys. Lett. B **441**, 313 (1998).
155. M. Sieber, diploma thesis, Tübingen University (1987).
156. J. Steinberger, Phys. Rev. **76**, 1180 (1949).
157. R. Tarrach, Phys. Lett. B **133**, 259 (1983).
158. J.S. Toll, *The Dispersion Relation for Light and its Application to the Problems Involving Electron Pairs*, Ph.D. Thesis, Princeton University (1952).
159. W. Tsai and T. Erber, Phys. Rev. D **10**, 492 (1974).

160. W. Tsai and T. Erber, Phys. Rev. D **12**, 1132 (1975).
161. W. Tsai and T. Erber, Acta Phys. Austriaca **45**, 245 (1976).
162. L.F. Urrutia, Phys. Rev. D **17**, 1977 (1978).
163. C. Vafa and E. Witten, Commun. Math. Phys. **95**, 257 (1984).
164. V. Weisskopf, K. Dan. Vidensk. Selsk. Mat. Fys. Medd. **14**, 1 (1936).
165. H.A. Weldon, Phys. Rev. D **26**, 1394 (1982).
166. K. Wittman, report, University of Graz (1971).
167. R. Pengo et al.; F. Nezrick; W.-T. Ni, in *Frontier Tests of QED and Physics of the Vacuum*, ed. by E. Zavattini, D. Bakalov and C. Rizzo, Heron Press, Sofia (1998).
168. V.Ch. Zhukovsky, T.L. Shoniya and P.A. Eminov, Zh. Eksp. Teor. Fiz. **107**, 299 (1995); J. Exp. Theor. Phys. **80**, 158 (1995).

Index

Springer Tracts in Modern Physics

Printing: Mercedes-Druck, Berlin
Binding: Stürtz AG, Würzburg